Special Functions in Fractional Calculus and Engineering

Special functions play a very important role in solving various families of ordinary and partial differential equations as well as their fractional-order analogs, which model real-life situations. Owing to the non-local nature and memory effect, fractional calculus is capable of modeling many situations, which arise in engineering. This book includes a collection of related topics associated with such equations and their relevance and significance in engineering.

Special Functions in Fractional Calculus and Engineering highlights the significance and applicability of special functions in solving fractional-order differential equations with engineering applications. This book focuses on the non-local nature and memory effect of fractional calculus in modeling relevant to engineering science and covers a variety of important and useful methods using special functions for solving various types of fractional-order models relevant to engineering science. This book also illustrates the applicability and usefulness of special functions by justifying their numerous and widespread occurrences in the solution of fractional-order differential, integral, and integrodifferential equations.

This book holds a wide variety of interconnected fundamental and advanced topics with interdisciplinary applications that combine applied mathematics and engineering sciences, which are useful to graduate students, Ph.D. scholars, researchers, and educators interested in special functions, fractional calculus, mathematical modeling, and engineering.

Mathematics and its Applications: Modelling, Engineering, and Social Sciences

Series Editor:
Hemen Dutta
Department of Mathematics, Gauhati University

Methods of Mathematical Modelling
Fractional Differential Equations
Edited by Harendra Singh, Devendra Kumar, and Dumitru Baleanu

Mathematical Methods in Engineering and Applied Sciences
Edited by Hemen Dutta

Sequence Spaces
Topics in Modern Summability Theory
Mohammad Mursaleen and Feyzi Basar

Fractional Calculus in Medical and Health Science
Devendra Kumar and Jagdev Singh

Topics in Contemporary Mathematical Analysis and Applications
Hemen Dutta

Sloshing in Upright Circular Containers
Theory, Analytical Solutions, and Applications
Alexander Timokha and Ihor Raynovskyy

Advanced Numerical Methods for Differential Equations
Applications in Science and Engineering
Edited by Harendra Singh, Jagdev Singh, S. D. Purohit, and Devendra Kumar

Concise Introduction to Logic and Set Theory
Edited by Iqbal H. Jebril, Hemen Dutta, and Ilwoo Cho

Integral Transforms and Engineering
Theory, Methods, and Applications
Abdon Atangana and Ali Akgül

Numerical Methods for Fractal-Fractional Differential Equations and Engineering
Simulations and Modeling
Muhammad Altaf Khan and Abdon Atangana

For more information about this series, please visit:
https://www.routledge.com/Mathematics-and-its-Applications/book-series/MES

ISSN (online): 2689-0224
ISSN (print): 2689-0232

Special Functions in Fractional Calculus and Engineering

Edited by
Harendra Singh
H. M. Srivastava
R. K. Pandey

CRC Press
Taylor & Francis Group
Boca Raton London New York

CRC Press is an imprint of the
Taylor & Francis Group, an **informa** business

First edition published 2023
by CRC Press
6000 Broken Sound Parkway NW, Suite 300, Boca Raton, FL 33487-2742

and by CRC Press
4 Park Square, Milton Park, Abingdon, Oxon, OX14 4RN

CRC Press is an imprint of Taylor & Francis Group, LLC

Library of Congress Cataloging-in-Publication Data
Names: Singh, Harendra, editor. | Srivastava, H. M., editor. | Pandey, R. K., editor.
Title: Special functions in fractional calculus and engineering / edited by Harendra Singh,
H. M. Srivastava, R. K. Pandey.
Description: Boca Raton : CRC Press, [2023] |
Series: Mathematics and its applications : modelling, engineering, and social sciences |
Includes bibliographical references and index.
Identifiers: LCCN 2023000426 (print) | LCCN 2023000427 (ebook) | ISBN 9781032435008 (hbk) |
ISBN 9781032436029 (pbk) | ISBN 9781003368069 (ebk)
Subjects: LCSH: Engineering mathematics. | Functions. | Fractional calculus.
Classification: LCC TA347.F86 S64 2023 (print) | LCC TA347.F86 (ebook) |
DDC 620.001/51–dc23/eng/20230130
LC record available at https://lccn.loc.gov/2023000426
LC ebook record available at https://lccn.loc.gov/2023000427

ISBN: 978-1-032-43500-8 (hbk)
ISBN: 978-1-032-43602-9 (pbk)
ISBN: 978-1-003-36806-9 (ebk)

DOI: 10.1201/9781003368069

Typeset in Times
by codeMantra

Contents

Preface

Chapter 1 presents a brief introductory overview and survey of some of the recent developments in the theory of various extensively studied special functions and their usages in fractional calculus and its applications. The origin of special functions, known also as mathematical functions and higher transcendental functions, can be traced back to several widespread areas such as mathematical physics, analytic number theory, applied mathematical sciences, and other fields. On the other hand, in the current literature, there are remarkably extensive usages of the operators of fractional calculus (i.e., fractional-order integrals and fractional-order derivatives) in the modeling and analysis of a significantly large variety of applied scientific and real-world problems in mathematical, physical, biological, engineering and statistical sciences, and in other scientific disciplines.

Chapter 2 considers the applications of the special functions in describing the exact analytical solutions of a class of fluid models described by the fractional operators. In the present investigations, the Caputo derivative has been used as an operator. The special functions used in this chapter are the exponential function, the Mittag-Leffler function, the Gaussian error function, the Bessel function, and others.

Chapter 3 presents a set of d-dimensional fractional diffusion equations coupled with the reaction terms, which can be related to an irreversible or reversible process depending on the choice of the reaction terms. The differential operator connected to the time variable is extended to an arbitrary integro differential operator, which may be related to different operators with singular (Caputo operator) and nonsingular (Caputo-Fabrizio and Atangana-Baleanu operators) kernels. For the spatial operator, we consider an operator which may be related to different cases, particularly the Levy distributions. This operator is a composition of the Levy distribution with the generalized Hankel transform, which enables us to consider in a unified way in different cases. The solutions for free boundary conditions and arbitrary initial conditions for this set of equations are obtained.

In Chapter 4, fractional kinetic equations (FKE) involving broad variance of special functions have proven advantages in describing and establishing many physical world problems in physics and astrophysics. In this chapter, the solution of FKE involving the product of incomplete $\aleph$-functions and $\mathbb{M}$-series using highly accurate and efficient J-transform. Owing to manifold generality of aforementioned FKE, numerous (new and known) corollaries of the main result are effectively recorded.

In Chapter 5, the collocation method is extended for a new class of fractional advection diffusion equation (GFADE). The numerical solution of such GFADE is derived through a collocation method using Legendre polynomials. Convergence and error analysis of polynomial approximation are provided. Numerical examples with homogeneous and nonhomogeneous boundary conditions are presented to confirm the theoretical findings.

Chapter 6 presents the family of the incomplete generalized Mittag-Leffler functions. This chapter systematically investigates several properties of these incomplete generalized Mittag-Leffler functions including, for example, certain basic properties and various integral transforms representations. Certain properties of the Riemann-Liouville fractional integrals and derivatives associated with the incomplete generalized Mittag-Leffler functions are investigated. Some interesting special cases of our main results are also pointed out. Finally, as an application, the solution to the kinetic equations was established.

In Chapter 7, a numerical technique is developed by constructing a particular type of neural network using a Fibonacci polynomial. Various degrees of the Fibonacci polynomial are being used as an activation function in the middle layer of the neural network. The efficiency of the developed method is verified by using it in five examples.

Chapter 8 presents a special class of reaction diffusion equations viz. the Klein-Gordon equation. The key work of this chapter is to propose a tool to find the numerical solution of fractional nonlinear Klein-Gordon equation (FNKGE) with some given boundary and initial conditions. In this numerical technique the operational matrix based on orthogonal Laguerre polynomials is applied to fractional nonlinear Klein-Gordon equations and converted this equation in to an algebraic systems which further simplified by Newton method and gives the desired numerical solution of our considered problem. The stability and convergence analysis of the proposed numerical algorithm is given to show the effectiveness of the scheme.

Chapter 9 presents the systematic study of the generalized Saigo's hypergeometric fractional calculus operators to establish a number of key results for the families of extended Hurwitz-Lerch Zeta function. The new (presumably) and useful (potentially) results that are obtained in this chapter have been expressed in terms of the I–function. A large number of special cases involving Riemann-Liouville and Erdelyi-Kober fractional integral and differential operators and Hurwitz-Lerch zeta function have been expressed in terms of I–function and H–function.

Chapter 10 investigates a solution to the differential equations with spatial and temporal fractional derivatives by using two novel finite difference schemes. A Crank-Nicolson difference method is used for the temporal fractional derivatives. Also, a compact difference scheme is used for the temporal fractional derivatives of the fractional oscillation motion equation with viscoelastic damping. This model is used to describe the mechanical oscillation mechanism in a viscoelastic medium. Analysis of the stability and convergence are rigorously discussed for the two schemes.

Chapter 11 examines the Dadras-Momeni system in the frame of the Caputo-Fabrizio fractional derivative. Theoretical aspects such as boundedness, existence, and uniqueness of solutions are presented. A detailed analysis is presented regarding the stability of points of equilibrium. To regulate chaos in this fractional-order system with unpredictable dynamics, a sliding mode controller is developed, and the global stability of the system with control law is established.

Chapter 12 investigates a fractional order model with the Mittag-Leffler Kernel. The existence and uniqueness of the solutions of the model are also discussed. The numerical simulations for different values of fractional order and parameters are demonstrated.

Editors

Dr. Harendra Singh is an Assistant Professor at the Department of Mathematics, Post-Graduate College, Ghazipur-233001, Uttar Pradesh, India, and has been listed in the top 2% scientists list published by Stanford University. He primarily teaches subjects such as real and complex analysis, functional analysis, abstract algebra, and measure theory in postgraduate level courses in mathematics. Dr. Singh has published 50 research papers in various journals of repute and has published three books from Taylor & Francis, one with Springer and one with Elsevier. He has attained a number of national and international conferences and presented several research papers. He is a reviewer of various journals, and his areas of interest are mathematical modeling, fractional differential equations, integral equations, calculus of variations, and analytical and numerical methods.

Dr. H. M. Srivastava is a Professor Emeritus, Department of Mathematics and Statistics, University of Victoria, British Columbia V8W 3R4, Canada. He earned his Ph.D. in 1965 while he was a full-time member of the teaching faculty at the Jai Narain Vyas University of Jodhpur in India (since 1963). Professor Srivastava has held (and continues to hold) numerous Visiting, Honorary and Chair Professorships at many universities and research institutes in different parts of the world. Having received several D.Sc. degrees as well as honorary memberships and fellowships of many scientific academies and scientific societies around the world, he is also actively associated editorially with numerous international scientific research journals as an Honorary or Advisory Editor or as an Editorial Board Member. He has also edited many special issues of scientific research journals as the Lead or Joint Guest Editor. He has published 36 books, monographs, and edited volumes, 36 books (and encyclopedia) chapters, 48 papers in international conference proceedings, and more than 1,350 peer-reviewed international scientific research journal articles, as well as Forewords and Prefaces to many books and journals. Dr. Srivastava's research interests include several areas of pure and applied mathematical sciences, such as real and complex analysis, fractional calculus and its applications, integral equations and transforms, higher transcendental functions and their applications, q-series and q-polynomials, analytic number theory, analytic and geometric inequalities, probability and statistics, and inventory modeling and optimization.

Dr. R. K. Pandey is an Associate Professor in the Department of Mathematical Sciences, at the Indian Institute of Technology (BHU), Varanasi, India. He is the recipient of the Indo-US fellowship and INSA Visiting Fellowship for 2012. He had been a visiting faculty member at the Southern Illinois University, Carbondale Illinois, IL, during 2012–2013, The University of Tokyo, Japan from November to December, 2010, Central South University China, and Shanghai University Shanghai, China in December 2020. He has also worked as an Assistant Professor in Mathematics at

BITS Pilani and the Indian Institute of Information Technology Design and Manufacturing Jabalpur from July 2009 to May 2014. Dr. Pandey received his Ph.D. from the Department of Applied Mathematics, Institute of Technology (BHU), Varanasi, India in 2009 and has published 60 research papers in refereed journals and guided 8 Ph.D. students. He has been a member of management committees for academic institutions and also a member of the advisory committees in various national and international conferences. His current areas of research include fractional derivatives, image processing, and numerical methods for integral and integrodifferential equations.

Contributors

Ali Akgül
Department of Mathematics
Arts and Science Faculty
Siirt University
Siirt, Turkey
and
Mathematics Research Center
Department of Mathematics
Near East Unversity
Nicosia, Turkey

T. S. Aleroev
Moscow State University of Civil Engineering
Moscow, Russia

Chandrali Baishya
Department of Studies and Research in Mathematics
Tumkur University
Tumkur, India

Manish Kumar Bansal
Department of Mathematics
Jaypee Institute of Information Technology
Noida, India

Purnima Chopra
Formerly Department of Mathematics
Marudhar Engineering College Bikaner
Bikaner, India

Kushal Dhar Dwivedi
Department of Mathematics
S. N. Government Post-Graduate College
Khandwa, India

A. M. Elsayed
Department of Mathematics
Faculty of Science
Zagazig University
Zagazig, Egypt
and
Moscow State University of Civil Engineering
Moscow, Russia

Nidhi Jolly
Department of Computer Applications
Maharaja Surajmal Institute
Janakpuri, India

Sandeep Kumar
Department of Mathematical Sciences
Indian Institute of Technology
Banaras Hindu University
Varanasi, India

Priya Kumari
Department of Mathematics
Institute of Science
Banaras Hindu University
Varanasi, India

E. K. Lenzi
Departamento de Física
Universidade Estadual de Ponta Grossa
Ponta Grossa, Brazil
and
National Institute of Science and Technology for Complex Systems
Centro Brasileiro de Pesquisas Físicas
Rio de Janeiro, Brazil

M. K. Lenzi
Departamento de Engenharia Quìmica
Universidade Federal do Paraná
Curitiba, Brazil

Prashant Pandey
Department of Mathematics
KS Saket PG College
Ayodhya, India

R. K. Pandey
Department of Mathematical Sciences
Indian Institute of Technology
Banaras Hindu University
Varanasi, India

Rakesh K. Parmar
Department of Mathematics
Ramanujan School of Mathematical Sciences
Pondicherry University
Puducherry, India

S. D. Purohit
Department of HEAS (Mathematics)
Rajasthan Technical University
Kota, India

Arjun K. Rathie
Department of Mathematics
Vedant College of Engineering and Technology
Rajasthan Technical University
Kota, India

Ndolane Sene
Department of Mathematics and Statistics
Institut des Politiques Publiques
Cheikh Anta Diop University
Dakar Fann, Senegal

Shiva Sharma
Department of Mathematics
K P College Murliganj
Madhepura, India

Harendra Singh
Department of Mathematics
Post-Graduate College
Uttar Pradesh, India

H. M. Srivastava
Department of Mathematics and Statistics
University of Victoria
Victoria, Canada
and
Department of Medical Research
China Medical University Hospital
China Medical University
Taichung, Taiwan
and
Department of Mathematics and Informatics
Azerbaijan University,
Baku, Azerbaijan
and
Section of Mathematics
International Telematic University Uninettuno
Rome, Italy

P. Veeresha
Center for Mathematical Needs
Department of Mathematics
CHRIST (Deemed to be University)
Bengaluru, India

1 An Introductory Overview of Special Functions and Their Associated Operators of Fractional Calculus

H. M. Srivastava
University of Victoria
China Medical University
Azerbaijan University
International Telematic University Uninettuno

CONTENTS

1.1 INTRODUCTION, DEFINITIONS AND PRELIMINARIES

In this chapter, we begin by introducing the following standard notations:

$$\mathbb{N} := \{1,2,3,\ldots\}, \quad \mathbb{N}_0 := \{0,1,2,3,\ldots\} = \mathbb{N} \cup \{0\}$$

and

$$\mathbb{Z}^- := \{-1,-2,-3,\ldots\} = \mathbb{Z}_0^- \setminus \{0\}.$$

Moreover, as usual, $\mathbb{Z}$ denotes the set of integers, $\mathbb{R}$ denotes the set of real numbers and $\mathbb{C}$ denotes the set of complex numbers.

Beyond any doubt, one of the most useful and the most fundamental special functions of mathematical analysis and applied mathematics is the familiar (Euler's) Gamma function $\Gamma(z)$. Essentially, it stemmed out of an attempt by the Swiss mathematician, physicist, astronomer, geographer, logician, and engineer, Leonard

DOI: 10.1201/9781003368069-1

Euler (1707–1783), to render a meaning to $x!$ when x is any positive real number. In 1729, Euler undertook the problem of interpolating $n!$ between the positive integer values of n. It led to what is widely known as the (Euler's) Gamma function $\Gamma(z)$ which is defined, for $z \in \mathbb{C} \setminus \mathbb{Z}_0^-$, by

$$\Gamma(z) = \begin{cases} \displaystyle\int_0^\infty e^{-t}\, t^{z-1}\, dt & (\Re(z) > 0) \\[2ex] \dfrac{\Gamma(z+n)}{\prod\limits_{j=0}^{n-1} (z+j)} & \left(z \in \mathbb{C} \setminus \mathbb{Z}_0^-;\ n \in \mathbb{N}\right). \end{cases} \tag{1.1}$$

Historically, of course, the origin of the Gamma function $\Gamma(z)$ defined by Eq. (7.1) can be traced back to two letters from Euler to the German mathematician, Christian Goldbach (1690–1764), elaborating upon a simple desire to extend factorials to values between the integers. It is regrettable to see that, in several recent amateurish-type publications, some authors have trivially changed the variable t of integration in the integral definition in Eq. (7.1) and have thereby claimed to have produced a "generalization" (called k-Gamma function) of this classical Gamma function $\Gamma(z)$ (see, for details, [102, Section 3, pp. 1505–1506]).

Among numerous special functions, which are related rather closely to the Gamma function, we choose to mention the incomplete Gamma functions, the Beta and the incomplete Beta functions, the Error functions and the Probability integral, the Digamma and Polygamma functions, and so on (see, for details, [1,7,26,63–65,77,126,127,129]).

In this chapter, we aim at presenting a brief introductory overview and survey of some of the recent developments in the theory of a large variety of extensively studied special functions and their usages in the theory and applications of fractional calculus (i.e., fractional integrals and fractional derivatives).

1.2 HYPERGEOMETRIC FUNCTIONS: EXTENSIONS AND MULTIVARIATE GENERALIZATIONS

We first recall the following second-order homogeneous linear ordinary differential equation (popularly known as the *Gauss hypergeometric equation*):

$$z(1-z)\frac{d^2w}{dz^2} + [c-(a+b+1)z]\frac{dw}{dz} - abw = 0, \tag{1.2}$$

with three singularities at $z = 0, 1, \infty$, each of which is a regular singular point of Eq. (1.2). It is indeed the most celebrated differential equation of the *Fuchsian class* consisting of differential equations whose only singularities (including the point at ∞) are regular singular points. The importance of the differential equation (1.2) arises from the following known observation in the theory of differential equations.

Every homogeneous linear differential equation of the second order, whose singularities (including the point at infinity) are regular and at most three in number, can be transformed into the hypergeometric equation (1.2).

In a neighborhood of the regular singular point at the origin ($z=0$), the general solution of the hypergeometric equation (1.2) can be expressed in terms of the Gauss hypergeometric function ${}_2F_1$, which is named after the famous German mathematician, Carl Friedrich Gauss (1777–1855), who introduced this function in 1812 and gave the F-notation for it. By using the general Pochhammer symbol or the *shifted factorial* $(\lambda)_\nu$, since

$$(1)_n = n! \qquad (n \in \mathbb{N}_0),$$

which is defined (for $\lambda, \nu \in \mathbb{C}$), in terms of the Gamma function in Eq. (7.1), by

$$(\lambda)_\nu := \frac{\Gamma(\lambda+\nu)}{\Gamma(\lambda)} = \begin{cases} 1 & (\nu = 0;\ \lambda \in \mathbb{C}\setminus\{0\}) \\ \lambda(\lambda+1)\ldots(\lambda+n-1) & (\nu = n \in \mathbb{N};\ \lambda \in \mathbb{C}), \end{cases} \tag{1.3}$$

where it is understood *conventionally* that $(0)_0 := 1$ and assumed *tacitly* that the Γ-quotient exists, the Gauss hypergeometric function ${}_2F_1$ is defined by

$$\begin{aligned} {}_2F_1(a,b;c;z) &:= 1 + \frac{ab}{1\cdot c}\,z + \frac{a(a+1)b(b+1)}{1\cdot 2\cdot c(c+1)}\,z^2 + \cdots \\ &= \sum_{n=0}^{\infty} \frac{(a)_n\,(b)_n}{(c)_n}\,\frac{z^n}{n!} \qquad (a,b \in \mathbb{C};\ c \in \mathbb{C}\setminus\mathbb{Z}_0^-) \end{aligned} \tag{1.4}$$

or by its analytic continuation for all $z \in \mathbb{C}\setminus[1,\infty)$.

In the general case ${}_pF_q$ $(p,q \in \mathbb{N}_0)$, which is usually attributed to *Bishop* Ernest William Barnes (1874–1953) of the Church of England in Birmingham, a generalized hypergeometric function with p numerator parameters $\alpha_j \in \mathbb{C}$ $(j=1,\ldots,p)$ and q denominator parameters $\beta_j \in \mathbb{C}\setminus\mathbb{Z}_0^-$ $(j=1,\ldots,q)$ is given by

$$\begin{aligned} {}_pF_q\left[\begin{matrix} \alpha_1,\ldots,\alpha_p; \\ \beta_1,\ldots,\beta_q; \end{matrix}\ z\right] &:= \sum_{n=0}^{\infty} \frac{(\alpha_1)_n\cdots(\alpha_p)_n}{(\beta_1)_n\cdots(\beta_q)_n}\,\frac{z^n}{n!} \\ &=: {}_pF_q(\alpha_1,\ldots,\alpha_p;\beta_1,\ldots,\beta_q;z), \end{aligned} \tag{1.5}$$

in which the infinite series
(i) converges absolutely for $|z| < \infty$ if $p \leqq q$,
(ii) converges absolutely for $|z| < 1$ if $p = q+1$, and
(iii) diverges for all z $(z \neq 0)$ if $p > q+1$.
Furthermore, if we set

$$\omega := \sum_{j=1}^{q} \beta_j - \sum_{j=1}^{p} \alpha_j,$$

then it is known that the generalized hypergeometric ${}_pF_q$ series in Eq. (1.5) (with $p = q+1$) is

I. absolutely convergent for $|z| = 1$ if $\Re(\omega) > 0$,
II. conditionally convergent for $|z| = 1$ $(z \neq 1)$ if $-1 < \Re(\omega) \leqq 0$, and
III. divergent for $|z| = 1$ if $\Re(\omega) \leqq -1$.

In particular, when $p - 1 = q = 2$, the function ${}_3F_2$ is known as the Clausen hypergeometric function in honor of the Danish mathematician and astronomer, Thomas Clausen (1801–1885) who, in 1828, established the following hypergeometric identity:

$$\left({}_2F_1 \left[\begin{array}{c} a, b; \\ \\ a+b+\frac{1}{2}; \end{array} z \right] \right)^2 = {}_3F_2 \left[\begin{array}{c} 2a, a+b, 2b; \\ \\ a+b+\frac{1}{2}, 2(a+b); \end{array} z \right]. \tag{1.6}$$

For $a, b, c, z \in \mathbb{R}$, the positivity of the function ${}_3F_2$ on the right-hand side of Clausen's identity (1.6) was instrumental in de Branges' proof of the 68-year-old Bieberbach conjecture of 1916 that

$$|a_n| \leqq n \qquad (n \in \mathbb{N} \setminus \{1\})$$

for functions f in the normalized analytic and univalent function class $\mathcal{S}$ given by

$$\mathcal{S} := \left\{ f : f \in \mathcal{A}, \; f(z) = z + \sum_{n=0}^{\infty} a_n z^n \;\; (z \in \mathbb{U}) \text{ and } f \text{ is univalent in } \mathbb{U} \right\},$$

where $\mathcal{A}$ denotes the class of normalized functions which are analytic in the open unit disk $\mathbb{U}$ (see, for details, [25]). In fact, de Branges' theorem of 1984 asserts the truth of the Milin conjecture of 1971, which implies the Robertson conjecture of 1936 and also the above-mentioned Bieberbach conjecture of 1916. The proof by Louis de Branges included Löwner's differential equation and a certain nonnegativity result which may readily be put in the following Clausenian hypergeometric form:

$$\frac{(\lambda+2)_n}{n!} \, {}_3F_2 \left[\begin{array}{c} -n, \lambda+n+2, \frac{1}{2}(\lambda+1); \\ \\ \lambda+1, \frac{1}{2}(\lambda+3); \end{array} x \right] \geqq 0 \tag{1.7}$$

$$(0 \leqq x < 1; \; \lambda \geqq -2; \; n \in \mathbb{N}_0),$$

in which the exceptional values of the parameter λ are to be handled as limit cases. We remark in passing that the theory of special functions (and, especially, the generalized hypergeometric functions) has thus far remained unavoidable in proving the aforementioned conjectures in Geometric Function Theory of Complex Analysis.

Almost all of the widely used special functions of mathematical physics and applied mathematics such as the Bessel, Legendre and Lommel functions, Whittaker functions, Elliptic integrals, and so on, are expressible in terms of the generalized hypergeometric function ${}_pF_q$. Moreover, the Jacobi, Laguerre, Hermite, and other associated orthogonal and non-orthogonal polynomials (e.g., the Bessel polynomials) are essentially hypergeometric functions ${}_pF_q$ $(p, q \in \mathbb{N}_0)$ in which one or more of the numerator parameters α_j $(j = 1, \ldots, p)$ is a negative integer. In particular, the special orthogonal polynomials $\{\mathscr{S}_n(x)\}_{n\in\mathbb{N}_0}$, which are related rather closely to the Bessel polynomials as well as the Laguerre polynomials, occurred in the investigation of energy spectral functions for a certain family of isotropic turbulence fields (see, for details, [92] and the references cited therein). Quite recently in Ref. [50], a modified form of these special orthogonal polynomials $\{\mathscr{S}_n(x)\}_{n\in\mathbb{N}_0}$ has provided a novel set of (orthogonal) basis functions along with some suitable collocation points in a certain matrix technique for computationally treating a class of multi-order fractional pantograph differential equations by using matrix techniques. Thus, clearly, the potential for the usefulness of the generalized hypergeometric function ${}_pF_q$ $(p,q\in\mathbb{N}_0)$ and its considerably wide variety of special cases cannot be overemphasized. Some of the books, monographs, and tables along these lines include those by Abramowitz and Stegun [1], Andrews [6], Andrews et al. [7], Carlson [18], Erdélyi et al. [26, Vols. I and II], Lebedev [61], Luke [63,64]), Magnus et al. [65], Miller [69], Srivastava et al. [126,127,129], Temme [150], Szegö [149], and others.

Various extensions and multivariate generalizations of hypergeometric functions have stemmed from, and are somewhat motivated by, the potential for their applications in solutions of systems of partial differential equations. Some of the multivariate generalizations of hypergeometric functions include the two-variable Appell and Kampé de Fériet functions and the Lauricella functions in several variables, the Srivastava-Daoust hypergeometric functions in two and more variables, and so on. For the theory and applications of many of these univariate and multivariate hypergeometric functions, the interested reader is referred to the monographs by Appell and Kampé de Fériet [9], Bailey [12], Erdélyi et al. [26, Vol. I], Slater [88], Seaborn [82], Srivastava et al. [126,129], and other authors. In particular, for the theory and applications of the substantially more general G and H functions in one, two and more variables, the reader is referred to the monographs by Mathai et al. [68] and Srivastava et al. (see [123,127,129]), and others. Historically, the H-function of several complex variables was first introduced and investigated by Srivastava and Panda [133,134]. It was used as the kernel of some general families of multidimensional integral transforms in Refs. [136,137] (see also [2,3,21,23,32,33,53,59,74,120–122,143,145,147] for other results involving the Srivastava-Panda multivariable H-function).

It should be remarked in passing that, as a nontrivial *and* non-vacuous generalization of the H-function encountered in a systematic study of a class of Feynman integrals, the $\overline{H}$-function was introduced and investigated rather widely and extensively (see, for details, [14,48,49,128,141]).

We choose to conclude this section by recalling that, as once remarked by Gian-Carlo Rota (1932–1999), *the above-mentioned univariate and multivariate hypergeometric functions are capable of encompassing just about everything in sight.*

1.3 THE ZETA AND RELATED FUNCTIONS OF ANALYTIC NUMBER THEORY

Many important and potentially useful functions in *Analytic Number Theory* include (for example) the Riemann Zeta function $\zeta(s)$ and the Hurwitz (or generalized) Zeta function $\zeta(s,a)$, which are defined, for $\Re(s) > 1$, by

$$\zeta(s) := \begin{cases} \sum\limits_{n=1}^{\infty} \dfrac{1}{n^s} = \dfrac{1}{1-2^{-s}} \sum\limits_{n=1}^{\infty} \dfrac{1}{(2n-1)^s} & (\Re(s) > 1) \\[2ex] \dfrac{1}{1-2^{1-s}} \sum\limits_{n=1}^{\infty} \dfrac{(-1)^{n-1}}{n^s} & (\Re(s) > 0;\ s \neq 1) \end{cases} \tag{1.8}$$

and

$$\zeta(s,a) := \sum_{n=0}^{\infty} \frac{1}{(n+a)^s} \qquad \left(\Re(s) > 1;\ a \in \mathbb{C} \setminus \mathbb{Z}_0^-\right), \tag{1.9}$$

and $\left(\text{for } \Re(s) \leqq 1; s \neq 1\right)$ by their *meromorphic* continuations (see, for details, the remarkable works by Titchmarsh [151] and Apostol [8] *as well as* the monumental treatise by Whittaker and Watson [154]; see also [1, Chapter 23] and [113, Chapter 2]).

A substantially more general Dirichlet-type series than those in Eqs. (1.8) and (1.9) happens to define the Hurwitz-Lerch Zeta function $\Phi(z,s,a)$ as follows (see, for example, [26, Vol. I, p. 27. Eq. 1.11 (1)]; see also [113, pp. 121 *et seq.*], [93] and [114, pp. 135 *et seq.*])

$$\Phi(z,s,a) := \sum_{n=0}^{\infty} \frac{z^n}{(n+a)^s} \tag{1.10}$$

$$\left(a \in \mathbb{C} \setminus \mathbb{Z}_0^-;\ s \in \mathbb{C} \text{ when } |z| < 1;\ \Re(s) > 1 \text{ when } |z| = 1\right).$$

In fact, just as in the cases of the Riemann Zeta function $\zeta(s)$ and the Hurwitz (or generalized) Zeta function $\zeta(s,a)$, the Hurwitz-Lerch Zeta function $\Phi(z,s,a)$ can be continued *meromorphically* to the whole complex s-plane, except for a simple pole at $s = 1$ with its residue 1 (see also the recent survey-cum-expository review articles [97,98] on various widely and extensively studied families of the Hurwitz-Lerch and related Zeta functions). It is also known that [26, Vol. I, p. 27, Equation 1.11 (3)]

$$\Phi(z,s,a) = \frac{1}{\Gamma(s)} \int_0^{\infty} \frac{t^{s-1}\, \mathrm{e}^{-at}}{1 - z\mathrm{e}^{-t}}\, \mathrm{d}t = \frac{1}{\Gamma(s)} \int_0^{\infty} \frac{t^{s-1}\, \mathrm{e}^{-(a-1)t}}{\mathrm{e}^t - z}\, \mathrm{d}t \tag{1.11}$$

$$\left(\Re(a) > 0;\ \Re(s) > 0 \text{ when } |z| \leqq 1;\ \Re(s) > 1 \text{ when } z = 1\right).$$

The above-mentioned *Polygamma functions* $\psi^{(n)}(z)$ $(n \in \mathbb{N}_0)$ defined by

$$\psi^{(n)}(z) := \frac{d^{n+1}}{dz^{n+1}} \{\log\Gamma(z)\} = \frac{d^n}{dz^n} \{\psi(z)\} \qquad \left(n \in \mathbb{N}_0;\ z \in \mathbb{C} \setminus \mathbb{Z}_0^-\right), \tag{1.12}$$

of which $\psi^{(0)}(z) = \psi(z)$ is called the Digamma function, satisfy the following relationship with the Hurwitz (or generalized) Zeta function $\zeta(s,a)$ defined by Eq. (1.9):

$$\psi^{(n)}(z) = (-1)^{n+1}\, n! \sum_{k=0}^{\infty} \frac{1}{(k+z)^{n+1}} = (-1)^{n+1}\, n!\, \zeta(n+1,z) \tag{1.13}$$

$$\left(n \in \mathbb{N};\ z \in \mathbb{C} \setminus \mathbb{Z}_0^-\right).$$

The relationship (1.13) may be used to deduce the properties of the Polygamma functions $\psi^{(n)}(z)$ $(n \in \mathbb{N})$ from those of the Hurwitz (or generalized) Zeta function $\zeta(s,z)$ $\quad (s = n+1;\ n \in \mathbb{N})$. The special case of the Hurwitz (or generalized) Zeta function $\zeta(s,z)$ when $z = 1$ is the familiar Riemann Zeta function $\zeta(s)$, which is involved in the following Taylor-Maclaurin series expansion of $\log\Gamma(1+z)$ about the origin $z = 0$:

$$\log\Gamma(1+z) = -\gamma z + \sum_{n=2}^{\infty} (-1)^n \zeta(n) \frac{z^n}{n} \qquad (|z| < 1), \tag{1.14}$$

where γ denotes the Euler-Mascheroni constant.

The general Hurwitz-Lerch Zeta function $\Phi(z,s,a)$ defined by Eq. (1.10) contains, as its *special* cases, not only the Riemann Zeta function $\zeta(s)$ and the Hurwitz (or generalized) Zeta function $\zeta(s,a)$ and the Lerch Zeta function $\ell_s(\xi)$ defined by

$$\ell_s(\xi) := \sum_{n=1}^{\infty} \frac{e^{2n\pi i\xi}}{n^s} = e^{2\pi i\xi}\, \Phi\left(e^{2\pi i\xi}, s, 1\right) \qquad \left(\xi \in \mathbb{R};\ \Re(s) > 1\right), \tag{1.15}$$

but also such other important functions of *Analytic Number Theory* as the Polylogarithmic function $\mathrm{Li}_s(z)$:

$$\mathrm{Li}_s(z) := \sum_{n=1}^{\infty} \frac{z^n}{n^s} = z\,\Phi(z,s,1) \tag{1.16}$$

$$\left(s \in \mathbb{C} \quad \text{when} \quad |z| < 1;\ \Re(s) > 1 \quad \text{when} \quad |z| = 1\right)$$

and the Lipschitz-Lerch Zeta function $\phi(\xi,a,s)$:

$$\phi(\xi,a,s) := \sum_{n=0}^{\infty} \frac{e^{2n\pi i\xi}}{(n+a)^s} = \Phi\left(e^{2\pi i\xi}, s, a\right) =: L(\xi,s,a) \tag{1.17}$$

$$\left(a \in \mathbb{C}\setminus\mathbb{Z}_0^-;\ \Re(s) > 0 \quad \text{when} \quad \xi \in \mathbb{R}\setminus\mathbb{Z};\ \Re(s) > 1 \quad \text{when} \quad \xi \in \mathbb{Z}\right),$$

which was first studied by Rudolf Lipschitz (1832–1903) and Matyás Lerch (1860–1922) in connection with Dirichlet's famous theorem on primes in arithmetic progressions (see, for details, [151]).

For a number of further extensions of the Hurwitz-Lerch Zeta function $\Phi(s,z,a)$, including its *multi-parameter* extensions, the reader is referred to the recent monograph by Srivastava and Choi [114] (see also [97,98,113]). In each of these two recent monographs [113,114], one can also find a detailed systematic presentation of the *double*, *triple* and *multiple* Gamma functions which were studied by Barnes and others (see also Whittaker and Watson [154, p. 264] and Gradshteyn and Ryzhik [34, p. 661, Entry 6.441(4); p. 937, Entry 8.333]).

An interesting and potentially useful family of the λ-generalized Hurwitz-Lerch Zeta functions, which *further* extend the multi-parameter Hurwitz-Lerch Zeta function

$$\Phi^{(\rho_1,\ldots,\rho_p;\sigma_1,\ldots,\sigma_q)}_{\lambda_1,\ldots,\lambda_p;\mu_1,\ldots,\mu_q}(z,s,\kappa)$$

defined by Ref. [142, p. 503, Equation (6.2)] (see also [94,114])

$$\Phi_{\lambda_1,\ldots,\lambda_p;\mu_1,\ldots,\mu_q}^{(\rho_1,\ldots,\rho_p,\sigma_1,\ldots,\sigma_q)}(z,s,\kappa) := \sum_{n=0}^{\infty} \frac{\prod_{j=1}^{p}(\lambda_j)_{n\rho_j}}{n!\cdot\prod_{j=1}^{q}(\mu_j)_{n\sigma_j}} \frac{z^n}{(n+\kappa)^s} \tag{1.18}$$

$$\Big(p,q\in\mathbb{N}_0;\ \lambda_j\in\mathbb{C}\ (j=1,\ldots,p);\ \kappa,\mu_j\in\mathbb{C}\setminus\mathbb{Z}_0^-\ (j=1,\ldots,q);$$

$$\rho_j,\sigma_k\in\mathbb{R}^+\ (j=1,\ldots,p;\ k=1,\ldots,q);\Delta>-1 \text{ when } s,z\in\mathbb{C};$$

$$\Delta=-1 \text{ and } s\in\mathbb{C} \text{ when } |z|<\nabla^*;$$

$$\Delta=-1 \text{ and } \Re(\Xi)>\frac{1}{2} \text{ when } |z|=\nabla^*\Big),$$

where, for convenience,

$$\Delta := \sum_{j=1}^{q}\sigma_j - \sum_{j=1}^{p}\rho_j \quad \text{and} \quad \Xi := s+\sum_{j=1}^{q}\mu_j - \sum_{j=1}^{p}\lambda_j + \frac{p-q}{2} \tag{1.19}$$

and

$$\nabla^* := \left(\prod_{j=1}^{p}\rho_j^{-\rho_j}\right)\cdot\left(\prod_{j=1}^{q}\sigma_j^{\sigma_j}\right). \tag{1.20}$$

It was introduced and investigated systematically in a recent paper by Srivastava [94], who also discussed their potential application in Number Theory by appropriately constructing a presumably new continuous analogue of Lippert's Hurwitz measure and, in addition, considered some other statistical applications of these families of the λ-generalized Hurwitz-Lerch Zeta functions in probability distribution theory (see also the references to several related earlier works cited by Srivastava [94]). For the convenience of the interested reader in pursuing some of the related open problems, we choose to reproduce here the definition of the λ-generalized Hurwitz-Lerch Zeta function whose investigation was initiated by Srivastava [94]:

$$\Phi_{\lambda_1,\ldots,\lambda_p;\mu_1,\ldots,\mu_q}^{(\rho_1,\ldots,\rho_p,\sigma_1,\ldots,\sigma_q)}(z,s,a;b,\lambda) := \frac{1}{\Gamma(s)}\int_0^\infty t^{s-1}\exp\left(-at-\frac{b}{t^\lambda}\right) \cdot {}_p\Psi_q^*\left[\begin{matrix}(\lambda_1,\rho_1),\ldots,(\lambda_p,\rho_p); \\ (\mu_1,\sigma_1),\ldots,(\mu_q,\sigma_q);\end{matrix}\ ze^{-t}\right]\mathrm{d}t \tag{1.21}$$

$$\left(\min\{\Re(a),\Re(s)\}>0;\ \Re(b)\geqq 0;\ \lambda\geqq 0\right),$$

so that, obviously, we have the following relationship with the multi-parameter Hurwitz-Lerch Zeta function defined by Eq. (1.18):

$$\begin{aligned}\Phi^{(\rho_1,\ldots,\rho_p,\sigma_1,\ldots,\sigma_q)}_{\lambda_1,\ldots,\lambda_p;\mu_1,\ldots,\mu_q}(z,s,a;0,\lambda) &= \Phi^{(\rho_1,\ldots,\rho_p,\sigma_1,\ldots,\sigma_q)}_{\lambda_1,\ldots,\lambda_p;\mu_1,\ldots,\mu_q}(z,s,a)\\ &= e^b\,\Phi^{(\rho_1,\ldots,\rho_p,\sigma_1,\ldots,\sigma_q)}_{\lambda_1,\ldots,\lambda_p;\mu_1,\ldots,\mu_q}(z,s,a;b,0). \qquad (1.22)\end{aligned}$$

The general Fox-Wright functions ${}_p\Psi_q$ and ${}_p\Psi^*_q$, occurring in the integral representation (1.21), are defined by (see, for details, [26, Vol. I, p. 183] and [126, p. 21]; see also [57, p. 56], [54, p. 65] and [123, p. 19])

$$\begin{aligned}{}_p\Psi^*_q\left[\begin{array}{c}(a_1,A_1),\ldots,(a_p,A_p);\\ \\ (b_1,B_1),\ldots,(b_q,B_q);\end{array} z\right] &:= \sum_{n=0}^{\infty}\frac{(a_1)_{A_1n}\cdots(a_p)_{A_pn}}{(b_1)_{B_1n}\cdots(b_q)_{B_qn}}\,\frac{z^n}{n!}\\ &=: \frac{\Gamma(b_1)\ldots\Gamma(b_q)}{\Gamma(a_1)\ldots\Gamma(a_p)}\,{}_p\Psi_q\left[\begin{array}{c}(a_1,A_1),\ldots,(a_p,A_p);\\ \\ (b_1,B_1),\ldots,(b_q,B_q);\end{array} z\right] \qquad (1.23)\end{aligned}$$

$$\left(\Re(A_j)>0\ (j=1,\ldots,p);\ \Re(B_j)>0\ (j=1,\ldots,q);\ \Re\left(\sum_{j=1}^{q}B_j-\sum_{j=1}^{p}A_j\right)\geqq -1\right),$$

where, and in what follows, $(\lambda)_\nu$ denotes the general Pochhammer symbol or the *shifted* factorial defined already by Eq. (1.3), and the equality in the convergence condition holds true only for suitably bounded values of $|z|$ given by

$$|z| < \nabla := \left(\prod_{j=1}^{p}A_j^{-A_j}\right)\cdot\left(\prod_{j=1}^{q}B_j^{B_j}\right). \qquad (1.24)$$

Clearly, the generalized hypergeometric function ${}_pF_q\ (p,q\in\mathbb{N}_0)$ in Eq. (1.25), with p numerator parameters $a_1,\ldots,a_p$ and q denominator parameters $b_1,\ldots,b_q$, is a widely and extensively investigated and potentially useful special case of the general Fox-Wright function ${}_p\Psi_q\ (p,q\in\mathbb{N}_0)$ when

$$A_j = 1 \quad (j=1,\ldots,p) \qquad \text{and} \qquad B_j = 1 \quad (j=1,\ldots,q),$$

given by

$$
{}_pF_q\left[\begin{array}{c} a_1,\ldots,a_p; \\ \\ b_1,\ldots,b_q; \end{array} z\right] := \sum_{n=0}^{\infty} \frac{(a_1)_n\cdots(a_p)_n}{(b_1)_n\cdots(b_q)_n}\,\frac{z^n}{n!}
$$

$$
= {}_p\Psi_q^*\left[\begin{array}{c} (a_1,1),\ldots,(a_p,1); \\ \\ (b_1,1),\ldots,(b_q,1); \end{array} z\right]
$$

$$
= \frac{\Gamma(b_1)\ldots\Gamma(b_q)}{\Gamma(a_1)\ldots\Gamma(a_p)}\;{}_p\Psi_q\left[\begin{array}{c} (a_1,1),\ldots,(a_p,1); \\ \\ (b_1,1),\ldots,(b_q,1); \end{array} z\right]. \tag{1.25}
$$

Indeed it was my proud privilege to have met many times and discussed mathematical researches, especially on various families of special functions and related topics, with my Canadian colleague, Charles Fox (1897–1977) of birth and education in England, both at McGill University and Sir George Williams University (*now* Concordia University) in Montréal, mainly during the 1970s (see, for details, [90]). Another remarkable mathematical scientist of modern times happens to be *Sir* Edward Maitland Wright (1906–2005), with whom I had the privilege to meet and discuss researches emerging from his publications on hypergeometric and related higher transcendental functions during my visit to the University of Aberdeen in Scotland in 1976.

Finally, in their recent work, for $\Re(\sigma) > 0$ and $\Re(\omega) \geqq 1 + \Re(\sigma)$, Srivastava et al. [112] introduced and investigated a family of multiple Hurwitz-Lerch Zeta functions, associated with the Srivastava-Daoust hypergeometric series in two and more variables by using the currently popular (simplified) notation given in Ref. [132] (see, for details, [19]; see also [20,38,126]), which is defined by

$$
{}^{\sigma}_{\omega}F^{A:B^{(1)};\ldots;B^{(n)}}_{C:D^{(1)};\ldots;D^{(n)}}(z_1,\ldots,z_n)
$$

$$
= {}^{\sigma}_{\omega}F^{A:B^{(1)};\ldots;B^{(n)}}_{C:D^{(1)};\ldots;D^{(n)}}
$$

$$
\left(\begin{array}{c} \left[(a):\theta^{(1)},\ldots,\theta^{(n)}\right]:\left[\left(b^{(1)}\right):\psi^{(1)}\right];\ldots;\left[\left(b^{(n)}\right):\psi^{(n)}\right]; \\ \\ \left[(c):\delta^{(1)},\ldots,\delta^{(n)}\right]:\left[\left(d^{(1)}\right):\phi^{(1)}\right];\ldots;\left[\left(d^{(n)}\right):\phi^{(n)}\right]; \end{array} z_1,\ldots,z_n\right)
$$

$$
:= \sum_{m_1,\ldots,m_n=0}^{\infty} \mathscr{K}^{A:B^{(1)};\ldots;B^{(n)}}_{C:D^{(1)};\ldots;D^{(n)}}(m_1,\ldots,m_n)
$$

$$
\cdot \frac{z_1^{m_1}}{m_1!}\cdots\frac{z_n^{m_n}}{m_n!}\,\frac{1}{(m_1+\ldots+m_n+\omega)^{\sigma}}, \tag{1.26}
$$

where, for convenience,

$$\mathcal{K}^{A:B^{(1)};\ldots;B^{(n)}}_{C:D^{(1)};\ldots;D^{(n)}}(m_1,\ldots,m_n)$$

$$:= \frac{\prod_{j=1}^{A}(a_j)_{\theta_j^{(1)}m_1+\ldots+\theta_j^{(n)}m_n}}{\prod_{j=1}^{C}(a_j)_{\delta_j^{(1)}m_1+\ldots+\delta_j^{(n)}m_n}} \frac{\prod_{j=1}^{B^{(1)}}\left(b_j^{(1)}\right)_{\psi_j^{(1)}m_1}}{\prod_{j=1}^{D^{(1)}}\left(d_j^{(1)}\right)_{\phi_j^{(1)}m_1}} \cdots \frac{\prod_{j=1}^{B^{(n)}}\left(b_j^{(n)}\right)_{\psi_j^{(n)}m_n}}{\prod_{j=1}^{D^{(n)}}\left(d_j^{(n)}\right)_{\phi_j^{(n)}m_n}}, \tag{1.27}$$

in which the multiple series converges for

$$|z_1| < 1,\ldots,|z_n| < 1,$$

provided that

$$1+\sum_{j=1}^{C}\delta_j^{(\ell)}+\sum_{j=1}^{D^{(\ell)}}\phi_j^{(\ell)}-\sum_{j=1}^{A}\theta_j^{(\ell)}-\sum_{j=1}^{B^{(\ell)}}\psi_j^{(\ell)}=0 \qquad (\forall\,\ell=1,\ldots,n). \tag{1.28}$$

By applying the methods and techniques, which were developed and used by Srivastava et al. [112], we can derive the following integral representation for the family of multiple Hurwitz-Lerch Zeta-type functions of this section.

Theorem 1.1. (see [112, p. 312, Theorem 4.1]) *Let* $\Re(\sigma) > 0$ *and* $\Re(\omega) \geqq 1+\Re(\sigma)$. *Suppose also that*

$$|z_1| < 1,\ldots,|z_n| < 1$$

and

$$1+\sum_{j=1}^{C}\delta_j^{(\ell)}+\sum_{j=1}^{D^{(\ell)}}\phi_j^{(\ell)}-\sum_{j=1}^{A}\theta_j^{(\ell)}-\sum_{j=1}^{B^{(\ell)}}\psi_j^{(\ell)}=0 \qquad (\forall\,\ell=1,\ldots,n).$$

Then the following integral representation holds true for the multiple Hurwitz-Lerch-type functions in Eq. (1.26):

$$ {}^{\sigma}_{\omega}F^{A:B^{(1)};\ldots;B^{(n)}}_{C:D^{(1)};\ldots;D^{(n)}}(z_1,\ldots,z_n) = \frac{1}{\Gamma(\sigma)}\int_0^{\infty} t^{\sigma-1}\,\mathrm{e}^{-\omega t}\, F^{A:B^{(1)};\ldots;B^{(n)}}_{C:D^{(1)};\ldots;D^{(n)}}$$
$$\left(\begin{matrix}\left[(a):\theta^{(1)},\ldots,\theta^{(n)}\right]:\left[\left(b^{(1)}\right):\psi^{(1)}\right];\ldots;\left[\left(b^{(n)}\right):\psi^{(n)}\right]; \\ \left[(c):\delta^{(1)},\ldots,\delta^{(n)}\right]:\left[\left(d^{(1)}\right):\phi^{(1)}\right];\ldots;\left[\left(d^{(n)}\right):\phi^{(n)}\right];\end{matrix}\; z_1\mathrm{e}^{-t},\ldots,z_n\mathrm{e}^{-t}\right)\mathrm{d}t, \tag{1.29}$$

provided that the integral in Eq. (1.29) *is convergent.*

1.4 EXTENSIONS AND GENERALIZATIONS OF THE MITTAG-LEFFLER-TYPE FUNCTIONS

The familiar Mittag-Leffler function $E_\alpha(z)$ and its two-parameter version $E_{\alpha,\beta}(z)$, which are defined, respectively, by Refs. [70,155,156]

$$E_\alpha(z) := \sum_{k=0}^{\infty} \frac{z^k}{\Gamma(\alpha k+1)} \quad \text{and} \quad E_{\alpha,\beta}(z) := \sum_{k=0}^{\infty} \frac{z^k}{\Gamma(\alpha k+\beta)} \tag{1.30}$$

$$\left(z, \alpha, \beta \in \mathbb{C};\ \Re(\alpha) > 0\right).$$

The one-parameter function $E_\alpha(z)$ was first considered by Magnus Gustaf (Gösta) Mittag-Leffler (1846–1927) in 1903 and its two-parameter version $E_{\alpha,\beta}(z)$ was introduced by Anders Wiman (1865–1959) in 1905 (see also [66,89]).

The Mittag-Leffler functions $E_\alpha(z)$ and $E_{\alpha,\beta}(z)$ are *natural* extensions of the exponential, hyperbolic, and trigonometric functions. In fact, it is easily observed that

$$E_1(z) = e^z, \ E_2\left(z^2\right) = \cosh z, \ E_2\left(-z^2\right) = \cos z,$$

$$E_{1,2}(z) = \frac{e^z - 1}{z} \quad \text{and} \quad E_{2,2}(z^2) = \frac{\sinh z}{z}.$$

For a reasonably detailed account of various properties, generalizations and applications of the Mittag-Leffler functions $E_\alpha(z)$ and $E_{\alpha,\beta}(z)$, the reader may refer to the recent works by Gorenflo et al. [31], Haubold et al. [40] and Kilbas et al. ([12], [13] and [57, Chapter 1]). The Mittag-Leffler function $E_\alpha(z)$ given by Eq. (1.30) and some of its various generalizations have only recently been calculated numerically in the whole complex plane (see, for example, [11,22]; see also [1,73]).

In many recent investigations, the interest in the families of Mittag-Leffler-type functions has grown considerably due mainly to their potential for applications in some reaction-diffusion problems and other problems in applied and engineering sciences. Moreover, their various extensions and generalizations appear in the solutions of fractional-order differential and integral equations (see, for example, [95]; see also [27,119]). In particular, a family of the multi-index Mittag-Leffler functions was considered and used as a kernel of some fractional-calculus operators by Srivastava et al. [109,110]; see also the references to the related earlier works which are cited in each of these papers). For a detailed presentation of some of these recent developments, one is referred to a recent survey-cum-expository review article [103].

Almost all of the above-mentioned multi-parameter generalizations and extensions of the Mittag-Leffler function $E_\alpha(z)$ in Eq. (1.30) are contained, as obvious special cases, of the general Fox-Wright function ${}_p\Psi_q$ $(p, q \in \mathbb{N}_0)$ or ${}_p\Psi_q^*$ $(p, q \in \mathbb{N}_0)$, defined by Eq. (1.23), with p numerator parameters $a_1, \ldots, a_p$ and q denominator parameters $b_1, \ldots, b_q$ such that

$$a_j \in \mathbb{C} \quad (j = 1, \ldots, p) \qquad \text{and} \qquad b_j \in \mathbb{C} \setminus \mathbb{Z}_0^- \quad (j = 1, \ldots, q).$$

The families of higher transcendental functions of the Mittag-Leffler and the Fox-Wright types are known to play important roles in the theory of fractional calculus and operational calculus as well as in their applications in the basic processes of evolution, relaxation, diffusion, oscillation, and wave propagation. Some other general families of the Mittag-Leffler-type functions were investigated and applied recently by Srivastava and Tomovski (see, for details, [24]; see also [103] and the references to the earlier works cited in each of these works).

We recall here a series of monumental works by Wright [157]– [159] in which he introduced and systematically studied the asymptotic expansion of the following Taylor-Maclaurin series (see [157, p. 424]):

$$\mathfrak{E}_{\alpha,\beta}(\phi;z) := \sum_{n=0}^{\infty} \frac{\phi(n)}{\Gamma(\alpha n+\beta)}\, z^n \qquad (\alpha,\beta \in \mathbb{C};\ \Re(\alpha) > 0), \tag{1.31}$$

where $\phi(t)$ is a function satisfying suitable conditions. Wright's work was motivated substantially by the earlier developments involving simpler cases, which were reported by Mittag-Leffler in 1905, Wiman in 1905, Ernest William Barnes (1874–1953) in 1906, Godfrey Harold Hardy (1877–1947) in 1905, George Neville Watson (1886–1965) in 1913, Fox in 1928, and many other authors. In particular, the aforementioned work [13] by *Bishop* Ernest William Barnes (1874–1953) of the Church of England in Birmingham considered the asymptotic expansions of functions in the class defined below:

$$E_{\alpha,\beta}^{(\kappa)}(s;z) := \sum_{n=0}^{\infty} \frac{z^n}{(n+\kappa)^s\, \Gamma(\alpha n+\beta)} \qquad (\alpha,\beta \in \mathbb{C};\ \Re(\alpha) > 0) \tag{1.32}$$

for suitably restricted parameters κ and s. It is easy to deduce, from the definition (1.32), the following relationships with the Mittag-Leffler-type function $E_{\alpha,\beta}^{(\kappa)}(s;z)$ of Barnes [13]:

$$E_\alpha(z) = \lim_{s\to 0}\left\{E_{\alpha,1}^{(\kappa)}(s;z)\right\} \qquad \text{and} \qquad E_{\alpha,\beta}(z) = \lim_{s\to 0}\left\{E_{\alpha,\beta}^{(\kappa)}(s;z)\right\}. \tag{1.33}$$

More interestingly, we also have the following relationship:

$$\lim_{\alpha\to 0}\left\{E_{\alpha,\beta}^{(\kappa)}(s;z)\right\} = \frac{1}{\Gamma(\beta)}\,\Phi(z,s,\kappa)$$

with the classical Lerch transcendent (or the Hurwitz-Lerch Zeta function) $\Phi(z,s,\kappa)$ defined by Eq. (1.10).

As we have indicated above, Barnes [13] systematically presented asymptotic expansions of many functions such as those in the class of the Mittag-Leffler-type function $E_{\alpha,\beta}^{(\kappa)}(s;z)$ defined by Eq. (1.32), and the classical Mittag-Leffler functions $E_\alpha(z)$ and $E_{\alpha,\beta}(z)$ defined by Eq. (1.30). On the other hand, the multi-parameter Hurwitz-Lerch Zeta function

$$\Phi_{\lambda_1,\ldots,\lambda_p;\mu_1,\ldots,\mu_q}^{(\rho_1,\ldots,\rho_p;\sigma_1,\ldots,\sigma_q)}(z,s,\kappa),$$

defined by Eq. (1.18), obviously provides a natural unification and generalization of the Fox-Wright function ${}_p\Psi_q^*$ defined by Eq. (1.23) as well as the Hurwitz-Lerch Zeta function $\Phi(z,s,\kappa)$ defined by Eq. (1.10).

We conclude this section by recalling the following interesting unification of the definitions in Eqs. (1.31) and (1.18) for suitably restricted function $\varphi(\tau)$ (see, for details, [102,103]):

$$\mathscr{E}_{\alpha,\beta}(\varphi;z,s,\kappa) := \sum_{n=0}^{\infty} \frac{\varphi(n)}{(n+\kappa)^s\,\Gamma(\alpha n+\beta)}\, z^n \qquad \big(\alpha,\beta \in \mathbb{C};\ \Re(\alpha) > 0\big), \quad (1.34)$$

where the parameters α, β, s, and κ are appropriately constrained as above. Such general families of special functions of the Mittag-Leffler and the Hurwitz-Lerch types were applied recently in Ref. [106] in connection with some double integrals stemming from the Boltzmann equation in the kinetic theory of gasses. Upon replacing the sequence $\{\varphi(n)\}_{n=0}^{\infty}$ in the definition (1.34) by the sequence $\{\phi(n)\}_{n=0}^{\infty}$, we readily observe that

$$\mathfrak{E}_{\alpha,\beta}(\phi;z) = \lim_{s\to 0}\big\{\mathscr{E}_{\alpha,\beta}(\varphi;z,s,\kappa)\big\}\big|_{\varphi\equiv\phi}. \quad (1.35)$$

Moreover, if we put $\alpha = \beta = 1$ and

$$\varphi(n) = \frac{\prod_{j=1}^{p}(\lambda_j)_{n\rho_j}}{\prod_{j=1}^{q}(\mu_j)_{n\sigma_j}} \qquad (n \in \mathbb{N}_0) \quad (1.36)$$

in definition (1.34), then definition (1.34) will immediately yield the definition (1.18) of the extended Hurwitz-Lerch Zeta function

$$\Phi^{(\rho_1,\ldots,\rho_p;\sigma_1,\ldots,\sigma_q)}_{\lambda_1,\ldots,\lambda_p;\mu_1,\ldots,\mu_q}(z,s,\kappa).$$

Alternatively, in the special case of Eq. (1.34) when $\alpha \to 0$, $\beta = 1$ and

$$\varphi(n) = \frac{\prod_{j=1}^{p}(\lambda_j)_{n\rho_j}}{n!\cdot\prod_{j=1}^{q}(\mu_j)_{n\sigma_j}} \qquad (n \in \mathbb{N}_0) \quad (1.37)$$

or, more simply, by setting

$$\varphi(n) = \frac{\Gamma(\alpha n+\beta)\ \prod_{j=1}^{p}(\lambda_j)_{n\rho_j}}{n!\cdot\prod_{j=1}^{q}(\mu_j)_{n\sigma_j}} \qquad (n \in \mathbb{N}_0), \quad (1.38)$$

we are led to the extended Hurwitz-Lerch Zeta function

$$\Phi^{(\rho_1,\ldots,\rho_p;\sigma_1,\ldots,\sigma_q)}_{\lambda_1,\ldots,\lambda_p;\mu_1,\ldots,\mu_q}(z,s,\kappa)$$

defined by Eq. (1.18).

1.5 FRACTIONAL CALCULUS AND ITS APPLICATIONS

The current scientific literature consists of a significantly large number of developments on fractional-order modeling and analysis of real-world and other problems, which arise in the mathematical, physical, biological, statistical, and engineering sciences. It is truly amazing to see how fast the subject of fractional calculus has grown in recent years, the concept of which is rooted essentially in a question raised in 1695 by Marquis de l'Hôpital (1661–1704) to Gottfried Wilhelm Leibniz (1646–1716), which sought the meaning of Leibniz's (currently popular) notation

$$\frac{d^n y}{dx^n}$$

for the derivative of order $n \in \mathbb{N}_0$ when $n = \frac{1}{2}$, and indeed also in Leibniz's reply dated 30 September 1695 to l'Hôpital as follows:

> "$\cdots$ *This is an apparent paradox from which, one day, useful consequences will be drawn.* $\cdots$"

In widespread applications of fractional calculus, the use is made of fractional-order derivatives of different (and, occasionally, *ad hoc*) kinds (see, for example, [5]– [10], [16,20,23,24,71,81,111,124,152]). It is fairly traditional to define the fractional-order integrals and fractional-order derivatives by means of the following right-sided Riemann-Liouville fractional integral operator ${}^{\mathrm{RL}}I^{\mu}_{a+}$ and the left-sided Riemann-Liouville fractional integral operator ${}^{\mathrm{RL}}I^{\mu}_{a-}$, and the corresponding Riemann-Liouville fractional derivative operators ${}^{\mathrm{RL}}D^{\mu}_{a+}$ and ${}^{\mathrm{RL}}D^{\mu}_{a-}$, as follows (see, for example, [26, Vol. II, Chapter 13], [57, pp. 69–70] and [79]; see also [22,107,148] for the Marichev-Saigo-Maeda generalization using Appell's third function F_3 in the kernel):

$$\left({}^{\mathrm{RL}}I^{\mu}_{a+}f\right)(x) = \frac{1}{\Gamma(\mu)}\int_a^x (x-t)^{\mu-1}\, f(t)\,dt \qquad \left(x > a;\ \Re(\mu) > 0\right), \tag{1.39}$$

$$\left({}^{\mathrm{RL}}I^{\mu}_{a-}f\right)(x) = \frac{1}{\Gamma(\mu)}\int_x^a (t-x)^{\mu-1}\, f(t)\,dt \qquad \left(x < a;\ \Re(\mu) > 0\right) \tag{1.40}$$

and

$$\left({}^{\mathrm{RL}}D^{\mu}_{a\pm}f\right)(x) = \left(\pm\frac{d}{dx}\right)^n \left(I^{n-\mu}_{a\pm}f\right)(x) \qquad \left(\Re(\mu) \geqq 0;\ n = [\Re(\mu)]+1\right), \tag{1.41}$$

where the function f is assumed to be at least locally integrable and $[\Re(\mu)]$ means the greatest integer in $\Re(\mu)$. It is worthwhile to remark in passing that, in the current literature, there is an unfortunate trend to trivially and inconsequentially translate known theory and known results based on the classical Riemann-Liouville and other types of fractional integrals and fractional derivatives in terms of the corresponding theory and the corresponding results by forcing in some obviously redundant (or superfluous) parameters and variables in the widely and extensively investigated definitions and results (see, for details, [102, Section 2, pp. 1504–1505]).

In a series of papers, Hilfer et al. [5]– [7]; [9,10,80,152] introduced and studied an interesting family of generalized Riemann-Liouville fractional derivatives of order μ $(0 < \mu < 1)$ and type ν $(0 \leqq \nu \leqq 1)$. The right-sided Hilfer fractional derivative operator ${}^{\mathrm{H}}D_{a+}^{\mu,\nu}$ and the left-sided Hilfer fractional derivative operator ${}^{\mathrm{H}}D_{a-}^{\mu,\nu}$ of order μ $(0 < \mu < 1)$ and type ν $(0 \leqq \nu \leqq 1)$ with respect to x are defined by

$$\left({}^{\mathrm{H}}D_{a\pm}^{\mu,\nu} f\right)(x) = \left(\pm\, {}^{\mathrm{H}}I_{a\pm}^{\nu(1-\mu)} \frac{d}{dx} \left({}^{\mathrm{H}}I_{a\pm}^{(1-\nu)(1-\mu)} f\right)\right)(x), \tag{1.42}$$

where it is tacitly assumed that the second member of Eq. (1.42) exists. The generalization (1.42) yields the classical Riemann-Liouville fractional derivative operator when $\nu = 0$. Moreover, for $\nu = 1$, it leads to the fractional derivative operator introduced by Liouville [62, p. 10], which is quite frequently attributed wrongly to Caputo [15] nowadays, but which should more appropriately be referred to as the *Liouville-Caputo fractional derivative*, giving due credits to Joseph Liouville (1809–1882) who considered such fractional derivatives many decades earlier in 1832 (see [62]). Many authors (see, for example, [16,24]) called the general operators in Eq. (1.42) the Hilfer fractional derivative operators. Several applications of the Hilfer fractional derivative operator $D_{a\pm}^{\mu,\nu}$ can indeed be found in Ref. [7] (see also [95,96]).

In the vast scientific literature on fractional calculus and its widespread applications, not only the Fox-Wright hypergeometric function ${}_p\Psi_q(z)$ defined by Eq. (1.23), but also much more general functions such as (for example) Meijer's G-function and Fox's H-function, have already been used as kernels of many different families of fractional-calculus operators (see, for details, [123]. [124,139]; see also the references cited in each of these earlier works). In fact, Srivastava et al. [124] used the Riemann-Liouville type fractional integrals with the Fox H-function and the Fox-Wright hypergeometric function ${}_p\Psi_q(z)$ as kernels and also applied their results to the substantially more general $\overline{H}$-function (see, for example, [14,128]).

The Wright function $\mathfrak{E}_{\alpha,\beta}(\varphi;z)$ in Eq. (1.31), which was introduced and studied in Ref. [157] as long ago as 1940, has appeared recently in Ref. [76] in connection with fractional calculus, but *without* giving due credits to Wright [157]. Closely following the recent works [102]– [104], the general right-sided fractional integral operator $\mathscr{I}_{a+}^{\mu}(\varphi;z,s,\kappa,\nu)$ and the general left-sided fractional integral operator $\mathscr{I}_{a-}^{\mu}(\varphi;z,s,\kappa,\nu)$, and the corresponding fractional derivative operators $\mathscr{D}_{a+}^{\mu}(\varphi;z,s,\kappa,\nu)$ and $\mathscr{D}_{a-}^{\mu}(\varphi;z,s,\kappa,\nu)$, each of the Riemann-Liouville type, are defined by

$$\left(\mathscr{I}_{a+}^{\mu}(\varphi;z,s,\kappa,\nu) f\right)(x) = \frac{1}{\Gamma(\mu)} \int_a^x (x-t)^{\mu-1}\, \mathscr{E}_{\alpha,\beta}\left(\varphi;z(x-t)^{\nu},s,\kappa\right) f(t)\, dt \tag{1.43}$$

$$\left(x > a;\ \Re(\mu) > 0\right),$$

$$\left(\mathscr{I}_{a-}^{\mu}(\varphi;z,s,\kappa,\nu)f\right)(x)=\frac{1}{\Gamma(\mu)}\int_x^a (t-x)^{\mu-1}\,\mathscr{E}_{\alpha,\beta}\big(\varphi;z(t-x)^{\nu},s,\kappa\big)\,f(t)\,dt$$
$$\big(x<a;\ \Re(\mu)>0\big) \tag{1.44}$$

and

$$\left(\mathscr{D}_{a\pm}^{\mu}(\varphi;z,s,\kappa,\nu)f\right)(x)=\left(\pm\frac{d}{dx}\right)^{n}\left(\mathscr{I}_{a\pm}^{n-\mu}(\varphi;z,s,\kappa,\nu)f\right)(x) \tag{1.45}$$
$$\big(\Re(\mu)\geqq 0;\ n=[\Re(\mu)]+1\big),$$

where the function f is in the space $L(\mathfrak{a},\mathfrak{b})$ of Lebesgue integrable functions on a finite closed interval $[\mathfrak{a},\mathfrak{b}]$ $(\mathfrak{b}>\mathfrak{a})$ of the real line $\mathbb{R}$ given by

$$L(\mathfrak{a},\mathfrak{b})=\left\{f:\ \|f\|_1=\int_{\mathfrak{a}}^{\mathfrak{b}}|f(x)|\,dx<\infty\right\} \tag{1.46}$$

and, moreover, it is *tacitly* assumed that, in situations such as those occurring in conjunction with the usages of the definitions in Eqs. (1.43)–(1.45), the point $\mathfrak{a}$ in all such function spaces as (for example) the function space $L(\mathfrak{a},\mathfrak{b})$ coincides precisely with the *lower* terminal a in the integrals involved in the definitions (1.43)–(1.45).

For potential applications based on the general fractional-calculus operators defined by Eqs. (1.43), (1.44) and (1.45), we list below several useful properties of the kernel $\mathscr{E}_{\alpha,\beta}(\varphi;zx^{\nu},s,\kappa)$ involved therein.

$$\frac{d^n}{dx^n}\left\{x^{\mu-1}\,\mathscr{E}_{\alpha,\beta}(\varphi;zx^{\nu},s,\kappa)\right\}$$
$$=x^{\mu-n-1}\sum_{j=0}^{\infty}\frac{\varphi(j)}{(j+\kappa)^s\,\Gamma(\alpha j+\beta)}\,\frac{\Gamma(\nu j+\mu)}{\Gamma(\nu j+\mu-n)}\,(zx^{\nu})^j \tag{1.47}$$
$$\big(n\in\mathbb{N}_0;\ \Re(\mu)>0;\ \Re(\nu)>0;\ \Re(\alpha)>0\big),$$

which, in the special case when $\mu=\beta$ and $\nu=\alpha$, yields

$$\frac{d^n}{dx^n}\left\{x^{\beta-1}\,\mathscr{E}_{\alpha,\beta}(\varphi;zx^{\alpha},s,\kappa)\right\}=x^{\beta-n-1}\,\mathscr{E}_{\alpha,\beta-n}(\varphi;zx^{\alpha},s,\kappa) \tag{1.48}$$
$$\big(n\in\mathbb{N}_0;\ \Re(\alpha)>0;\ \Re(\beta)>0\big),$$

provided that each member of equations (1.47) and (1.48) exists.

$$\mathfrak{I}\left\{t^{\mu-1}\,\mathscr{E}_{\alpha,\beta}(\varphi;zt^{\nu},s,\kappa)\right\}(x)$$
$$:=\int_0^x t^{\mu-1}\,\mathscr{E}_{\alpha,\beta}(\varphi;zt^{\nu},s,\kappa)\,dt$$
$$=x^{\mu}\sum_{j=0}^{\infty}\frac{\varphi(j)}{(j+\kappa)^s\,\Gamma(\alpha j+\beta)}\,\frac{\Gamma(\nu j+\mu)}{\Gamma(\nu j+\mu+1)}\,(zx^{\nu})^j \tag{1.49}$$
$$\big(\Re(\mu)>0;\ \Re(\nu)>0;\ \Re(\alpha)>0\big),$$

provided that the integral exists. More generally, we have

$$\begin{aligned}\mathfrak{I}^n \left\{t^{\mu-1}\, \mathscr{E}_{\alpha,\beta}\left(\varphi; zt^{\nu}, s, \kappa\right)\right\}(x) \\ = x^{\mu+n-1} \sum_{j=0}^{\infty} \frac{\varphi(j)}{(j+\kappa)^s\, \Gamma(\alpha j+\beta)} \frac{\Gamma(\nu j+\mu+n-1)}{\Gamma(\nu j+\mu+n)} \left(zx^{\nu}\right)^j\end{aligned} \tag{1.50}$$

$$\left(n \in \mathbb{N};\ \Re(\mu) > 0;\ \Re(\nu) > 0;\ \Re(\alpha) > 0\right),$$

which, in the special case when $\mu = \beta$ and $\nu = \alpha$, yields

$$\mathfrak{I}^n \left\{t^{\beta-1}\, \mathscr{E}_{\alpha,\beta}\left(\varphi; zt^{\alpha}, s, \kappa\right)\right\}(x) = x^{\beta+n-1}\, \mathscr{E}_{\alpha,\beta+n}\left(\varphi; zx^{\alpha}, s, \kappa\right) \tag{1.51}$$

$$\left(n \in \mathbb{N};\ \Re(\alpha) > 0;\ \Re(\beta) > 0\right),$$

provided that each member of equations (1.50) and (1.51) exists.

For the operator $\mathscr{L}$ of the Laplace transform given by

$$\mathscr{L}\{f(\tau) : \mathfrak{s}\} := \int_0^{\infty} \mathrm{e}^{-\mathfrak{s}\tau} f(\tau)\, d\tau =: F(\mathfrak{s}) \qquad \left(\Re(\mathfrak{s}) > 0\right), \tag{1.52}$$

where the function $f(\tau)$ is so constrained that the integral exists, we get

$$\mathscr{L}\left\{\tau^{\mu-1}\, \mathscr{E}_{\alpha,\beta}\left(\varphi; z\tau^{\nu}, s, \kappa\right) : \mathfrak{s}\right\} = \frac{1}{\mathfrak{s}^{\mu}} \sum_{j=0}^{\infty} \frac{\varphi(j)\, \Gamma(\nu j+\mu)}{(j+\kappa)^s\, \Gamma(\alpha j+\beta)} \left(\frac{z}{\mathfrak{s}^{\nu}}\right)^j \tag{1.53}$$

$$\left(\Re(\mathfrak{s}) > 0;\ \Re(\mu) > 0;\ \Re(\nu) > 0;\ \Re(\alpha) > 0\right),$$

provided that each member of Eq. (1.53) exists.

The $\mathfrak{s}$-multiplied Laplace transform, also known as the Laplace-Carson transform, is defined by

$$\mathscr{L}\mathscr{C}\{f(\tau) : \mathfrak{s}\} := \mathfrak{s} \int_0^{\infty} \mathrm{e}^{-\mathfrak{s}\tau} f(\tau)\, d\tau =: F(\mathfrak{s}) \qquad \left(\Re(\mathfrak{s}) > 0\right), \tag{1.54}$$

provided that the integral exists.

For $\mu = \beta$ and $\nu = \alpha$, the Laplace transform formula (1.53) assumes the following simpler form:

$$\mathscr{L}\left\{\tau^{\beta-1}\, \mathscr{E}_{\alpha,\beta}\left(\varphi; z\tau^{\alpha}, s, \kappa\right) : \mathfrak{s}\right\} = \frac{1}{\mathfrak{s}^{\mu}} \sum_{j=0}^{\infty} \frac{\varphi(j)}{(j+\kappa)^s} \left(\frac{z}{\mathfrak{s}^{\alpha}}\right)^j \tag{1.55}$$

$$\left(\Re(\mathfrak{s}) > 0;\ \Re(\alpha) > 0;\ \Re(\beta) > 0\right).$$

Unfortunately, by trivially changing, in the definition of the classical Laplace transform (1.52), the index $\mathfrak{s}$ or the integration variable t or both the index $\mathfrak{s}$ *and* the integration variable t, many mainly amateurish-type researchers have made and continue to make the obviously false claim to have "generalized" the classical Laplace transform (1.52) itself. Some of the examples in this connection include, but are not limited to, the so-called Sumudu transform, the so-called natural transform, the so-called

Shehu transform, the so-called $\mathscr{P}_\delta$-transform, the so-called k-Laplace transform, the so-called Elzaki transform, and so on (see, for details, [102, Section 4, pp. 1508–1510]; see also [105, pp. 36–38]). So far there are no convincing arguments and evidence as to why all these rather trivial and inconsequential parameters and argument variations of the classical Laplace transform (1.52) or the Laplace-Carson transform (1.54) are preferable in any way to the classical Laplace transform in Eq. (1.52) or the Laplace-Carson transform (1.54) themselves.

Various special cases and consequences of the above key results, involving simpler functions of the types which we have considered in this and the preceding sections, can be deduced fairly easily. We turn instead to the following examples of applications involving fractional-order derivatives of one kind or the other. In this connection, we note the following relationship between the Riemann-Liouville fractional derivative ${}^{\mathrm{RL}}D^{\mu}_{0+}$ and the Liouville-Caputo fractional derivative ${}^{\mathrm{LC}}D^{\mu}_{0+}$ of order μ (see, for example, [57, p. 91, Eq. (2.4.1)] *with* $a=0$):

$$\left({}^{\mathrm{LC}}D^{\mu}_{0+}f\right)(x)=\left({}^{\mathrm{RL}}D^{\mu}_{0+}\left\{f(t)-\sum_{k=0}^{n-1}\frac{f^{(k)}(0)}{k!}\,t^k\right\}\right)(x), \tag{1.56}$$

where n is given by

$$n=\begin{cases}[\Re(\mu)]+1 & (\mu\neq\mathbb{N}_0)\\[2ex] \mu & (\mu\in\mathbb{N}_0).\end{cases} \tag{1.57}$$

Equivalently, since

$$\left({}^{\mathrm{RL}}D^{\mu}_{0+}\left\{t^{\lambda-1}\right\}\right)(x)=\frac{\Gamma(\lambda)}{\Gamma(\lambda-\mu)}\,x^{\lambda-\mu-1} \qquad \big(\Re(\lambda)>0;\ \Re(\mu)\geqq 0\big), \tag{1.58}$$

the relationship (1.56) can be written as follows:

$$\left({}^{\mathrm{LC}}D^{\mu}_{0+}f\right)(x)=\left({}^{\mathrm{RL}}D^{\mu}_{0+}f\right)(x)-\sum_{k=0}^{n-1}\frac{f^{(k)}(0)}{\Gamma(k-\mu+1)}\,x^{k-\mu}, \tag{1.59}$$

where n is given, as before in Eq. (1.56), by Eq. (1.57).

I. The basic processes of relaxation, diffusion, oscillations, and wave propagation were revisited by Gorenflo et al. [31] who introduced the Liouville-Caputo-type fractional-order derivatives in the governing (ordinary or partial) differential equations and considered each of the following fractional differential equations:

$$\frac{d^{\alpha}u}{dt^{\alpha}}+c^{\alpha}\,u(t;\alpha)=0 \qquad (c>0;\ 0<\alpha\leqq 2) \tag{1.60}$$

and

$$\frac{\partial^{2\beta}u}{\partial t^{2\beta}}=k\,\frac{\partial^2 u}{\partial x^2} \qquad (k>0;\ -\infty<x<\infty;\ 0<\beta\leqq 1), \tag{1.61}$$

where the aforementioned Liuoville-Caputo fractional derivative of order $\mu > 0$ of a *causal* function $f(t)$, that is,

$$f(t) = 0 \qquad (t < 0),$$

is given by

$$\frac{d^{\mu}}{dx^{\mu}}\{f(x)\} = \left({}^{\mathrm{LC}}D^{\mu}_{0+}f\right)(x) := \begin{cases} f^{(n)}(x) & (\mu = n \in \mathbb{N}_0) \\ \dfrac{1}{\Gamma(n-\mu)}\displaystyle\int_0^x \frac{f^{(n)}(t)}{(x-t)^{\mu-n+1}}\,dt & (n-1 < \Re(\mu) < n;\ n \in \mathbb{N}). \end{cases} \tag{1.62}$$

Here, as usual, n is given by Eq. (1.57) and $f^{(n)}(t)$ denotes the ordinary derivative of $f(t)$ of order n.

Equation (1.60) represents fractional relaxation when $0 < \alpha \leqq 1$ under the initial condition $u(0+;\alpha) = u_0$. Moreover, it can be viewed as fractional oscillation when $1 < \alpha \leqq 2$ under the following initial conditions:

$$u(0+;\alpha) = u_0 \qquad \text{and} \qquad \dot{u}(0+;\alpha) = v_0,$$

where $v_0 \equiv 0$ for continuous dependence of the solution on the parameter α also in the transition from $\alpha = 1-$ to $\alpha = 1+$, and $\dot{u}$ is the time-derivative of u.

In equation (1.61), $u = u(x,t;\beta)$ is assumed to be a *causal* function of time t $(t > 0)$ such that

$$u(\mp\infty, t;\beta) = 0.$$

Clearly, equation (1.61) represents fractional diffusion when $0 < \beta \leqq \frac{1}{2}$ under the initial condition $u(x,0+;\beta) = f(x)$. It can also be viewed as fractional wave equation when $\frac{1}{2} < \beta \leqq 1$ under the following initial conditions:

$$u(x,0+;\beta) = f(x) \qquad \text{and} \qquad \dot{u}(x,0+;\beta) = g(x),$$

where $g(x) \equiv 0$ for continuous dependence of the solution on the parameter β also in the transition from $\beta = \frac{1}{2}-$ to $\beta = \frac{1}{2}+$.

In terms of the Mittag-Leffler function $E_{\alpha}(z)$ defined by Eq. (1.30), the explicit solution of the initial-value problem involving the fractional differential equation (1.60) is given by

$$u(t;\alpha) = u_0\, E_{\alpha}\left(-(ct)^{\alpha}\right). \tag{1.63}$$

On the other hand, the explicit solution of the initial-value problem involving the fractional differential equation (1.61) can be expressed as follows:

$$u(x,t;\beta) = \int_{-\infty}^{\infty} \mathscr{G}_c(\xi,t;\beta)\, f(x-\xi)\, d\xi, \tag{1.64}$$

where $\mathscr{G}_c(x,t;\beta)$ denotes the Green function given by

$$|x|\ \mathscr{G}_c(x,t;\beta) = \frac{z}{2} \sum_{n=0}^{\infty} \frac{(-z)^n}{n!\,\Gamma(1-\beta-\beta n)} \qquad \left(z = \frac{|x|}{\sqrt{kt^{\beta}}};\ 0<\beta<1\right), \tag{1.65}$$

which, in turn, can be easily expressed in terms of Wright's generalized Bessel function or the Bessel-Wright function $J_{\nu}^{\mu}(z)$ defined by (see, for example, [127, p. 42, Equation II.5 (22)])

$$J_{\nu}^{\mu}(z) := \sum_{n=0}^{\infty} \frac{(-z)^n}{n!\,\Gamma(\mu n+\nu+1)} = {}_0\Psi_2 \left[\begin{array}{c} \overline{\qquad\qquad\qquad}\,; \\ (1,1),(\nu+1,\mu); \end{array} -z \right]. \tag{1.66}$$

II. In recent years, many different forms of kinetic equations of fractional order have been widely investigated, especially in the modeling and analysis of a number of important problems of physics and astrophysics (see, for details, [39]). Particularly, in the past decade or so, kinetic equations of fractional order have apparently gained popularity, mainly because of the discovery of their relation with the theory of CTRW (Continuous-Time Random Walks) (see [9]). These equations have been and are being investigated with the aim to first determine and then interpret certain physical phenomena which are known to govern processes such as diffusion in porous media, reaction and relaxation in complex systems, anomalous diffusion, and so on (see also [6,58,100]).

Theorems 1.2–1.4 below, each of which was established in Ref. [103] (see also [104]), are sufficiently general key results, which are capable of being appropriately and suitably specialized with a view to including solutions of the corresponding (known or new) fractional-order kinetic equations associated with a large variety of simpler functions than those involved herein.

Theorem 1.2. *Let* $c,\mu,\nu,\rho,\sigma \in \mathbb{R}^+$. *Suppose also that the general function* $\mathscr{E}_{\alpha,\beta}(\varphi;z,s,\kappa)$, *defined by Eq.* (1.34), *exists. Then the solution of the following generalized fractional kinetic equation*:

$$N(t) - N_0\, t^{\mu-1}\, \mathscr{E}_{\alpha,\beta}(\varphi; zt^{\nu}, s, \kappa) = -c^{\rho}\left({}^{\mathrm{RL}}I_{0+}^{\sigma} N\right)(t) \tag{1.67}$$

is given by

$$N(t) = N_0\, t^{\mu-1} \sum_{r=0}^{\infty} (-c^{\rho} t^{\sigma})^r \cdot \sum_{j=0}^{\infty} \frac{\varphi(j)\,\Gamma(\nu j+\mu)}{(j+\kappa)^s\,\Gamma(\alpha j+\beta)\,\Gamma(\nu j+\sigma r+\mu)} (zt^{\nu})^j \qquad (t>0), \tag{1.68}$$

provided that the right-hand side of the solution asserted by Eq. (1.68) *exists.*

Theorem 1.3. *Let* $c,\mu,\nu,\rho,\sigma \in \mathbb{R}^+$. *Suppose also that the general function* $\mathfrak{E}_{\alpha,\beta}(\phi;z)$, *defined by Eq.* (1.31), *exists. Then the solution of the following generalized fractional kinetic equation*:

$$N(t) - N_0\, t^{\mu-1}\, \mathfrak{E}_{\alpha,\beta}(\phi; zt^{\nu}) = -c^{\rho}\left({}^{\mathrm{RL}}I_{0+}^{\sigma} N\right)(t) \tag{1.69}$$

is given by

$$N(t) = N_0 t^{\mu-1} \sum_{r=0}^{\infty} \left(-c^{\rho} t^{\sigma}\right)^r \cdot \sum_{j=0}^{\infty} \frac{\phi(j)\,\Gamma(\nu j+\mu)}{\Gamma(\alpha j+\beta)\,\Gamma(\nu j+\sigma r+\mu)} \left(zt^{\nu}\right)^j \qquad (t>0), \tag{1.70}$$

provided that the right-hand side of the solution asserted by Eq. (1.70) *exists.*

Theorem 1.4. *For* $c,\mu,\nu,\rho,\sigma \in \mathbb{R}^+$, *let the extended Hurwitz-Lerch Zeta function*:

$$\Phi_{\lambda_1,\ldots,\lambda_p;\mu_1,\ldots,\mu_q}^{(\rho_1,\ldots,\rho_p;\sigma_1,\ldots,\sigma_q)}(z,s,\kappa),$$

defined by Eq. (1.18), *exist. Then the solution of the following generalized fractional kinetic equation*:

$$N(t) - N_0\, t^{\mu-1}\, \Phi_{\lambda_1,\ldots,\lambda_p;\mu_1,\ldots,\mu_q}^{(\rho_1,\ldots,\rho_p;\sigma_1,\ldots,\sigma_q)}\left(zt^{\nu},s,\kappa\right) = -c^{\rho}\left({}^{\mathrm{RL}}I_{0+}^{\sigma} N\right)(t) \tag{1.71}$$

is given by

$$N(t) = N_0 t^{\mu-1} \sum_{r=0}^{\infty} \left(-c^{\rho} t^{\sigma}\right)^r \frac{\Gamma(\mu)}{\Gamma(\sigma r+\mu)} \cdot \Phi_{\mu,\lambda_1,\ldots,\lambda_p;\sigma r+\mu,\mu_1,\ldots,\mu_q}^{(\nu,\rho_1,\ldots,\rho_p;\nu,\sigma_1,\ldots,\sigma_q)}\left(zt^{\nu},s,\kappa\right) \qquad (t>0), \tag{1.72}$$

provided that the right-hand side of the solution asserted by Eq. (1.72) *exists.*

Remarkably, the distinct advantage of using the general function $\mathscr{E}_{\alpha,\beta}(\varphi;z,s,\kappa)$, defined by Eq. (1.34), in the non-homogeneous term of the fractional-order kinetic equation (1.67) lies in its generality so that solutions of other kinetic equations involving relatively simpler non-homogeneous terms can be derived by appropriately specializing the solution (1.68) asserted by Theorem 1.2. Similar remarks would apply equally strongly to the results (1.70) and (1.72), which are provided by Theorems 1.3 and 1.4, respectively.

III. The unified fractional derivative operator ${}^{\mathrm{H}}D_{0+}^{\alpha,\beta}$ of order α $(0<\alpha<1)$ and type β $(0 \leqq \beta \leqq 1)$, defined by equation (1.42), was considered by Hilfer [6] in order to derive the solution of the following general fractional differential equation:

$$\left({}^{\mathrm{H}}D_{0+}^{\alpha,\beta} f\right)(x) = \lambda f(x) \qquad (x>0) \tag{1.73}$$

under the initial condition given, in terms of the corresponding *two-parameter* fractional integral operator ${}^{\mathrm{H}}I_{0+}^{\alpha,\beta}$, by

$$\left({}^{\mathrm{H}}I_{0+}^{(1-\beta)(1-\alpha)}f\right)(0+) = c_0, \tag{1.74}$$

where it is tacitly assumed that

$$\left({}^{\mathrm{H}}I_{0+}^{(1-\beta)(1-\alpha)}f\right)(0+) := \lim_{x\to 0+}\left\{\left({}^{\mathrm{H}}I_{0+}^{(1-\beta)(1-\alpha)}f\right)(x)\right\},$$

c_0 is a given constant and the parameter λ is the eigenvalue. Hilfer's solution of the above initial-value problem is given by (see [6, p. 115, Eq. (124)]):

$$f(x) = c_0\, x^{(1-\beta)(\alpha-1)}\, E_{\alpha,\alpha+\beta(1-\alpha)}\left(\lambda x^{\alpha}\right), \tag{1.75}$$

where $E_{\alpha,\beta}(z)$ denotes the two-parameter Mittag-Leffler function defined by Eq. (1.30).

If we put $\beta = 0$ and $c_0 = 1$ in Hilfer's solution (1.74), we can deduce the corrected version of the claimed solution (see [152, p. 802, Eq. (3.1)]; see also [96]) of the following initial-value problem:

$$\left({}^{\mathrm{RL}}D_{0+}^{\alpha}f\right)(x) = \lambda f(x) \qquad (x > 0), \tag{1.76}$$

under the initial condition given by

$$\left({}^{\mathrm{RL}}I_{0+}^{1-\alpha}f\right)(0+) = 1, \tag{1.77}$$

where, as also in equation (1.74),

$$\left({}^{\mathrm{RL}}I_{0+}^{1-\alpha}f\right)(0+) :- \lim_{x\to 0+}\left\{\left({}^{\mathrm{RL}}I_{0+}^{1-\alpha}f\right)(x)\right\}$$

in the form given by

$$f(x) = x^{\alpha-1}\, E_{\alpha,\alpha}\left(\lambda x^{\alpha}\right), \tag{1.78}$$

in terms of the two-parameter Mittag-Leffler function defined by Eq. (1.30).

In concluding this section, we remark that, in recent years, various real-world problems and issues in many areas of mathematical, physical, and engineering sciences have been modeled and analyzed by appealing to several powerful tools, one of which involves applications of the operators of fractional calculus (see, for example, a recent development [138] involving applications of the operators of fractional calculus to model and analyze some cobweb economic models). Notably, a number of important and potentially useful definitions have been introduced and used for fractional-order derivatives. These include, among others, the fractional-order derivative operators which stem from the Riemann-Liouville, the Grünwald-Letnikov, the Liouville-Caputo, the Caputo-Fabrizio and the Atangana-Baleanu fractional-order derivatives (see, for example, [10,16,19,57,75,160]).

The Riemann-Liouville fractional derivative is known to involve the convolution of a given function and a power-law kernel (see, for details, [57,75]). On the other hand, the Liouville-Caputo (LC) fractional derivative involves the convolution of the local derivative of a given function with a power-law function [17]. The fractional-order derivatives proposed by Caputo and Fabrizio [16] and Atangana and Baleanu [10] are based on the exponential decay law which is a generalized power-law function. On the other hand, the Caputo-Fabrizio (CFC) fractional-order derivative as well as the Atangana-Baleanu (ABC) fractional-order derivative allow us to describe complex physical problems that follow, at the same time, the power law and the exponential decay law. Such items of information as those presented in many of these works are believed to have the potential to generate further developments on fractional-order modeling and analysis of interesting applied problems. Many experiments and theories have shown that a fairly large number of abnormal phenomena, which occur in the engineering and applied sciences, can be well described by using discrete fractional calculus (see, for example, [30]; see also the recent work [36]). In particular, fractional difference equations have been found to provide powerful tools in the modeling and analysis of various phenomena in many different fields of science and engineering including those in mathematical physics, fluid mechanics, heat conduction, and so on.

1.6 CONCLUDING REMARKS AND OBSERVATIONS

The origin of many members of the remarkably vast family of higher transcendental functions, which are popularly known as special functions or mathematical functions, can be traced back to such widespread areas as (for example) mathematical physics, analytic number theory and applied mathematical sciences. Here, in this chapter, we have presented a brief introductory overview and survey of some of the recent developments in the theory of several extensively studied families of higher transcendental functions (or, more popularly, special functions or mathematical functions) and their potential applications in (for example) mathematical physics, analytic number theory and applied mathematical sciences. For further reading and researching by those who are interested in pursuing this subject, we have chosen to provide references to various useful monographs and textbooks on the theory and applications of higher transcendental functions. We have also considered several operators of fractional calculus, which are associated with higher transcendental functions, briefly indicating some of their applications as well.

The interested reader can find in (for example) [102,103,105] a considerably large number of recently published journal articles which have dealt with the extensively investigated subject of fractional calculus and its widespread applications. In fact, judging by the ongoing contributions to the theory and applications of fractional calculus, which are continually appearing in some of the leading journals of mathematical, physical, statistical and engineering sciences, the importance of the subject matter dealt with in this review article cannot be over-emphasized. For readers, who are interested in researching these topics, we have chosen to include citations of some of the recent developments on fractional calculus and its widespread applications.

There is a considerable literature investigating and applying the quantum or basic (or $\mathfrak{q}$-) calculus not only in the area of higher transcendental functions (see, for example, [91] in which the $\mathfrak{q}$-extension of the aforementioned general family of Srivastava-Daoust multivariable hypergeometric functions [19] was introduced and studied systematically; see also [29,126]), which we have presented in Section 1.2, and in Geometric Function Theory of Complex Analysis (see, for a detailed historical and introductory overview, the recently published survey-cum-expository review article [99]), but also in the modeling and analysis of applied problems as well as in extending the well-established theory and applications of various rather classical mathematical functions and mathematical inequalities.

The so-called $(\mathfrak{p},\mathfrak{q})$-variation of the widely investigated $\mathfrak{q}$-analysis was contained in a 1991 Letter to the Editor by Chakrabarti and Jagannathan [20] in which the already known development of $\mathfrak{q}$-deformed oscillators was more-or-less trivially translated to what they called the $(\mathfrak{p},\mathfrak{q})$-deformed oscillators with an extra (albeit inconsequential) parameter $\mathfrak{p}$. Apparently, in addition to [20], many works on attempts to involve one extra redundant (or superfluous) parameter of the type $\mathfrak{p}$ as in (for example) [52,72,78] have been misconstrued by a large number of mostly amateurish-type researches to believe that they can "generalize" the known $\mathfrak{q}$-theory and the known $\mathfrak{q}$-results by trivially and inconsequentially following the same route as indicated above.

Historically and traditionally, the classical $\mathfrak{q}$-number $[n]_\mathfrak{q}$ is given (for $0 < \mathfrak{q} < 1$) by

$$[n]_\mathfrak{q} := \begin{cases} \dfrac{1-\mathfrak{q}^n}{1-\mathfrak{q}} & (n \in \mathbb{N} := \{1,2,3,\ldots\}) \\[2ex] 0 & (n=0), \end{cases} \tag{1.79}$$

so that, for the classical $\mathfrak{q}$-number $[n]_\mathfrak{q}$, we have

$$[n]_\mathfrak{q} := \frac{1-\mathfrak{q}^n}{1-\mathfrak{q}} = 1+\mathfrak{q}+\mathfrak{q}^2+\ldots+\mathfrak{q}^{n-1} \qquad (n \in \mathbb{N}). \tag{1.80}$$

Furthermore, the so-called $\mathfrak{q}$-derivative or the so-called $\mathfrak{q}$-difference of a suitably constrained function $f(z)$ is denoted by $(D_\mathfrak{q} f)(z)$ and defined, in a given subset of $\mathbb{C}$, by

$$(D_\mathfrak{q} f)(z) := \begin{cases} \dfrac{f(z)-f(\mathfrak{q}z)}{(1-\mathfrak{q})z} & (z \in \mathbb{C}\setminus\{0\};\ 0<\mathfrak{q}<1) \\[2ex] f'(0) & (z=0;\ 0<\mathfrak{q}<1). \end{cases} \tag{1.81}$$

On the other hand, the so-called $(\mathfrak{p},\mathfrak{q})$-number $[n]_{\mathfrak{p},\mathfrak{q}}$ is given (for $0<\mathfrak{q}<\mathfrak{p}\leqq 1$) by

$$\begin{aligned} [n]_{\mathfrak{p},\mathfrak{q}} &:= \begin{cases} \dfrac{\mathfrak{p}^n-\mathfrak{q}^n}{\mathfrak{p}-\mathfrak{q}} & (n \in \mathbb{N} := \{1,2,3,\ldots\}) \\[2ex] 0 & (n=0) \end{cases} \\ &=: \mathfrak{p}^{n-1}\,[n]_{\frac{\mathfrak{q}}{\mathfrak{p}}}, \end{aligned} \tag{1.82}$$

where, for the classical q-number $[n]_{\mathfrak{q}}$, we have

$$[n]_{\mathfrak{q}} := \frac{1-\mathfrak{q}^n}{1-\mathfrak{q}} = \mathfrak{p}^{1-n}\left(\frac{\mathfrak{p}^n-(\mathfrak{p}\mathfrak{q})^n}{\mathfrak{p}-(\mathfrak{p}\mathfrak{q})}\right) = \mathfrak{p}^{1-n}\,[n]_{\mathfrak{p},\mathfrak{p}\mathfrak{q}}. \tag{1.83}$$

Furthermore, the so-called $(\mathfrak{p},\mathfrak{q})$-derivative or the so-called $(\mathfrak{p},\mathfrak{q})$-difference of a suitably constrained function $f(z)$ is denoted by $(D_{\mathfrak{p},\mathfrak{q}}\,f)(z)$ and defined, in a given subset of $\mathbb{C}$, by

$$(D_{\mathfrak{p},\mathfrak{q}}\,f)(z) = \begin{cases} \dfrac{f(\mathfrak{p}z)-f(\mathfrak{q}z)}{(\mathfrak{p}-\mathfrak{q})z} & (z\in\mathbb{C}\setminus\{0\};\ 0<\mathfrak{q}<\mathfrak{p}\leqq 1) \\[2ex] f'(0) & (z=0;\ 0<\mathfrak{q}<\mathfrak{p}\leqq 1). \end{cases} \tag{1.84}$$

Thus, clearly, we have the following simple connection of the so-called $(\mathfrak{p},\mathfrak{q})$-derivative or the $(\mathfrak{p},\mathfrak{q})$-difference $(D_{\mathfrak{q}}f)(z)$ with the familiar and widely investigated q-derivative or the q-difference $(D_{\mathfrak{q}}f)(z)$ given by the definition (1.81):

$$(D_{\mathfrak{p},\mathfrak{q}}\,f)(z) = \left(D_{\frac{\mathfrak{q}}{\mathfrak{p}}}\,f\right)(\mathfrak{p}z) \qquad \text{and} \qquad (D_{\mathfrak{q}}\,f)(z) = (D_{\mathfrak{p},\mathfrak{p}\mathfrak{q}}\,f)\left(\frac{z}{\mathfrak{p}}\right) \tag{1.85}$$

$$(z\in\mathbb{C};\ 0<\mathfrak{q}<\mathfrak{p}\leqq 1).$$

These last equations (1.82)–(1.85) show rather clearly that, in most cases, the already considered and readily accessible q-analogues for $0<\mathfrak{q}<1$ can easily (and possibly trivially) be translated into the corresponding $(\mathfrak{p},\mathfrak{q})$-analogues (with $0<\mathfrak{q}<\mathfrak{p}\leqq 1$) by applying some straightforward parametric and argument variations of the types indicated above, because the additional parameter $\mathfrak{p}$ is redundant or superfluous.

It is regretful, therefore, to see that a large number of mostly amateurish-type researchers on these and other related topics continue to produce and publish obvious and inconsequential variations and straightforward translations of the known q-results in terms of the so-called $(\mathfrak{p},\mathfrak{q})$-calculus by unnecessarily and inconsequentially forcing-in an obviously superfluous (or redundant) parameter $\mathfrak{p}$ into the classical q-calculus and thereby falsely claiming "generalization" (see [99, p. 340] and [102, Section 5, pp. 1511–1512]). Since, obviously, there is hardly any advantage or any innovation in straightforward translations of known q-theory and known q-results in terms of the corresponding $(\mathfrak{p},\mathfrak{q})$-theory and the corresponding $(\mathfrak{p},\mathfrak{q})$-results, such tendencies to produce and flood the literature with trivialities and inconsequential variations of known q-theory and known q-results should be discouraged by all means.

CONFLICTS OF INTEREST

The author declares that there are no conflicts of interest.

Bibliography

1. M. Abramowitz and I. A. Stegun (Editors), *Handbook of Mathematical Functions with Formulas, Graphs, and Mathematical Tables*, Applied Mathematics Series **55**, National Bureau of Standards, Washington, D.C. (1964); Reprinted by Dover Publications, New York (1965) (see also [73]).
2. P. Agarwal, P. Harjule and R. Jain, A general Volterra-type integral equation associated with an integral operator involving the product of general class of polynomials and multivariable H-function in the kernel, *Proc. Jangjeon Math. Soc.* **20** (2017), 391–396.
3. P. Agarwal, S. V. Rogosin, E. T. Karimov and M. Chand, Generalized fractional integral operators and the multivariable H-function, *J. Inequal. Appl.* **2015** (2015), 1–17.
4. B. Ahmad, J. Henderson and R. Luca, *Boundary Value Problems for Fractional Differential Equations and Systems*, Trends in Abstract and Applied Analysis, Vol. **9**, World Scientific Publishing Company, Singapore, New Jersey, London and Hong Kong (2021).
5. G. A. Anastassiou, *Generalized Fractional Calculus: New Advancements and Applications*, Studies in Systems, Decision and Control, Vol. **305**, Springer, Cham (2021).
6. L. C. Andrews, *Special Functions for Engineers and Applied Mathematicians*, Macmillan Company, New York (1984).
7. G. E. Andrews, R. Askey and R. Roy, *Special Functions*, Encyclopedia of Mathematics and Its Applications, Vol. **71**, Cambridge University Press, Cambridge, London and New York (1999).
8. T. M. Apostol, *Introduction to Analytic Number Theory*, Springer-Verlag, New York, Heidelberg and Berlin (1976).
9. P. Appell and J. Kampé de Fériet, *Fonctions Hypergéométriques et Hypersphériques; Polynβmes d'Hermite*, Gauthier-Villars, Paris (1926).
10. A. Atangana and D. Baleanu, New fractional derivatives with nonlocal and non-singular kernel: Theory and application to heat transfer model, *Therm. Sci.* **20** (2016), 763–769.
11. A. Badík and M. Feckan, Applying fractional calculus to analyze final consumption and gross investment influence on GDP, *J. Appl. Math. Statist. Inform.* **17** (1) (2021), 65–72.
12. W. N. Bailey, *Generalized Hypergeometric Series*, Cambridge Tracts in Mathematics and Mathematical Physics, Vol. **32**, Cambridge University Press, Cambridge, London and New York (1935); Reprinted by Stechert-Hafner Service Agency, New York and London (1964).
13. E. W. Barnes, The asymptotic expansion of integral functions defined by Taylor's series, *Philos. Trans. Roy. Soc. London Ser. A Math. Phys. Sci.* **206** (1906), 249–297.
14. R. G. Buschman and H. M. Srivastava, The $\overline{H}$ function associated with a certain class of Feynman integrals, *J. Phys. A: Math. Gen.* **23** (1990), 4707–4710.
15. M. Caputo, *Elasticitáe Dissipazionne*, Zanichelli, Bologna (1969).
16. M. Caputo and M. Fabrizio, A new definition of fractional derivative without singular kernel, *Progr. Fract. Differ. Appl.* **1** (2015), 73–85.
17. M. Caputo and F. Mainardi, A new dissipation model based on memory mechanism, *Pure Appl. Geophys.* **91** (1971), 134–147.
18. B. C. Carlson, *Special Functions of Applied Mathematics*, Academic Press, New York, San Francisco and London (1977).
19. C. Cattani, H. M. Srivastava and X.-J. Yang (Editors), *Fractional Dynamics*, Emerging Science Publishers (De Gruyter Open), Berlin and Warsaw (2015).
20. R. Chakrabarti and R. Jagannathan, A (p,q)-oscillator realization of two-parameter quantum algebras, *J. Phys. A Math. Gen.* **24** (1991), L711–L718.

21. V. B. L. Chaurasia, Expansion theorem for the multivariable H-function, *Bull. Math. Soc. Sci. Math. Roumanie* (Nouvelle Sér) **36** (84) (1992), 109–112.
22. J. Choi, J. Daiya, D. Kumar and R. K. Saxena, Fractional differentiation of the product of Appell function F_3 and multivariable H-functions, *Commun. Korean Math. Soc.* **31** (2016), 115–129.
23. J. Daiya, J. Ram and D. Kumar, The multivariable H-function and the general class of Srivastava polynomials involving the generalized Mellin-Barnes contour integrals, *Filomat* **30** (2016), 1457–1464.
24. P. Debnath, H. M. Srivastava, P. Kumam and B. Hazarika (Editors), *Fixed Point Theory and Fractional Calculus: Recent Advances and Applications*, Forum for Interdisciplinary Mathematics, Springer Nature Singapore Private Limited, Singapore (2022).
25. L. de Branges, A proof of the Bieberbach conjecture, *Acta Math.* **154** (1985), 137–152.
26. A. Erdélyi, W. Magnus, F. Oberhettinger and F. G. Tricomi, *Higher Transcendental Functions*, Vols. **I**, **II** and **III**, McGraw-Hill Book Company, New York, Toronto and London, (1953), (1953) and (1955).
27. A. Fernandez, D. Baleanu and H. M. Srivastava, Series representations for fractional-calculus operators involving generalised Mittag-Leffler functions, *Commun. Nonlinear Sci. Numer. Simulat.* **67** (2019), 517–527; see also Corrigendum, *Commun. Nonlinear Sci. Numer. Simulat.* **82** (2020), 1–1.
28. R. Garrappa, A. Giusti and F. Mainardi, Variable-order fractional calculus: A change of perspective, *Commun. Nonlinear Sci. Numer. Simul.* **102** (2021), 1–16.
29. G. Gasper and M. Rahman, *Basic Hypergeometric Series* (with a Foreword by Richard Askey), Encyclopedia of Mathematics and Its Applications, Vol. **35**, Cambridge University Press, Cambridge, New York, Port Chester, Melbourne and Sydney (1990); Second edition, Encyclopedia of Mathematics and Its Applications, Vol. **96**, Cambridge University Press, Cambridge, London and New York (2004).
30. C. Goodrich and A. C. Peterson, *Discrete Fractional Calculus*, Springer, Berlin (2015).
31. R. Gorenflo, F. Mainardi and H. M. Srivastava, Special functions in fractional relaxation-oscillation and fractional diffusion-wave phenomena, in *Proceedings of the Eighth International Colloquium on Differential Equations* (Plovdiv, Bulgaria; August 18–23, 1997) (D. Bainov, Editor), pp. 195–202, VSP Publishers, Utrecht and Tokyo (1998).
32. S. Gaboury and R. Tremblay, An expansion theorem involving H-function of several complex variables, *Internat. J. Anal.* **2013** (2013), 1–7.
33. S. P. Goyal and R. Goyal, A general theorem for the generalized Weyl fractional integral operator involving the multivariable H-function, *Taiwanese J. Math.* **8** (2004), 559–568.
34. I. S. Gradshteyn and I. M. Ryzhik, *Tables of Integrals, Series, and Products* (Corrected and Enlarged edition prepared by A. Jeffrey), Academic Press, New York (1980); Sixth edition (2000).
35. E. Guariglia, Fractional calculus, Zeta functions and Shannon entropy, *Open Math.* **19** (2021), 87–100.
36. J. L. G. Guirao, P. O. Mohammed, H. M. Srivastava, D. Baleanu and M. S. Abualrub, Relationships between the discrete Riemann-Liouville and Liouville-Caputo fractional differences and their associated convexity results, *AIMS Math.* **7** (2022), 18127–18141.
37. A. R. Hadhoud, H. M. Srivastava and A. A. M. Rageh, Non-polynomial B-spline and shifted Jacobi spectral collocation techniques to solve time-fractional nonlinear coupled Burgers' equations numerically, *Adv. Differ. Equ.* **2021** (2021), 1–28.
38. N. T. Hài, O. I. Marichev and H. M. Srivastava, A note on the convergence of certain families of multiple hypergeometric series, *J. Math. Anal. Appl.* **164** (1992), 104–115.

39. H. J. Haubold and A. M. Mathai, The fractional kinetic equation and thermonuclear functions, *Astrophys. Space Sci.* **273** (2000), 53–63.
40. H. J. Haubold, A. M. Mathai and R. K. Saxena, Mittag-Leffler functions and their applications, *J. Appl. Math.* **2011** (2011), 1–51.
41. R. Hilfer (Editor), *Applications of Fractional Calculus in Physics*, World Scientific Publishing Company, Singapore, New Jersey, London and Hong Kong (2000).
42. R. Hilfer, Fractional time evolution, in *Applications of Fractional Calculus in Physics* (R. Hilfer, Editor), pp. 87–130, World Scientific Publishing Company, Singapore, New Jersey, London and Hong Kong (2000).
43. R. Hilfer, Experimental evidence for fractional time evolution in glass forming materials, *J. Chem. Phys.* **284** (2002), 399–408.
44. R. Hilfer, Threefold introduction to fractional derivatives, in *Anomalous Transport: Foundations and Applications* (R. Klages, G. Radons and I. M. Sokolov, Editors), pp. 17–73, Wiley-VCH Verlag, Weinheim (2008).
45. R. Hilfer and L. Anton, Fractional master equations and fractal time random walks, *Phys. Rev. E* **51** (1995), R848–R851.
46. R. Hilfer, Y. Luchko and Ž. Tomovski, Operational method for solution of the fractional differential equations with the generalized Riemann-Liouville fractional derivatives, *Fract. Calc. Appl. Anal.* **12** (2009), 299–318.
47. R. Hilfer and H. J. Seybold, Computation of the generalized Mittag-Leffler function and its inverse in the complex plane, *Integral Transforms Spec. Funct.* **17** (2006), 637–652.
48. A. A. Inayat-Hussain, New properties of hypergeometric series derivable from Feynman integrals. I: Transformation and reduction formulae, *J. Phys. A Math. Gen.* **20** (1987), 4109–4118.
49. A. A. Inayat-Hussain, New properties of hypergeometric series derivable from Feynman integrals. II: A generalisation of the H-function, *J. Phys. A Math. Gen.* **20** (1987), 4119–4128.
50. M. Izadi and H. M. Srivastava, A novel matrix technique for multi-order pantograph differential equations of fractional order, *Proc. Roy. Soc. London Ser. A Math. Phys. Engrg. Sci.* **477** (2021), 1–21.
51. M. Izadi and H. M. Srivastava, Numerical approximations to the nonlinear fractional-order logistic population model with fractional-order Bessel and Legendre bases, *Chaos Solitons Fractals* **145** (2021), 1–11.
52. R. Jagannathan and J. Van der Jeugt, Finite dimensional representations of the quantum group $\mathrm{GL}_{p,q}(2)$ using the exponential map from $\mathrm{U}_{p,q}\big(\mathrm{gl}(2)\big)$, *J. Phys. A Math. Gen.* **28** (1995), 2819–2831.
53. C. M. Joshi and N. L. Joshi, Successive derivatives and finite expansions involving the H-function of one and more variables, *Ann. Polon. Math.* **67** (1997), 15–29.
54. A. A. Kilbas and M. Saigo, *H-Transforms: Theory and Applications*, Analytical Methods and Special Functions: An International Series of Monographs in Mathematics, Vol. **9**, Chapman and Hall (A CRC Press Company), Boca Raton, FL, London and New York (2004).
55. A. A. Kilbas, M. Saigo and R. K. Saxena, Generalized Mittag-Leffler function and generalized fractional calculus operators, *Integral Transforms Spec. Funct.* **15** (2004), 31–49.
56. A. A. Kilbas, M. Saigo and R. K. Saxena, Solution of Volterra integro-differential equations with generalized Mittag-Leffler functions in the kernels, *J. Integral Equations Appl.* **14** (2002), 377–396.

57. A. A. Kilbas, H. M. Srivastava and J. J. Trujillo, *Theory and Applications of Fractional Differential Equations*, North-Holland Mathematical Studies, Vol. **204**, Elsevier (North-Holland) Science Publishers, Amsterdam, London and New York (2006).
58. D. Kumar, J. Choi and H. M. Srivastava, Solution of a general family of kinetic equations associated with the Mittag-Leffler function, *Nonlinear Funct. Anal. Appl.* **23** (2018), 455–471.
59. D. Kumar, S. D. Purohit and J. Choi, Generalized fractional integrals involving product of multivariable H-function and a general class of polynomials, *J. Nonlinear Sci. Appl.* **9** (2016), 8–21.
60. S. Kumar, R. K. Pandey, H. M. Srivastava and G. N. Singh, A convergent collocation approach for generalized fractional integro-differential equations using Jacobi polyfractonomials, *Mathematics* **9** (2021), 1–17.
61. N. N. Lebedev, *Special Functions and Their Applications*, Translated from the Russian by R. A. Silverman, Prentice-Hall, Englewood Cliffs, New Jersey (1965).
62. J. Liouville, Mémoire sur quelques quéstions de géometrie et de mécanique, et sur un nouveau genre de calcul pour résoudre ces quéstions, *J. École Polytech.* **13** (21) (1832), 1–69.
63. Y. L. Luke, *The Special Functions and Their Approximations*, Vols. **I** and **II**, Mathematics in Science and Engineering, Vols. **53-I** and **53-II**, A Series of Monographs and Textbooks, Academic Press, New York and London (1969).
64. Y. L. Luke, *Mathematical Functions and Their Approximations*, Academic Press, New York, San Francisco and London (1975).
65. W. Magnus, F. Oberhettinger and R. P. Soni, *Formulas and Theorems for the Special Functions of Mathematical Physics*, Third Enlarged Edition, Die Grundlehren der Mathematischen Wissenschaften in Einzeldarstellungen mit besonderer Berücksichtingung der Anwendungsgebiete, Bd. **52**, Springer-Verlag, Berlin, Heidelberg and New York (1966).
66. F. Mainardi, Why the Mittag-Leffler function can be considered the queen function of the fractional calculus? *Entropy* **22** (2020), 1–29.
67. F. Mainardi and R. Gorenflo, Time-fractional derivatives in relaxation processes: A tutorial survey, *Fract. Calc. Appl. Anal.* **10** (2007), 269–308.
68. A. M. Mathai, R. K. Saxena and H. J. Haubold, *The H-Function: Theory and Applications*, Springer, New York, Dordrecht, Heidelberg and London (2010).
69. W. Miller Jr., *Lie Theory and Special Functions*, Mathematics in Science and Engineering, Vol. **43**, A Series of Monographs and Textbooks, Academic Press, London and New York (1968).
70. G. M. Mittag-Leffler, Sur la nouvelle fonction $E_\alpha(x)$, *C. R. Acad. Sci. Paris* **137** (1903), 554–558.
71. P. Mokhtary, F. Ghoreishi and H. M. Srivastava, The Müntz-Legendre Tau method for fractional differential equations, *Appl. Math. Model.* **40** (2016), 671–684.
72. M. Nishizawa, $\mathrm{U}_{r,s}(gl4)$-symmetry for (r,s)-hypergeometric series, *J. Comput. Appl. Math.* **160** (2003), 233–239.
73. F. W. J. Olver, D. W. Lozier, R. F. Boisvert and C. W. Clark (Editors), *NIST Handbook of Mathematical Functions* [With 1 CD-ROM (Windows, Macintosh and UNIX)], U. S. Department of Commerce, National Institute of Standards and Technology, Washington, D. C. (2010); Cambridge University Press, Cambridge, London and New York (2010).
74. R. Panda, On a multiple integral involving the H-function of several variables, *Indian J. Math.* **19** (1977), 157–162.

75. I. Podlubny, *Fractional Differential Equations: An Introduction to Fractional Derivatives, Fractional Differential Equations, to Methods of Their Solution and Some of Their Applications*, Mathematics in Science and Engineering, Vol. **198**, Academic Press, New York, London, Sydney, Tokyo and Toronto (1999).
76. R. K. Raina, On generalized Wright's hypergeometric functions and fractional calculus operators, *East Asian Math. J.* **21** (2005), 191–203.
77. E. D. Rainville, *Special Functions*, Macmillan Company, New York, 1960; Reprinted by Chelsea Publishing Company, Bronx, New York (1971).
78. V. Sahai and S. Yadav, Representations of two-parameter quantum algebras and (p,q)-special functions, *J. Math. Anal. Appl.* **335** (2007), 268–279.
79. S. G. Samko, A. A. Kilbas and O. I. Marichev, *Fractional Integrals and Derivatives: Theory and Applications*, Gordon and Breach Science Publishers, Yverdon, Switzerland (1993).
80. T. Sandev and Ž. Tomovski, General time fractional wave equation for a vibrating string, *J. Phys. A Math. Theoret.* **43** (2010), 1–12.
81. R. K. Saxena and S. L. Kalla, Solution of Volterra-type integro-differential equations with a generalized Lauricella confluent hypergeometric function in the kernels, *Internat. J. Math. Math. Sci.* **2005** (2005), 1155–1170.
82. J. B. Seaborn, *Hypergeometric Functions and Their Applications*, Springer-Verlag, Berlin, Heidelberg and New York (1991).
83. H. J. Seybold and R. Hilfer, Numerical results for the generalized Mittag-Leffler function, *Fract. Calc. Appl. Anal.* **8** (2005), 127–139.
84. E. Shishkina and S. Sitnik, *Transmutations, Singular and Fractional Differential Equations with Applications to Mathematical Physics*, Mathematics in Science and Engineering, Academic Press (Elsevier Science Publishers), New York, London and Toronto (2020).
85. H. Singh, H. M. Srivastava and D. Baleanu (Editors), *Methods of Mathematical Modeling: Infectious Desease*, Academic Press (Elsevier Science BV), San Diego, London, Cambridge and Oxford (2022).
86. H. Singh, H. M. Srivastava and J. J. Nieto (Editors), *Handbook of Fractional Calculus for Engineering and Science*, Series on Advances in Applied Mathematics, Chapman and Hall/CRC Press (Taylor & Francis), Baton Roca, FL (2022).
87. H. Singh, H. M. Srivastava, Z. Hammouch and K. S. Nisar, Numerical simulation and stability analysis for the fractional-order dynamics of COVID-19, *Results Phys.* **20** (2021), 1–8.
88. L. J. Slater, *Generalized Hypergeometric Functions*, Cambridge University Press, Cambridge, London and New York (1966).
89. H. M. Srivastava, On an extension of the Mittag-Leffler function, *Yokohama Math. J.* **16** (1968), 77–88.
90. H. M. Srivastava, Charles Fox, *Bull. London Math. Soc.* **12** (1980), 67–70.
91. H. M. Srivastava, Certain q-polynomial expansions for functions of several variables I and II, *IMA J. Appl. Math.* **30** (1983), 315–323; *ibid.* **33** (1984), 205–209.
92. H. M. Srivastava, Some orthogonal polynomials representing the energy spectral functions for a family of isotropic turbulence fields, *Zeitschr. Angew. Math. Mech.* **64** (1984), 255–257.
93. H. M. Srivastava, Some formulas for the Bernoulli and Euler polynomials at rational arguments, *Math. Proc. Cambridge Philos. Soc.* **129** (2000), 77–84.

94. H. M. Srivastava, A new family of the λ-generalized Hurwitz-Lerch Zeta functions with applications, *Appl. Math. Inform. Sci.* **8** (2014), 1485–1500.
95. H. M. Srivastava, Some families of Mittag-Leffler type functions and associated operators of fractional calculus, *TWMS J. Pure Appl. Math.* **7** (2016), 123–145.
96. H. M. Srivastava, Remarks on some fractional-order differential equations, *Integral Transforms Spec. Funct.* **28** (2017), 560–564.
97. H. M. Srivastava, The Zeta and related functions: Recent developments, *J. Adv. Engrg. Comput.* **3** (2019), 329–354.
98. H. M. Srivastava, Some general families of the Hurwitz-Lerch Zeta functions and their applications: Recent developments and directions for further researches, *Proc. Inst. Math. Mech. Nat. Acad. Sci. Azerbaijan* **45** (2019), 234–269.
99. H. M. Srivastava, Operators of basic (or q-) calculus and fractional q-calculus and their applications in geometric function theory of complex analysis, *Iran. J. Sci. Technol. Trans. A Sci.* **44** (2020), 327–344.
100. H. M. Srivastava, Fractional-order derivatives and integrals: Introductory overview and recent developments, *Kyungpook Math. J.* **60** (2020), 73–116.
101. H. M. Srivastava, Diabetes and its resulting complications: Mathematical modeling via fractional calculus, *Public Health Open Access* **4** (3) (2020), 1–5.
102. H. M. Srivastava, Some parametric and argument variations of the operators of fractional calculus and related special functions and integral transformations, *J. Nonlinear Convex Anal.* **22** (2021), 1501–1520.
103. H. M. Srivastava, An introductory overview of fractional-calculus operators based upon the Fox-Wright and related higher transcendental functions, *J. Adv. Engrg. Comput.* **5** (2021), 135–166.
104. H. M. Srivastava, A survey of some recent developments on higher transcendental functions of analytic number theory and applied mathematics, *Symmetry* **13** (2021), 1–22.
105. H. M. Srivastava, Some general families of integral transformations and related results, *Appl. Math. Comput. Sci.* **6** (2022), 27–41.
106. H. M. Srivastava, Some double integrals stemming from the Boltzmann equation in the kinetic theory of gasses, *Eur. J. Pure Appl. Math.* **15** (2022), 810–820.
107. H. M. Srivastava, E. S. A. AbuJarad, F. Jarad, G. Srivastava and M. H. A. AbuJarad, The Marichev-Saigo-Maeda fractional-calculus operators involving the (p,q)-extended Bessel and Bessel-Wright functions, *Fractal Fract.* **5** (2021), 1–15.
108. H. M. Srivastava, A.-K. N. Alomari, K. M. Saad and W. M. Hamanah, Some dynamical models involving fractional-order derivatives with the Mittag-Leffler type kernels and their applications based upon the Legendre spectral collocation method, *Fractal Fract.* **5** (2021), 1–13.
109. H. M. Srivastava, M. K. Bansal and P. Harjule, A study of fractional integral operators involving a certain generalized multi-index Mittag-Leffler function, *Math. Methods Appl. Sci.* **41** (2018), 6108–6121.
110. H. M. Srivastava, M. K. Bansal and P. Harjule, A class of fractional integral operators involving a certain general multi-index Mittag-Leffler function, *Ukrainian Math. J.* (**2020**) (In Press).
111. H. M. Srivastava, S. V. Bedre, S. M. Khairnar and B. S. Desale, Krasnosel'skii type hybrid fixed point theorems and their applications to fractional integral equations, *Abstr. Appl. Anal.* **2014** (2014), 1–9; see also Corrigendum, *Abstr. Appl. Anal.* **2015** (2015), 1–2.

112. H. M. Srivastava, R. C. S. Chandel and H. Kumar, Some general Hurwitz-Lerch type Zeta functions associated with the Srivastava-Daoust multiple hypergeometric functions, *J. Nonlinear Var. Anal.* **6** (2022), 299–315.
113. H. M. Srivastava and J. Choi, *Series Associated with the Zeta and Related Functions*, Kluwer Acedemic Publishers, Dordrecht, Boston and London (2001).
114. H. M. Srivastava and J. Choi, *Zeta and q-Zeta Functions and Associated Series and Integrals*, Elsevier Science Publishers, Amsterdam, London and New York (2012).
115. H. M. Srivastava and J. V. da Costa Sousa, Multiplicity of solutions for fractional-order differential equations via the $\kappa(x)$-Laplacian operator and the Genus theory, *Fractal Fract.* **6** (2022), 1–27.
116. H. M. Srivastava and M. C. Daoust, Certain generalized Neumann expansions associated with the Kampé de Fériet function, *Nederl. Akad. Wetensch. Proc. Ser. A 72 = Indag. Math.* **31** (1969), 449–457.
117. H. M. Srivastava and M. C. Daoust, A note on the convergence of Kampé de Fériet's double hypergeometric series, *Math. Nachr.* **53** (1972), 151–159.
118. H. M. Srivastava, S. Deniz and K. M. Saad, An efficient semi-analytical method for solving the generalized regularized long wave equations with a new fractional derivative operator, *J. King Saud Univ. Sci.* **33** (2021), 1–7.
119. H. M. Srivastava, A. Fernandez and D. Baleanu, Some new fractional-calculus connections between Mittag-Leffler functions, *Mathematics* **7** (2019), 1–10.
120. H. M. Srivastava and M. Garg, Some integrals involving a general class of polynomials and the multivariable H-function, *Rev. Roumaine Phys.* **32** (1987), 685–692.
121. H. M. Srivastava and S. P. Goyal, Fractional derivatives of the H-function of several variables, *J. Math. Anal. Appl.* **112** (1985), 641–652.
122. H. M. Srivastava, S. P. Goyal and R. K. Agrawal, Some multiple integral relations for the H-function of several variables, *Bull. Inst. Math. Acad. Sinica* **9** (1981), 261–277.
123. H. M. Srivastava, K. C. Gupta and S. P. Goyal, *The H-Functions of One and Two Variables with Applications*, South Asian Publishers, New Delhi and Madras (1982).
124. H. M. Srivastava, P. Harjule and R. Jain, A general fractional differential equation associated with an integral operator with the H-function in the kernel, *Russian J. Math. Phys.* **22** (2015), 112–126.
125. H. M. Srivastava, R. Jan, A. Jan, W. Deebani and M. Shutaywi, Fractional-calculus analysis of the transmission dynamics of the dengue infection, *Chaos* 31 (2021), 1–18.
126. H. M. Srivastava and P. W. Karlsson, *Multiple Gaussian Hypergeometric Series*, Halsted Press (Ellis Horwood Limited, Chichester), John Wiley & Sons, New York, Chichester, Brisbane and Toronto (1985).
127. H. M. Srivastava and B. R. K. Kashyap, *Special Functions in Queuing Theory and Related Stochastic Processes*, Academic Press, New York and London (1982).
128. H. M. Srivastava, S.-D. Lin and P.-Y. Wang, Some fractional-calculus results for the $\overline{H}$-function associated with a class of Feynman integrals, *Russian J. Math. Phys.* **13** (2006), 94–100.
129. H. M. Srivastava and H. L. Manocha, *A Treatise on Generating Functions*, Halsted Press (Ellis Horwood Limited, Chichester), John Wiley and Sons, New York, Chichester, Brisbane and Toronto (1984).
130. H. M. Srivastava and P. O. Mohammed, A correlation between solutions of uncertain fractional forward difference equations and their paths, *Frontiers Phys.* **8** (2020), 1–10.
131. H. M. Srivastava, P. O. Mohammed, C. S. Ryoo and Y. S. Hamed, Existence and uniqueness of a class of uncertain Liouville-Caputo fractional difference equations, *J. King Saud Univ. Sci.* **33** (2021), 1–7.

132. H. M. Srivastava and R. Panda, An integral representation for the product of two Jacobi polynomials, *J. London Math. Soc.* (*Ser.* 2) **12** (1976), 419–425.
133. H. M. Srivastava and R. Panda, Some bilateral generating functions for a class of generalized hypergeometric polynomials, *J. Reine Angew. Math.* **283/284** (1976), 265–274.
134. H. M. Srivastava and R. Panda, Expansion theorems for the H function of several complex variables, *J. Reine Angew. Math.* **288** (1976), 129–145.
135. H. M. Srivastava and R. Panda, Some expansion theorems and generating relations for the H function of several complex variables. I and II, *Comment. Math. Univ. St. Paul.* **24** (2) (1975), 119–137; *ibid.* **25** (2) (1976), 167–197.
136. H. M. Srivastava and R. Panda, Certain multidimensional integral transformations. I and II, *Nederl. Akad. Wetensch. Proc. Ser. A 81* = *Indag. Math.* **40** (1978), 118–131; *ibid.* **40** (1978), 132–144.
137. H. M. Srivastava and R. Panda, Some multiple integral transformations involving the H-function of several variables, *Nederl. Akad. Wetensch. Proc. Ser. A 82* = *Indag. Math.* **41** (1979), 353–362.
138. H. M. Srivastava, D. Raghavan and S. Nagarajan, A comparative study of the stability of some fractional-order cobweb economic models, *Rev. Real Acad. Cienc. Exactas Fís. Natur. Ser. A Mat.* (*RACSAM*) **116** (2022), 1–20.
139. H. M. Srivastava and R. K. Saxena, Operators of fractional integration and applications, *Appl. Math. Comput.* **118** (2001), 1–52.
140. H. M. Srivastava and R. K. Saxena, Some Volterra-type fractional integro-differential equations with a multivariable confluent hypergeometric function as their kernel, *J. Integral Equations Appl.* **17** (2005), 199–217.
141. H. M. Srivastava, R. K. Saxena and R. K. Parmar, Some families of the incomplete H-functions and the incomplete $\bar{H}$-functions and associated integral transforms and operators of fractional calculus with applications, *Russ. J. Math. Phys.* **25** (2018), 116–138.
142. H. M. Srivastava, R. K. Saxena, T. K. Pogány and R. Saxena, Integral and computational representations of the extended Hurwitz-Lerch Zeta function, *Integral Transforms Spec. Funct.* **22** (2011), 487–506.
143. H. M. Srivastava and N. P. Singh, The integration of certain products of the multivariable H-function with a general class of polynomials, *Rend. Circ. Mat. Palermo* (Ser. 2) **32** (1983), 157–187.
144. H. M. Srivastava and Ž. Tomovski, Fractional calculus with an integral operator containing a generalized Mittag-Leffler function in the kernel, *Appl. Math. Comput.* **211** (2009), 198–210.
145. R. Srivastava, Definite integrals associated with the H-function of several variables, *Comment. Math. Univ. St. Paul.* **30** (2) (1981), 125–129.
146. R. Srivastava, A simplified overview of certain relations among infinite series that arose in the context of fractional calculus, *J. Math. Anal. Appl.* **162** (1991), 152–158.
147. D. L. Suther and P. Agarwal, Generalized Mittag-Leffler function and the multivariable H-function involving the generalized Mellin-Barnes contour integrals, *Commun. Numer. Anal.* **2017** (2017), 25–33.
148. D. L. Suthar, B. Debalkie and M. Andualem, Modified Saigo fractional integral operators involving multivariable H-function and general class of multivariable polynomials, *Adv. Differ. Equ.* **2019** (2019), 1–11.
149. G. Szegö, *Orthogonal Polynomials*, Fourth edition, Amererican Mathematical Society Colloquium Publications, Vol. **23**, American Mathematical Society, Providence, Rhode Island (1975).

150. N. M. Temme, *Special Functions: An Introduction to Classical Functions of Mathematical Physics*, A Wiley-Interscience Publication, John Wiley & Sons, New York, Chichester, Brisbane and Toronto (1996).
151. E. C. Titchmarsh, *The Theory of the Riemann Zeta-Function*, Clarendon (Oxford University) Press, Oxford and London (1951); Second Edition (Revised by D. R. Heath-Brown) (1986).
152. Ž. Tomovski, R. Hilfer and H. M. Srivastava, Fractional and operational calculus with generalized fractional derivative operators and Mittag-Leffler type functions, *Integral Transforms Spec. Funct.* **21** (2010), 797–814.
153. S. K. Verma, R. K. Vats, H. K. Nashine and H. M. Srivastava, Existence results for a fractional differential inclusion of arbitrary order with three-point boundary conditions, *Kragujevac J. Math.* **47** (2023), 935–945.
154. E. T. Whittaker and G. N. Watson, *A Course of Modern Analysis: An Introduction to the General Theory of Infinite Processes and of Analytic Functions; with an Account of the Principal Transcendental Functions*, Fourth Edition, Cambridge University Press, Cambridge, London and New York (1927).
155. A. Wiman, Über den Fundamentalsatz in der Theorie der Funcktionen $E_\alpha(x)$, *Acta Math.* **29** (1905), 191–201.
156. A. Wiman, Über die Nullstellen der Funktionen $E_\alpha(x)$, *Acta Math.* **29** (1905), 217–234.
157. E. M. Wright, The asymptotic expansion of integral functions defined by Taylor series. I, *Philos. Trans. Roy. Soc. London Ser. A Math. Phys. Sci.* **238** (1940), 423–451.
158. E. M. Wright, The asymptotic expansion of integral functions defined by Taylor series. II, *Philos. Trans. Roy. Soc. London Ser. A Math. Phys. Sci.* **239** (1941), 217–232.
159. E. M. Wright, The asymptotic expansion of integral functions and of the coefficients in their Taylor series, *Trans. Amer. Math. Soc.* **64** (1948), 409–438.
160. X.-J. Yang, D. Baleanu and H. M. Srivastava, *Local Fractional Integral Transforms and Their Applications*, Academic Press (Elsevier Science Publishers), Amsterdam, Heidelberg, London and New York (2016).

2 Analytical Solutions for the Fluid Model Described by Fractional Derivative Operators Using Special Functions in Fractional Calculus

Ndolane Sene
Cheikh Anta Diop University

CONTENTS

2.1 INTRODUCTION

The problems consisting of modeling of the fluid models through the Caputo derivative, Mittag-Leffler fractional derivative, through Caputo-Fabrizio derivative, and other operators grow much interest this last decade. Many papers in the literature can be cited for illustrations [1–3]. This interest is because fractional operators find many applications in a real-world problem, which can be enumerated as the applications in mathematical physics [4–7], the applications in modeling biological phenomena [8–12], the applications in modeling physics models [4,13,14], and many

DOI: 10.1201/9781003368069-2

other fields where differential equations can be used. Note that the fractional operators can be used in all phenomena that can be modeled using differential equations. In fractional calculus, the memory effect is the fundamental property that explains the importance of modeling with fractional operators. As used fractional operators in the literature can cite the following operators: Caputo-Fabrizio derivative [15] known as the derivative with the exponential kernel, the Atangana-Baleanu fractional operator known as the derivative with Mittag-Leffler kernel [16], this derivative finds much success in the literature due to its generalized form and many applications it can find in modeling epidemic models. Note that the Caputo fractional operator [17,18] is the most used operator in the literature because it is the first derivative correcting the inconvenience found with the Riemann-Liouville fractional derivative. Many other fractional operators and their generalizations exist in the literature. For more information related to fractional operators, see the following investigations [19–21].

We recall the literature review about the fluid models already investigated in the field of fractional calculus. In Ref. [3], Vieru et al. have proposed the first investigation on modeling fluid models using the fractional operators and have determined the exact analytical solutions of the considered fluid model using the application of the Laplace transform. The method utilized in this chapter is due to Vieru et al.'s work previously cited. In Ref. [1], Saqib et al. have proposed an investigation using the Laplace transform to establish the solution of the free convection flow of generalized Jeffrey fluid using a derivative with an exponential kernel namely the Caputo-Fabrizio derivative. In Ref. [22], the authors modeled fluid using the Caputo-Fabrio-derivative and the Atangana-Baleanu derivative; they proposed the analytical solutions using the Laplace transform. In this investigation, the authors make a comparative study between these two new derivatives recently introduced in fractional calculus. In Ref. [23], Shah et al. have proposed the solutions of the second-grade fluid model over an oscillating vertical plate by using the Caputo-Fabrizio derivative, and the solution of the proposed model has been addressed with the Laplace transformation method. In Ref. [24], Khan et al. have used the recent derivative with exponential and Mittag-Lellfer kernel to model the so-called heat and mass transfer of second-grade fluids over a vertical plate. In this work, the Laplace transform method is also used for the determination of the exact solutions, and the authors have particularly analyzed the similarities and the differences between modeling with Caputo-Fabrizio derivative and modeling with Atangana-Baleanu derivative. In modeling fluid model with Caputo-Fabrizio derivative and Atangana-Baleanu derivative, the problem consists to get the exact solutions, see in Refs. [25,26]. In Ref. [2], Imran et al. have proposed work on the determination of the exact solution of the fractional fluid model using the Laplace method. In Ref. [27], the author has proposed the solutions through the Fourier sine method of the class of fractional fluid model. In Ref. [28], Sene has proposed the analytical method to obtain the solution of convection flow of an incompressible viscous fluid under Newtonian heating and mass diffusion. The authors used in their investigations the Laplace transform including the Laplace transform of the Caputo derivative. The representations of the solutions have been made through special functions utilized in fractional calculus. In Ref. [29], Yavuz et al. have proposed the solution of the fractional fluid model and have analyzed the influence of the parameter of the model in physical terms. There used the classical method to arrive at their findings, which means the utilization of the Laplace transform method.

The importance of this chapter can be described in the next lines. The first part is to introduce the use of the special functions used in fractional calculus as the Mittag-Leffler functions, the exponential function, the Bessel function, the Wright function, and others to represent the exact analytical solution of the fluid model. In this chapter, the Caputo fractional operator has been used in the modeling of the fluid model. The second importance is the capture of the memory effect known to be contained in the fractional operators. We will also obtain the exact representations of the solutions of the classical fluid model by choosing the order of the fractional operator converging to 1. The third importance of this chapter is to study the sensitivity of the fractional fluid models to the initial and boundary conditions. In this part, we consider the initial condition switching between two conditions described later in this chapter. The fourth objective of this chapter will be to analyze the influence of the parameters on modeling the fluid model and to interpret the influence from the physical points of view. Note that for the determination of the exact solutions, as in the existing literature the Laplace transform method will be our used tool.

The structure of this chapter is as follows. In Section 2.2, we recall and discuss the special functions object of our investigation and the fractional operators used in this chapter. In Section 2.3, we present the fluid model under investigation and give the motivations for the use of this model. In Section 2.4, we use the Laplace transform method to determine the exact analytical solution of our fluid model under investigation. In this section, we express the solution using particular special functions. In Section 2.5, we open the discussion section and analyze the behaviors of the fluid model graphically. In Section 2.6, we provide the final remark and discussion about the findings of the present investigation.

2.2 FRACTIONAL CALCULUS OPERATORS AND SPECIAL FUNCTIONS

This section is the part where we recall some special functions necessary in our investigation. We also give their Laplace transforms as possible, which are fundamental to get the exact analytical solutions of the considered fluid model in this investigation. The functions considered in this chapter are the Mittag-Leffler function, the Wright function, the Gaussian error function, the Bessel-Wright function, and others. The part of this section will concern defining the fractional operator used in the present investigations. We mean the Caputo derivative, the Riemann-Liouville derivative, the Riemann-Liouville integral, the Atangana-Baleanu derivative, and the fractional operator with the exponential kernel.

We begin with the function introduced in this chapter used to express the exact analytical solution. The Mittag-Leffler function as reported in the literature is described in the following definition. Many definitions reported in this chapter can be found also in Refs. [30,31].

Definition 1. *[30,31] We symbolize the Mittag-Leffler function with two parameters by the function defined by the following representation*

$$E_{\alpha,\beta}(z)=\sum_{k=0}^{\infty}\frac{z^k}{\Gamma(\alpha k+\beta)}, \tag{2.1}$$

with the parameters satisfying the conditions that $\alpha>0$, $\beta\in\mathbb{R}$ *and* $z\in\mathbb{C}$.

We continue the generalization of the Mittag-Leffler function reported in the literature, see in Ref. [31] for more pieces of information. We define a possible generalization in the following definition. Note that there are many types of generalizations existing in the literature for the Mittag-Leffler function.

Definition 2. *[30] The Mittag-Leffler function with three parameters can be defined by the function represented by the following description*

$$E_{\alpha,\beta}^{(\kappa)}(z)=\sum_{n=0}^{\infty}\frac{z^n}{(\Gamma(\alpha n+\beta))^{\kappa}}, \tag{2.2}$$

with the parameters satisfying the condition that $\alpha,\beta,\kappa>0$, *and* $z\in\mathbb{C}$.

The first interesting remark from the Mittag-Leffler function is that when the order $\alpha=\beta=1$, we get the classical exponential function that is $E_{1,1}(z)=\exp(z)$. This function has also some interesting properties. The first one concerns the representation of some known functions using the Mittag-Leffler function. There exist many other relations between the Mittag-Leffler function and the exponential function. We cite one of them. We have the following relationship

$$E_{1,n}(z)=\frac{1}{z^{n-1}}\left(e^z-\sum_{k=0}^{n-2}\frac{z^k}{k!}\right), \tag{2.3}$$

the proof of the above relationship can be found in the literature and is not a new relation. We can also find a relation between the Mittag-Leffler function and the Gaussian error function, and then we have that

$$E_{1/2,1}(z)=e^{z^2}erfc(-z), \tag{2.4}$$

where the complementary of the Gaussian error function used in our investigation is defined by the following expression

$$erfc(-z)=\frac{2}{\sqrt{\pi}}\int_0^z e^{-t^2}dt. \tag{2.5}$$

Another property established in the literature is the relationship between the Mittag-Leffler function and the Bessel-Wright function introduced in the literature, see in Ref. [31], we have the following relationship

Definition 3. *[31] We symbolize the Bessel-Wright function with two parameters by the function defined by the following representation*

$$J_{\nu}^{\mu}(z)=\sum_{n=0}^{\infty}\frac{(-z)^{n}}{n!\Gamma(n\mu+\nu+1)}, \tag{2.6}$$

with the parameters satisfying the conditions $\mu>0$, $\nu\in\mathbb{R}^{+}$ *and* $z\in\mathbb{C}$.

The relationship between the Bessel-Wright function and the Mittag-Leffler function with three parameters is represented in the following form

$$J_{\nu}^{\mu}(z)=E_{(1,1),(\mu,\nu+1)}^{(2)}(z). \tag{2.7}$$

We continue by recalling the Wright function as reported in the literature, which can also be found in the recent Srivastava et al.'s investigation [31]. We have the following definition.

Definition 4. *[31] The generalized Wright function with two variables and a function* ϕ *is defined by the following form*

$$\varepsilon_{\alpha,\beta}(\phi,z)=\sum_{n=0}^{\infty}\frac{\phi(n)}{\Gamma(\alpha n+\beta)}z^{n}, \tag{2.8}$$

where $\alpha,\beta\in\mathbb{C}$ *and* $z\in\mathbb{C}$.

As it will be observed, this function will particularly play an important role in the investigation notably in expressing the form of the solutions got for our fluid model. According to the explanation in Ref. [31], Gerhold and Garra, and Polito have proposed a new Mittag-Lellfer function with three parameters represented in the following definition.

Definition 5. *[31] We defined a new Mittag-Leffler function with three parameters by the function defined as the following representation*

$$E_{\alpha,\beta}^{(\gamma)}(z)=\sum_{n=0}^{\infty}\frac{z^{n}}{[\Gamma(\alpha_{k}n+\beta_{k})]^{\gamma}}, \tag{2.9}$$

with the parameters satisfying the conditions that $\alpha,\beta,\gamma>0$, *and* $z\in\mathbb{C}$.

For information, the so-called Roy function [31] is obtained when the parameters satisfy the condition that $\alpha=\beta=1$, there is $R_{\gamma}(z)=E_{1,1}^{(\gamma)}(z)$. The section on special functions will be ended by recalling some relationships between the special function including the utilization of the Mittag-Leffler function, we have the following relations which proof can be found in the literature

$$E_{\alpha,\beta}^{1}(z)=E_{\alpha,\beta}(z), \qquad E_{2,2}^{1}(z)=\frac{\sinh\sqrt{z}}{z}, \tag{2.10}$$

$$E_{2,1}^{1}(z)=\cosh\sqrt{z}, \qquad E_{1,1}^{2}(z)=J_{0}(2\sqrt{z}), \tag{2.11}$$

$$E_{1,1}^{\alpha+1}(z)=e_{\alpha}(z), \qquad E_{1,1}^{\nu}(z)=Z(\lambda,z). \tag{2.12}$$

where $J_v(...) = J_v^1(...)$ is Wright-Bessel function of order v [31], $e_\alpha(...)$ is the α L-exponential function and $Z(.,.)$ is the Com-Poisson renormalization constant [31]. For the Mittag-Leffler function, the positivity of the constant parameters generates the convergence of the series representation used to describe this kind of function. The Laplace transforms of different types of Mittag-Leffler functions are summarized in the following representations, they are

$$\mathscr{L}\left\{t^{\beta-1}E_{\alpha,\beta}\left(\pm\lambda t^{\alpha}\right)\right\} = \frac{s^{\alpha-\beta}}{s^{\alpha}\pm\lambda}, \quad \mathscr{L}\left\{t^{n\alpha+\beta-1}E_{\alpha,\beta}^{(n)}\left(\pm\lambda t^{\alpha}\right)\right\} = \frac{n!s^{\alpha-\beta}}{\left(s^{\alpha}\pm\lambda\right)^{n+1}}.$$

We now continue by recalling the fractional operators used in this chapter. Before more investigation, we recall the definition of the Riemann-Liouville integral. We have the following.

Definition 6. *We suppose the function $v : [0,+\infty] \longrightarrow \mathbb{R}$, then we denote the Riemann-Liouville integral [17,18] of the considered function as the form*

$$\left(I^{\alpha}v\right)(t) = \frac{1}{\Gamma(\alpha)}\int_0^t (t-s)^{\alpha-1}v(s)ds, \tag{2.13}$$

where the function $\Gamma(...)$ represents the Gamma Euler function and the order α verifies the condition described by the form $\alpha > 0$.

The classical integral follows when the order of the Riemann-Liouville integral converges to 1, we recover that

$$\left(I^{1}v\right)(t) = \int_0^t v(s)ds. \tag{2.14}$$

This remark in Eq. (2.14) proves that the Riemann-Liouville integral generalizes the classical integral to non-integer order integration. The next definition concerns the derivative called the Riemann-Liouville derivative, which comes from the previous integral, we have the following definition.

Definition 7. *[17,18] We suppose the function $v : [0,+\infty] \longrightarrow \mathbb{R}$, then we denote the Riemann-Liouville fractional derivative of the considered function as the following description*

$$D^{\alpha}v(t) = \frac{1}{\Gamma(1-\alpha)}\frac{d}{dt}\int_0^t v(s)(t-s)^{-\alpha}ds, \tag{2.15}$$

where $t > 0$, and the order of the operator satisfies the condition that $\alpha \in (0,1)$ and the function $\Gamma(...)$ represents the Gamma Euler function.

We now define the Caputo derivative, and this derivative is motivated by the fact that with the Riemann-Liouville derivative, the derivative of a constant function does not give zero. This is a serious inconvenience in fractional calculus, and then the Caputo derivative was provided in the first part to correct this inconvenience. We have the following definition.

Definition 8. *[17,18] We suppose the function* $v : [0, +\infty] \longrightarrow \mathbb{R}$*, then we denote the Riemann-Liouville fractional derivative of the considered function as the following description*

$$D^{\alpha} v(t) = \frac{1}{\Gamma(1-\alpha)} \int_0^t (t-s)^{-\alpha} v'(s) ds, \tag{2.16}$$

where $t > 0$*, and the order of the operator satisfies the condition that* $\alpha \in (0,1)$ *and where the function* $\Gamma(...)$ *represents the Gamma Euler function.*

The Laplace transform will be used to determine the exact analytical solution using the special function recalled in this section as the Mittag-Leffler function, the Wright function, and the Gaussian error function. Thus, we recall for this work the Laplace transform of the Caputo derivative when the order is into (0,1), we have the following relationship

$$\mathscr{L}\{(D_c^{\alpha} v)(t)\} = s^{\alpha} \mathscr{L}\{v(t)\} - s^{\alpha-1} v(0). \tag{2.17}$$

2.3 FRACTIONAL MODEL UNDER CONSIDERATION

We present the fractional differential equations, which we will consider in the present investigation. In this section, we consider the fluid model known as the free convection flow of the generalized Jeffrey fluid type. The model is described by the following differential equations

$$D_t^{\alpha} v = \frac{1}{1+\kappa} \frac{\partial^2 u}{\partial x^2} + \frac{\beta}{1+\kappa} D^{\alpha} \left(\frac{\partial^2 v}{\partial x^2} \right) + Grw, \tag{2.18}$$

$$D_t^{\alpha} w = \frac{1}{Pr} \frac{\partial^2 w}{\partial x^2}. \tag{2.19}$$

For the velocity and the temperature distribution, the initial conditions are imposed to verify the expression represented in the following equations, which are

$$v(x,0) = 0, \tag{2.20}$$

$$w(x,0) = 0. \tag{2.21}$$

Furthermore, we set the velocity v and the temperature w to satisfy the boundary conditions represented in the following relationships

$$v(0,t) = 0, \tag{2.22}$$

$$w(0,t) = 1 \text{ or } t. \tag{2.23}$$

The novel findings of this chapter are to give the form of the exact solutions using some special functions such as the Mittag-Leffler function and the Gaussian error function, and other functions described in the previous section. The second novelty of this chapter is the initial and boundary conditions considered in Eqs. (2.20)–(2.23). Note that here the boundary condition for the temperature distribution switch between two values $w(0,t) = 1$ or $w(0,t) = t$. The influence of the parameter of the

Table 2.1
Parameters of the Model

Parameters	Descriptions
Pr	Prandtl Number
κ	The first Jeffrey fluid parameter
Gr	Thermal Grashof number
β	The second Jeffrey fluid parameter

model represented in Table 2.1 will be focused on and discussed in the present investigations. The physical meanings of the behaviors of the dynamics will be provided as possible in the present investigations. It is important to mention that the classical fluid model in this chapter is obtained when the order of the fractional operator α converges to 1. In other words, the model described in Eqs. (2.18) and (2.19) becomes the following form

$$\frac{\partial v}{\partial t} = \frac{1}{1+\kappa}\frac{\partial^2 u}{\partial x^2} + \frac{\beta}{1+\kappa}D^{\alpha}\left(\frac{\partial^2 v}{\partial x^2}\right) + Grw, \tag{2.24}$$

$$\frac{\partial w}{\partial t} = \frac{1}{Pr}\frac{\partial^2 w}{\partial x^2}. \tag{2.25}$$

It will be interesting to get the exact solution through the special function considered in this chapter. In the present investigation, this special model (2.24) and (2.25) will also be analyzed, and the influence of the parameters of the model will be analyzed and interpreted from physical viewpoints.

2.4 SOLUTIONS PROCEDURES

This section will be used to get the exact analytical solutions of the fluid model considered in this chapter. As mentioned previously, we use the Caputo derivative to model the free convection flow of generalized Jeffrey fluid. The particularity of this section will be the use of some of the special functions defined in Section 2.4.2 of the present investigations. We begin this investigation by determining the exact solution of the temperature distribution.

2.4.1 SOLUTIONS PROCEDURES WITH TEMPERATURE DISTRBUTION

In this part, we consider the fractional differential equation represented in Eq. (2.19), with the initial and boundary condition defined by the following relations

$$w(0,t) = 1, \tag{2.26}$$

$$w(x,0) = 0. \tag{2.27}$$

With the application of the Laplace transform according to the time, we get the calculations described in the following lines, they are

$$s^{\alpha}\bar{w} - s^{\alpha-1}w(x,0) = \frac{1}{Pr}\frac{\partial^2\bar{w}}{\partial x^2},$$
$$s^{\alpha}\bar{w} = \frac{1}{Pr}\frac{\partial^2\bar{w}}{\partial x^2},$$
$$\frac{\partial^2\bar{w}}{\partial x^2} - Prs^{\alpha}\bar{w} = 0. \tag{2.28}$$

The next in the procedure consists to apply the Laplace on the boundary condition given in our present investigation by the description that $w(0,s) = 1/s$. In the final remark, we have to solve the second-order differential equation defined by the following representation

$$\frac{\partial^2\bar{w}}{\partial x^2} - Prs^{\alpha}\bar{w} = 0. \tag{2.29}$$

under the boundary condition represented by the form in the equation that

$$w(0,s) = 1/s. \tag{2.30}$$

The solution of the second-order differential equation according to the classical resolution method is given with the utilization of the exponential function by the form presented as the form

$$\bar{w}(x,s) = \frac{\exp\left[-x\sqrt{Prs^{\alpha}}\right]}{s}. \tag{2.31}$$

According to the Laplace transform, we recognize that the Laplace transform includes the Laplace of the Wright function. Finally, the exact solution can be represented with the use of the Wright function, and then the exact solution of the fractional heat equation defined in Eq. (2.19) is presented as follows:

$$w(x,t) = \varepsilon_{\frac{-\alpha}{2},1}\left(\phi(n), -x\sqrt{Pr}t^{-\frac{\alpha}{2}}\right), \tag{2.32}$$

where $\phi(n) = \frac{1}{\Gamma(n+1)}$. The advantage of this form is that the implementation in MATLAB uses series representations and is useful. For much other software, we have codes to compute the Wright function. The present solution is an exact solution of Eq. (2.19), and this is the second advantage of the previous investigations.

The second part of this section should be to consider the special case provided in Eq. (2.19) under the initial and boundary conditions described in Eqs. (2.26) and (2.27). Here we will use another function to express the exact solution. The procedure of the method to get the exact solution does not change, now we apply the Laplace transform to Eq. (2.19) under $\alpha = 1$, we have

$$s\bar{w} - w(x,0) = \frac{1}{Pr}\frac{\partial^2 \bar{w}}{\partial x^2},$$
$$s\bar{w} = \frac{1}{Pr}\frac{\partial^2 \bar{w}}{\partial x^2},$$
$$\frac{\partial^2 \bar{w}}{\partial x^2} - Prs\bar{w} = 0. \tag{2.33}$$

Here also we have to solve the second-order differential equation represented by Eq. (2.29) with $\alpha = 1$ under the boundary condition defined by the form

$$w(0,s) = 1/s. \tag{2.34}$$

Adopting the method of the resolution applied to the second differential order equation, we get as a solution the equation defined by the form that

$$\bar{w}(x,s) = \frac{\exp\left[-x\sqrt{Prs}\right]}{s}. \tag{2.35}$$

Here the importance of the exact analytical solution is that the solution uses the special function known as the Gaussian error function, which can also be expressed in terms of the Mittag-Leffler function. The inverse of the Laplace transform of Eq. (2.35) can be obtained by the relationship

$$w(x,t) = erfc\left(\frac{x\sqrt{Pr}}{2\sqrt{t}}\right). \tag{2.36}$$

According to the scope of this chapter, when we use the transformation represented in Eq. (2.4), we can rewrite Eq. (2.36) including the Mittag-Leffler function, and then we get

$$w(x,t) = \frac{E_{1/2,1}\left(-\frac{x\sqrt{Pr}}{2\sqrt{t}}\right)}{\exp\left(\left(\frac{x\sqrt{Pr}}{2\sqrt{t}}\right)^2\right)}. \tag{2.37}$$

The use of Eq. (2.37) in MATLAB needs to know how to implement the Mittag-Leffler function, which is already established in the literature by Garrapa's works.

The second part of this chapter consists to determine the exact analytical solution with the initial and boundary condition defined by the following form

$$w(0,t) = t, \tag{2.38}$$
$$w(x,0) = 0. \tag{2.39}$$

The Laplace transform of Eq. (2.38) is given as the form $w(0,s) = 1/s^2$. And then in this part, we have to solve the fractional differential equation defined by the following equation

$$\frac{\partial^2 \bar{w}}{\partial x^2} - Prs^{\alpha}\bar{w} = 0. \tag{2.40}$$

under the boundary condition represented by the form that

$$w(0,s) = 1/s^2. \tag{2.41}$$

In the previous consideration, the exact analytical solution of the second-order differential equation (40) in terms of the Laplace transform is described in the following form that

$$\bar{w}(x,s) = \frac{\exp\left[-x\sqrt{Prs^{\alpha}}\right]}{s^2}. \tag{2.42}$$

The exact solution of the fluid model considered with initial and boundaries conditions (38)-(39) in this section can be represented using the Wright function defined in the previous sections, we have the following form for the solution

$$w(x,t) = t\varepsilon_{\frac{-\alpha}{2},2}\left(\phi(n), -x\sqrt{Pr}t^{-\frac{\alpha}{2}}\right). \tag{2.43}$$

Here also $\phi(n) = \frac{1}{\Gamma(n+1)}$. We can observe that Eq. (2.43) is expressed by using the Wright function. We can also express the exact analytical solution in the case of the classical model, we mean that $\alpha = 1$. In this case, the equation with the Laplace transform Eq. (2.42), its solution becomes that

$$\bar{w}(x,s) = \frac{\exp\left[-x\sqrt{Prs}\right]}{s^2}. \tag{2.44}$$

The inverse of the Laplace transform for the exact solution of the fractional equation Eq. (2.44) under initial and boundary conditions (2.38) and (2.39) can be expressed in the following form

$$w(x,t) = \left(\frac{x^2 Pr}{2} + t\right)\frac{E_{1/2,1}\left(-\frac{x\sqrt{Pr}}{2\sqrt{t}}\right)}{\exp\left(\left(\frac{x\sqrt{Pr}}{2\sqrt{t}}\right)^2\right)} - \frac{x\sqrt{Prt}}{2\sqrt{\pi}}E_{1,1}\left(-\frac{x^2 Pr}{4t}\right). \tag{2.45}$$

Here also we notice the importance of the special function, which has permitted us to express the analytical solution as well.

2.4.2 SOLUTIONS PROCEDURES WITH VELOCITY DISTRIBUTION

In this part, we establish the analytical solution of the fractional differential equation (2.18) using the special functions under investigation in this chapter. For this section, the considered initial and boundary condition is defined by the following form

$$u(x,0) = 0, \tag{2.46}$$

$$u(0,t) = 0. \tag{2.47}$$

We repeat the same procedure as adopted with the temperature distribution. We consider two cases because the temperature distribution impacts the velocity of the considered fluid. The first case is the case where we have that $w(0,t) = 1$. We apply the Laplace transform to Eq. (2.18), and then we obtain the relationships defined by the forms

$$s^{\alpha}\bar{u}-s^{\alpha-1}u(x,0)=\left(\frac{1}{1+\kappa}+\frac{\beta}{1+\kappa}s^{\alpha}\right)\frac{\partial^2\bar{u}}{\partial x^2}+Gr\bar{w},$$

$$s^{\alpha}\bar{u}=\left(\frac{1}{1+\kappa}+\frac{\beta}{1+\kappa}s^{\alpha}\right)\frac{\partial^2\bar{u}}{\partial x^2}+Gr\bar{w},$$

$$\frac{\partial^2\bar{u}}{\partial x^2}-\frac{(1+\kappa)s^{\alpha}}{1+\beta s^{\alpha}}\bar{u}=-\frac{Gr(1+\kappa)}{1+\beta s^{\alpha}}\frac{\exp\left[-x\sqrt{Prs^{\alpha}}\right]}{s}. \tag{2.48}$$

For simplification in the manipulations of the equation, let $\rho=1+\kappa$, we have to solve the second-order differential equation given by the following form

$$\frac{\partial^2\bar{u}}{\partial x^2}-\frac{\rho s^{\alpha}}{1+\beta s^{\alpha}}\bar{u}=-\frac{Gr\rho}{1+\beta s^{\alpha}}\frac{\exp\left[-x\sqrt{Prs^{\alpha}}\right]}{s}. \tag{2.49}$$

We solve the second-order differential equation, we need to obtain both the particular and the homogeneous solution. We get a solution of Eq. (2.49), defined by the following form

$$\bar{u}(x,s)=A\exp\left[-x\sqrt{\frac{\rho s^{\alpha}}{1+\beta s^{\alpha}}}\right]+B\exp\left[-x\sqrt{Prs^{\alpha}}\right], \tag{2.50}$$

Using the initial and the boundary condition described in Eqs. (2.46) and (2.47) with the utilization of Eq. (2.50), we get the following values $A=-B$ and explicitly, we have the expression given as

$$A=\frac{Gr\rho}{Pr\beta}\frac{s^{\alpha-(1+2\alpha)}}{s^{\alpha}+\frac{Pr-\rho}{Pr\beta}}. \tag{2.51}$$

For the inverse of the Laplace transform, we have the product of two Laplace transforms, and then we use integral transformation to obtain the exact analytical solutions. Let's consider the first form

$$\bar{u}_1(x,s)=B\exp\left[-x\sqrt{Prs^{\alpha}}\right], \tag{2.52}$$

$$=-\frac{Gr\rho}{Pr\beta}\frac{s^{\alpha-2\alpha}}{s^{\alpha}+\frac{Pr-\rho}{Pr\beta}}\frac{\exp\left[-x\sqrt{Prs^{\alpha}}\right]}{s}. \tag{2.53}$$

The inverse of the Laplace transform of Eq. (2.53) can be written as the following form, which is defined by

$$u(x,t)=\int_0^t a(x,s)b(t-s)d\tau, \tag{2.54}$$

In our context, the function a is obtained by inverting the second term of the Laplace transform given in Eq. (2.53), there is

$$a(x,t)=\varepsilon_{\frac{-\alpha}{2},1}\left(\phi(n),-x\sqrt{Pr}t^{-\frac{\alpha}{2}}\right), \tag{2.55}$$

and the second part will include the Mittag-Leffler function, we have the following inverse of Laplace transform from Eq. (2.53), there is

$$b(t) = -\frac{Gr\rho}{Pr\beta} E_{\alpha,2\alpha}\left(-\frac{Pr-\rho}{Pr\beta} t^{\alpha}\right). \tag{2.56}$$

Here, we notice the importance of the Mittag-Leffler function in expressing the exact analytical solutions of the considered fluid model. We continue the inversion process. We now consider inverting the following equation that is defined by the following form:

$$\bar{u}_2(x,s) = A\exp\left[-x\sqrt{\frac{\rho s^{\alpha}}{1+\beta s^{\alpha}}}\right], \tag{2.57}$$

$$= \frac{Gr\rho}{Pr\beta} \frac{s^{\alpha-(1+\alpha)}}{s^{\alpha}+\frac{Pr-\rho}{Pr\beta}} \frac{\exp\left[-x\sqrt{\frac{\rho s^{\alpha}}{1+\beta s^{\alpha}}}\right]}{s^{\alpha}}, \tag{2.58}$$

In the inverse of the Laplace transform, we repeat the same reasoning as in the previous section and then we get the following form

$$u(x,t) = \int_0^t c(x,s)d(t-s)d\tau. \tag{2.59}$$

Note that, in Eq. (2.58), we consider the Laplace transform form represented by the following function that

$$\bar{d}(s) = \frac{Gr\rho}{Pr\beta} \frac{s^{\alpha-(1+\alpha)}}{s^{\alpha}+\frac{Pr-\rho}{Pr\beta}}, \tag{2.60}$$

and then the inverse including the Mittag-Leffler function is characterized by the following form

$$d(t) = \frac{Gr\rho}{Pr\beta} E_{\alpha,1+\alpha}\left(-\frac{Pr-\rho}{Pr\beta} t^{\alpha}\right). \tag{2.61}$$

Furthermore, note that in Eq. (2.58) we consider the Laplace transform form represented by the following function that

$$\bar{c}(x,s) = \frac{\exp\left[-x\sqrt{\frac{\rho s^{\alpha}}{1+\beta s^{\alpha}}}\right]}{s^{\alpha}}. \tag{2.62}$$

We now proceed to the inverse of the Laplace transform with Eq. (2.62), it is defined by the following functions, we include the special function in the following form:

$$c(x,t) = \int_0^{\infty} k(x\sqrt{\rho},t)g(v,t)dv, \tag{2.63}$$

where the function k can be expressed using an exponential function, the Wright-Bessel function, and integral transformation, we have the following form

$$k(x\sqrt{\rho},t)=\frac{\rho e^{-\frac{t}{\beta}}}{\beta}\int_0^{\infty} erfc\left(\frac{x}{2\sqrt{x}}\right)e^{-\frac{\rho x}{m}}J_0\left(\frac{2}{\beta}\sqrt{\rho tx}\right)dx, \tag{2.64}$$

where $m=\frac{1-\alpha}{\alpha}$, J_0 is the Bessel function described in the preliminaries section and then the function g can be represented by the following form

$$g(v,t)=t^{-1}\varepsilon_{-\alpha,0}\left(\phi\left(n\right),-vt^{-\alpha}\right). \tag{2.65}$$

The method for the inverse of the Laplace transform described for Eq. (2.62) can be found in the classical literature. Here, we adapted the formula with our context. Finally, the exact solution is obtained by summing Eqs. (2.54) and (2.59). We get that

$$u(x,t)=\int_0^t a(x,s)b(t-s)d\tau,+\int_0^t c(x,s)d(t-s)d\tau. \tag{2.66}$$

We complete this section by considering the second initial and boundary conditions. For this part, the considered initial and boundary conditions are defined by the following form

$$v(x,0)=0, \tag{2.67}$$
$$v(0,t)=t. \tag{2.68}$$

We repeat the same procedure with a change in the Laplace transform, here we will use the form defined after the application of the Laplace transform on Eq. (2.68), it is defined by the equation $\bar{u}(0,s)=1/s^2$. We have the following relationships in the calculations

$$s^{\alpha}\bar{u}-s^{\alpha-1}u(x,0)=\left(\frac{1}{1+\kappa}+\frac{\beta}{1+\kappa}s^{\alpha}\right)\frac{\partial^2\bar{u}}{\partial x^2}+Gr\bar{w},$$

$$s^{\alpha}\bar{u}=\left(\frac{1}{1+\kappa}+\frac{\beta}{1+\kappa}s^{\alpha}\right)\frac{\partial^2\bar{u}}{\partial x^2}+Gr\bar{w},$$

$$\frac{\partial^2\bar{u}}{\partial x^2}-\frac{(1+\kappa)s^{\alpha}}{1+\beta s^{\alpha}}\bar{u}=-\frac{Gr(1+\kappa)}{1+\beta s^{\alpha}}\frac{\exp\left[-x\sqrt{Prs^{\alpha}}\right]}{s^2}. \tag{2.69}$$

Here also for the simplification of the calculations, we use a new variable, we suppose as in the previous part that $\rho=1+\kappa$, and then the second-order differential equation under investigation will be defined by the form

$$\frac{\partial^2\bar{u}}{\partial x^2}-\frac{\rho s^{\alpha}}{1+\beta s^{\alpha}}\bar{u}=-\frac{Gr\rho}{1+\beta s^{\alpha}}\frac{\exp\left[-x\sqrt{Prs^{\alpha}}\right]}{s^2}. \tag{2.70}$$

Solving Eq. (2.70) under initial and boundary conditions according to the classical method used to solve the second-order differential, we get the equation as the form that

$$\bar{u}(x,s) = A\exp\left[-x\sqrt{\frac{\rho s^{\alpha}}{1+\beta s^{\alpha}}}\right] + B\exp\left[-x\sqrt{Prs^{\alpha}}\right]. \tag{2.71}$$

The relation between A and B according to the initial and boundary conditions is given by $A = -B$, using the initial and boundary conditions, we get the following expression for the parameter A, there is

$$A = \frac{Gr\rho}{Pr\beta}\frac{s^{\alpha-(2+2\alpha)}}{s^{\alpha}+\frac{Pr-\rho}{Pr\beta}}. \tag{2.72}$$

Let's calculate the inverse of the Laplace transform by including the special function considered in this present investigation. Let the equation represented by the following form

$$\bar{u}_1(x,s) = B\exp\left[-x\sqrt{Prs^{\alpha}}\right], \tag{2.73}$$

$$= -\frac{Gr\rho}{Pr\beta}\frac{s^{\alpha-2\alpha}}{s^{\alpha}+\frac{Pr-\rho}{Pr\beta}}\frac{\exp\left[-x\sqrt{Prs^{\alpha}}\right]}{s^{2}}. \tag{2.74}$$

Using the convolution rule, the inverse of the Laplace transform of the function in Eq. (2.74) can be described as follows:

$$u(x,t) = \int_0^t a(x,s)b(t-s)d\tau. \tag{2.75}$$

In the present context, the function a is obtained by the inverse of the Laplace transform and it is given by the following form

$$a(x,t) = \varepsilon_{\frac{-\alpha}{2},2}\left(\phi(n), -x\sqrt{Pr}t^{-\frac{\alpha}{2}}\right), \tag{2.76}$$

and the second part will include the Mittag-Leffler function as in the previous sections, we have the following inverse of Laplace transform from the first part of Eq. (2.74) as the form

$$b(t) = -\frac{Gr\rho}{Pr\beta}E_{\alpha,2\alpha}\left(-\frac{Pr-\rho}{Pr\beta}t^{\alpha}\right). \tag{2.77}$$

We continue by searching the inverse of the Laplace transform in the first part of Eq. (2.71), in other words, we consider the function represented by the equation that

$$\bar{u}_2(x,s) = A\exp\left[-x\sqrt{\frac{\rho s^{\alpha}}{1+\beta s^{\alpha}}}\right], \tag{2.78}$$

$$= \frac{Gr\rho}{Pr\beta}\frac{s^{\alpha-(2+\alpha)}}{s^{\alpha}+\frac{Pr-\rho}{Pr\beta}}\frac{\exp\left[-x\sqrt{\frac{\rho s^{\alpha}}{1+\beta s^{\alpha}}}\right]}{s^{\alpha}}. \tag{2.79}$$

The method of inversion used in the previous part will also be used for the exact analytical solution, we have

$$u(x,t) = \int_0^t c(x,s)d(t-s)d\tau. \tag{2.80}$$

For simplification, we consider the first function to inverse defined by the function that

$$\bar{d}(s) = \frac{Gr\rho}{Pr\beta} \frac{s^{\alpha-(2+\alpha)}}{s^{\alpha} + \frac{Pr-\rho}{Pr\beta}}. \tag{2.81}$$

Note that its inverse includes one special function called the Mittag-Leffler function and is represented by the form that

$$d(t) = \frac{Gr\rho}{Pr\beta} E_{\alpha,2+\alpha}\left(-\frac{Pr-\rho}{Pr\beta}t^{\alpha}\right). \tag{2.82}$$

Furthermore, we end this part by calculating the inverse of the Laplace transform of the function defined in the next line. Note that in Eq. (2.79) we consider the Laplace transform form represented by the following function that

$$\bar{c}(x,s) = \frac{\exp\left[-x\sqrt{\frac{\rho s^{\alpha}}{1+\beta s^{\alpha}}}\right]}{s^{\alpha}}. \tag{2.83}$$

We repeat the previous parts, we now proceed to the inverse of the Laplace transform with Eq. (2.83), we include the special functions in our reasoning, and we have

$$c(x,t) = \int_0^{\infty} k(x\sqrt{\rho},t)g(v,t)dv, \tag{2.84}$$

where the function k can be expressed through exponential function, the Bessel function, and the integral transformation, we have the following form

$$k(x\sqrt{\rho},t) = \frac{\rho e^{-\frac{t}{\beta}}}{\beta} \int_0^{\infty} erfc\left(\frac{x}{2\sqrt{x}}\right) e^{-\frac{\rho x}{m}} J_0\left(\frac{2}{\beta}\sqrt{\rho t x}\right) dx, \tag{2.85}$$

where $m = \frac{1-\alpha}{\alpha}$, J_0 is the Bessel function described in the preliminaries section, and then the function g can be represented by the following form

$$g(v,t) = t^{-1}\varepsilon_{-\alpha,0}\left(\phi(n), -vt^{-\alpha}\right). \tag{2.86}$$

The method for the inverse of the Laplace transform described for Eq. (2.83) can be found in the literature. Here we adapted the formula with our context. Finally, the exact solution is obtained by summing Eqs. (2.75) and (2.80). We get that

$$u(x,t) = \int_0^t a(x,s)b(t-s)d\tau, + \int_0^t c(x,s)d(t-s)d\tau. \tag{2.87}$$

2.5 RESULTS AND DISCUSSION

In this section, the first objective will be to draw the graphics obtained for our present model under investigation. We begin the graphics with the concentration distribution of our considered model. In this first part, the influence of the Prandtl number will be analyzed, and the impact of the order of the fractional operator will also be analyzed. We begin with the case that the initial and boundary conditions are given by Eqs. (2.26) and (2.27). For the first graphics, we fix that $Pr = 20$, and the time $t = 8$, and the order of the Caputo operator will vary. Note that we use Eq. (2.32) in the representations. We have the following graphics for our models (2.18) and (2.19), Figures 2.1 and 2.2.

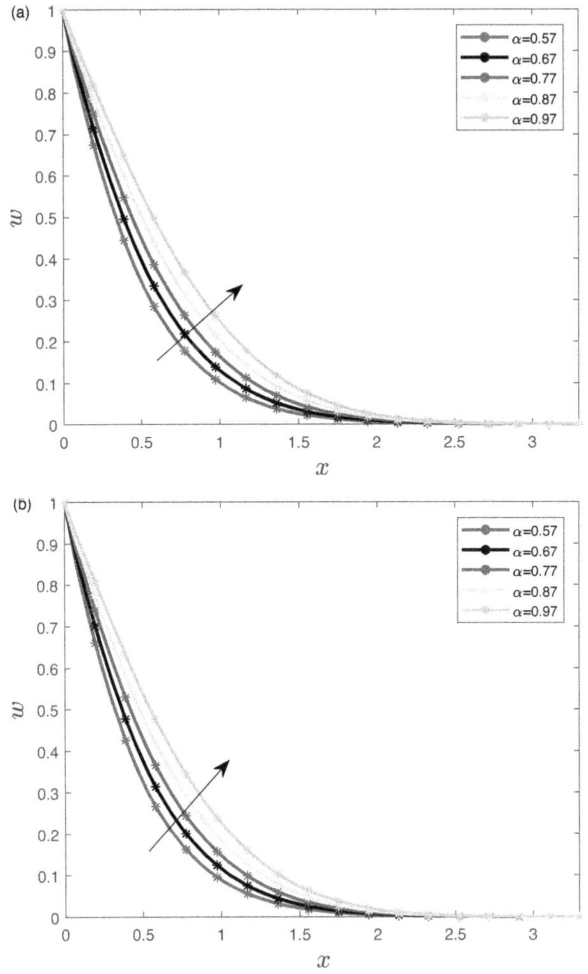

Figure 2.1 Temperature for different values of α with $Pr = 20$ (a) $Pr = 22$ (b).

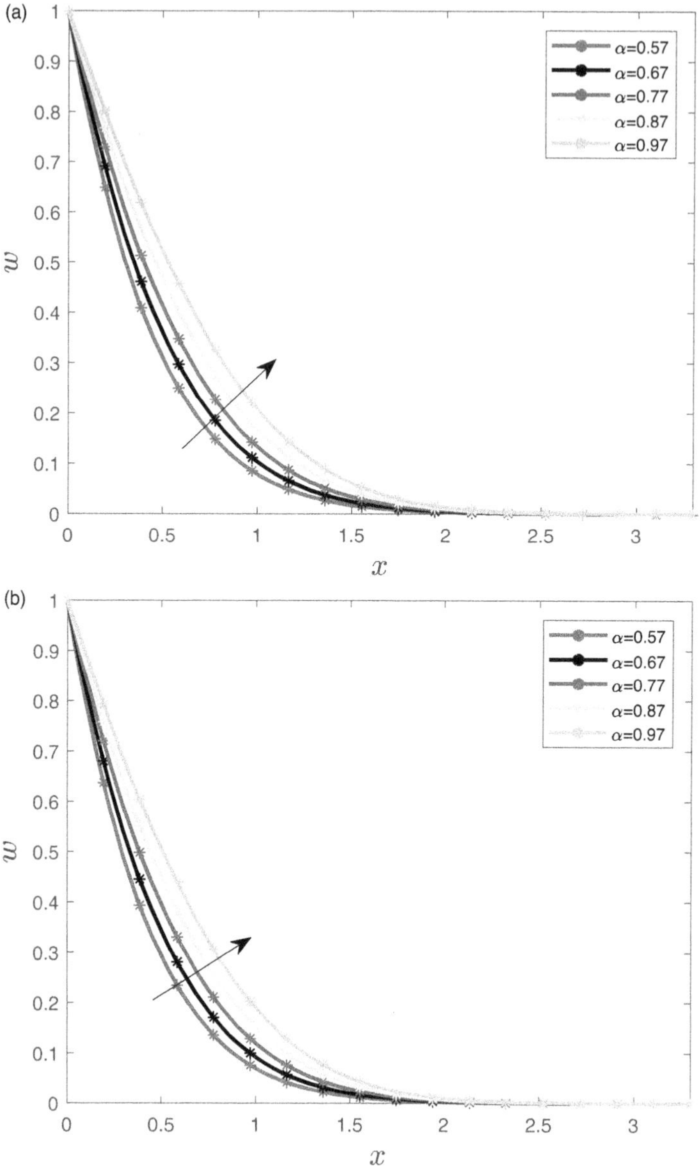

Figure 2.2 Temperature for different values of α with $Pr = 24$ (a) $Pr = 26$ (b).

We note that the increase in the order of the Caputo derivative implies the increase in the temperature of the considered fluid model (2.18) and (2.19). The impact is explained by the order of the fractional operator due to the accumulation of the memory impacts the diffusivity. More clearly, the increase of the order of the fractional operator increases the diffusivity, which in turn generates an increase in the temperature distribution as indicated by the arrows in Figures 2.1 and 2.2. By increasing the order the thermal diffusivity increases, and then we obtain an increase in the temperature distribution. We now consider the time $t = 8$, we consider the variation of the Prandtl number Pr. We have the following Figures 2.3 and 2.4.

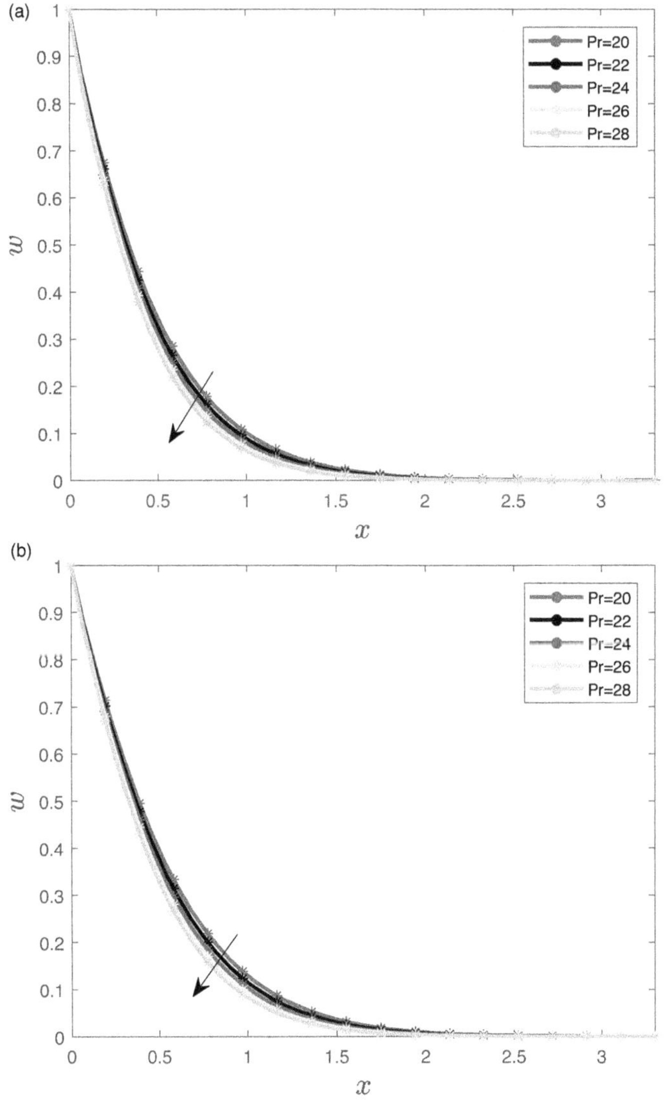

Figure 2.3 Temperature for different values of Prandtl number with $\alpha = 0.57$ (a) $\alpha = 0.67$(b).

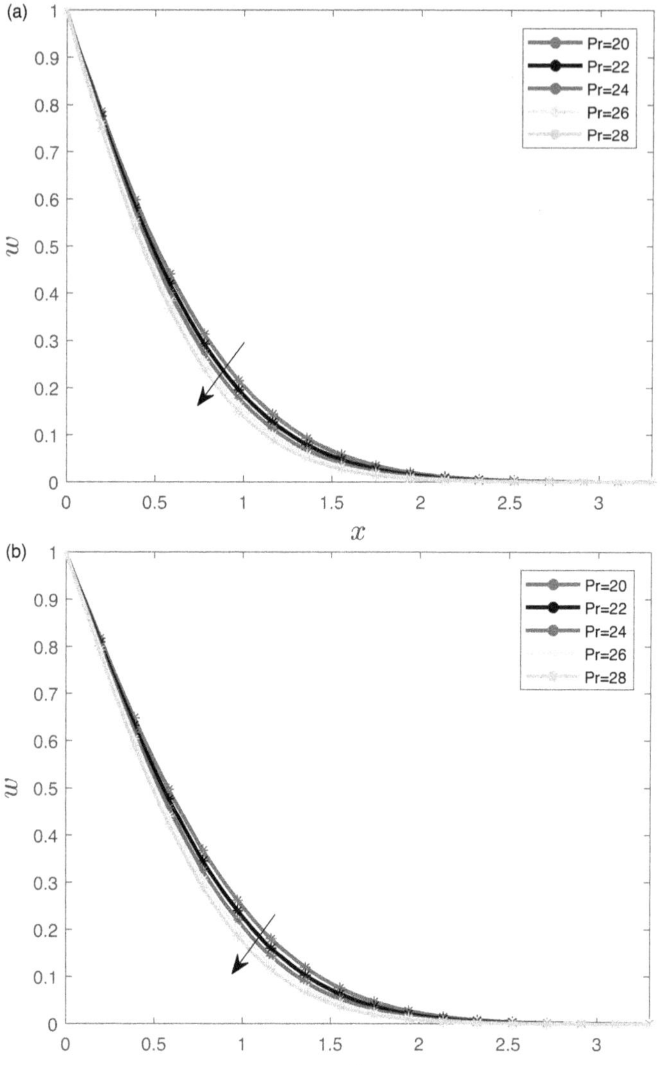

Figure 2.4 Temperature for different values of Prandtl number with $\alpha = 0.87$ (a) $\alpha = 0.97$(b).

We notice in all figures that the increase of the *Pr* number decreases the temperature distribution. This finding explains that the increase in the Prandtl number reduces the thermal boundary layer thickness, which in turn implies a decrease in the temperature distribution. Not here the role of the order of the fractional operator is to accelerate the diffusion process. There is that the fast decrease in the temperature distribution depends on the values of the considered fractional-order derivative.

We continue with the graphics of the temperature distribution. Here we consider the case of the initial and boundaries condition described by Eqs. (2.38) and (2.39). We begin by analyzing the impact of the order of the fractional operator on the dynamics of the temperature. We fix $Pr = 20$ and the time to $t = 8$. The graphics with different values of the fractional operator are described in Figures 2.5 and 2.6.

Carefully checking Figures 2.5 and 2.6, we observe that the increase of the order of the fractional operator generates high accumulation in the memories and affects

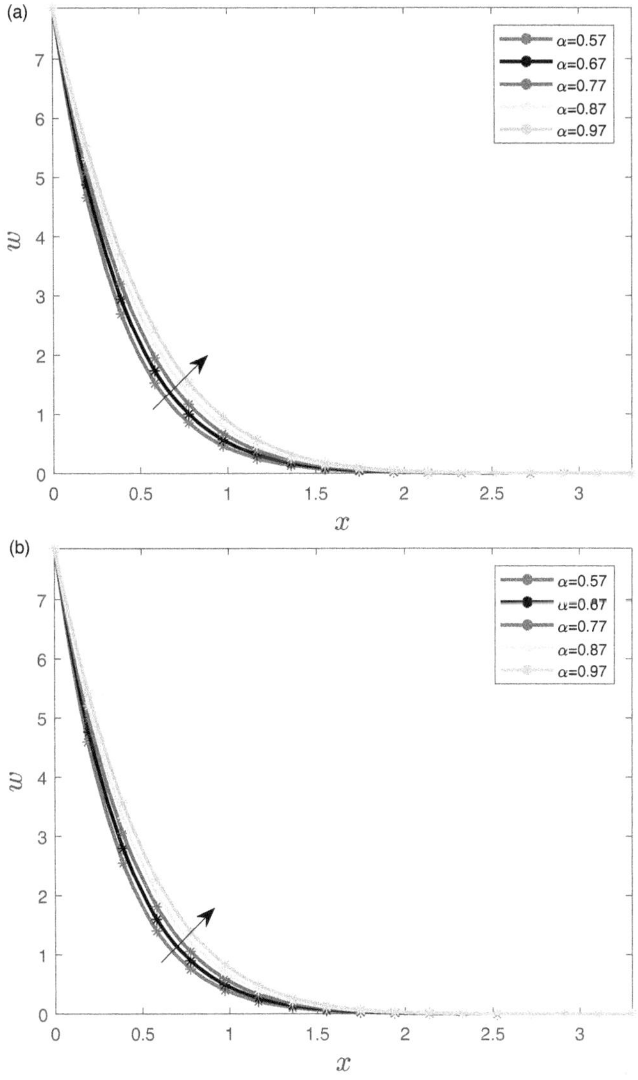

Figure 2.5 Temperature for different values of α with $Pr = 20$ (a) $Pr = 22$ (b).

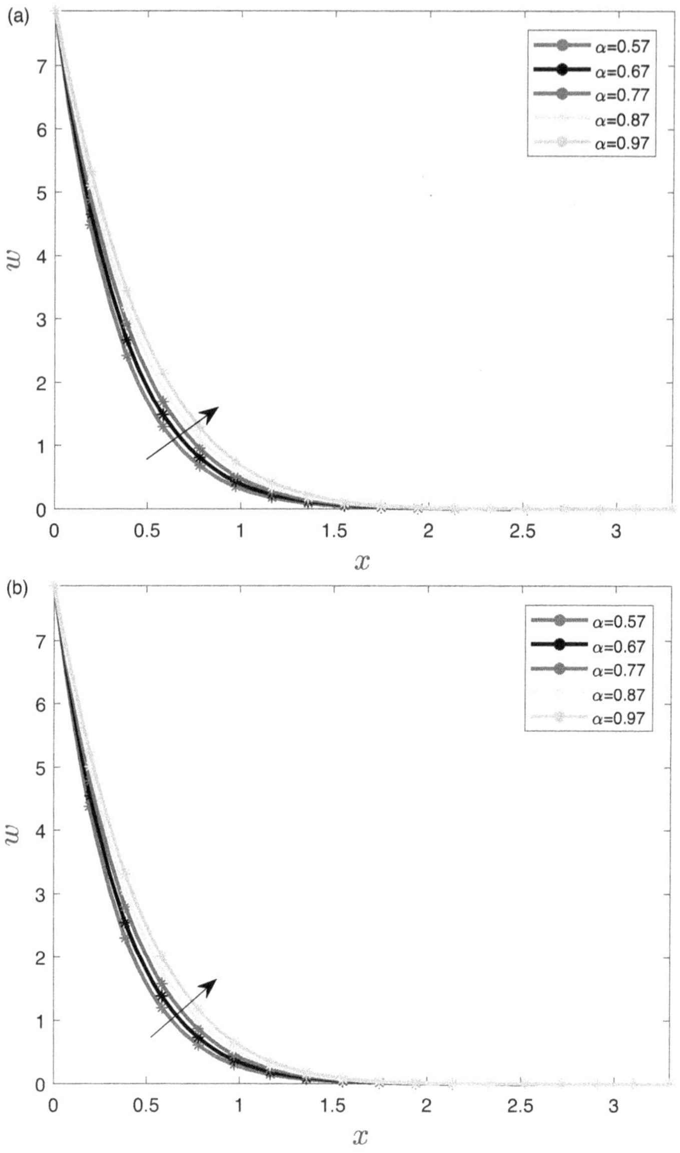

Figure 2.6 Temperature for different values of α with $Pr = 24$ (a) $Pr = 26$ (b).

the thermal condition. Thus, we notice an increase in the dynamics of the temperature distribution. As in the previous part, we finish the analysis of the thermal condition by analyzing again the influence of the Prandtl number in the new initial and boundary conditions. Here, we consider the variation of the Prandtl number, we have Figures 2.7 and 2.8.

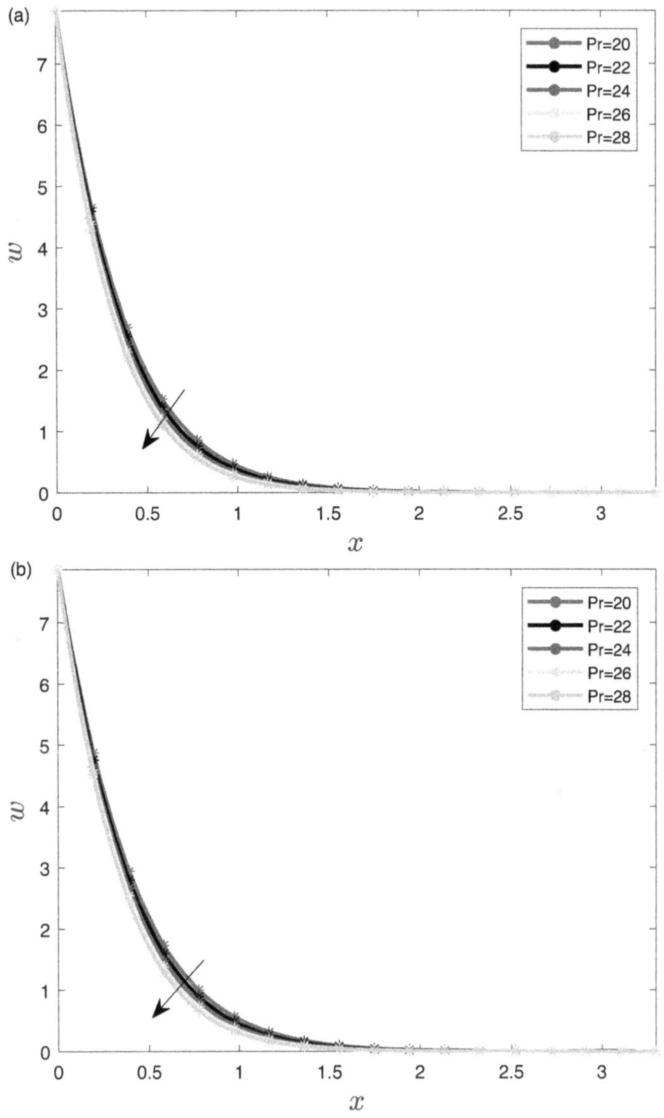

Figure 2.7 Temperature for different values of Prandtl number with $\alpha = 0.57$ (a) $\alpha = 0.67$(b).

The same behavior is noticed; we observe a decrease in the temperature distribution when the Prandtl number increases. As previously mentioned, the decrease in the dynamics is explained by the reduction of thermal diffusivity.

The second part of this section analyzes the behaviors of the velocity of the considered fluid according to the variation of the values of the Caputo derivative and the influence of the parameters of the model we mean that Prandtl number (Pr), thermal

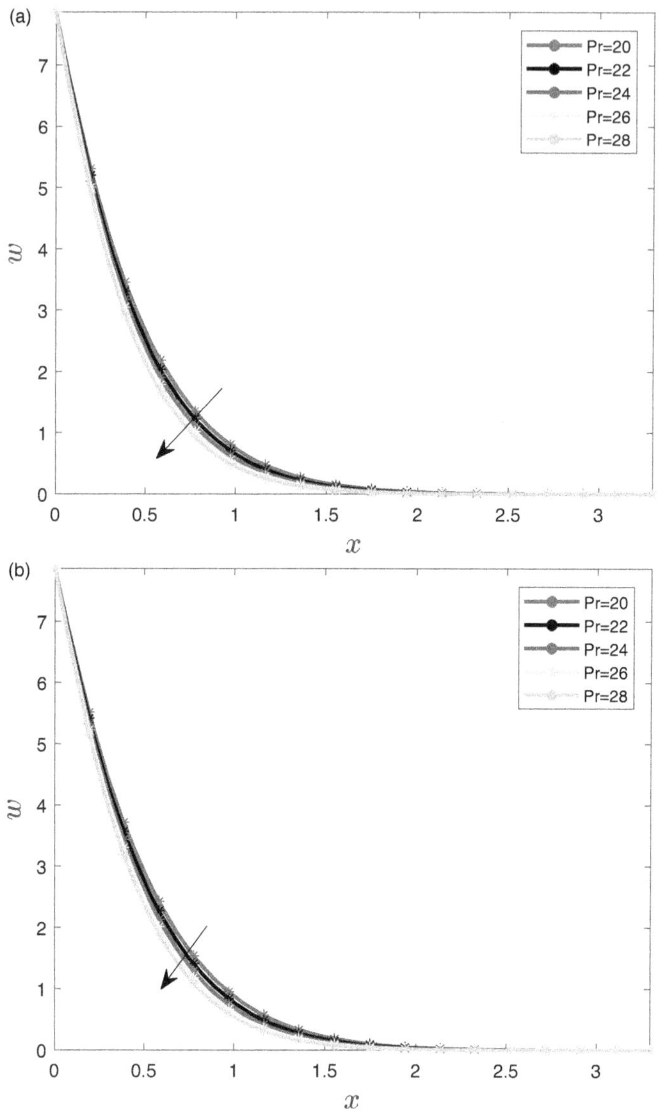

Figure 2.8 Temperature distribution with different values of Prandtl number with $\alpha = 0.87$ (a) $\alpha = 0.97$(b).

Grashof number (Gr), and the Jeffrey fluid parameter (β). We begin by analyzing the impact of the order of the Caputo derivative. We consider that $Gr = 15$, $\beta = 0.5$ and $\rho = 1.5$. The graphics are as follows (Figures 2.9 and 2.10):

In Figures 2.9 and 2.10, we can observe two conclusions. The first finding is that the increase of the order of the fractional operator gives an increase in the velocity of the considered fluid. The second finding is that the increase in the Prandtl number

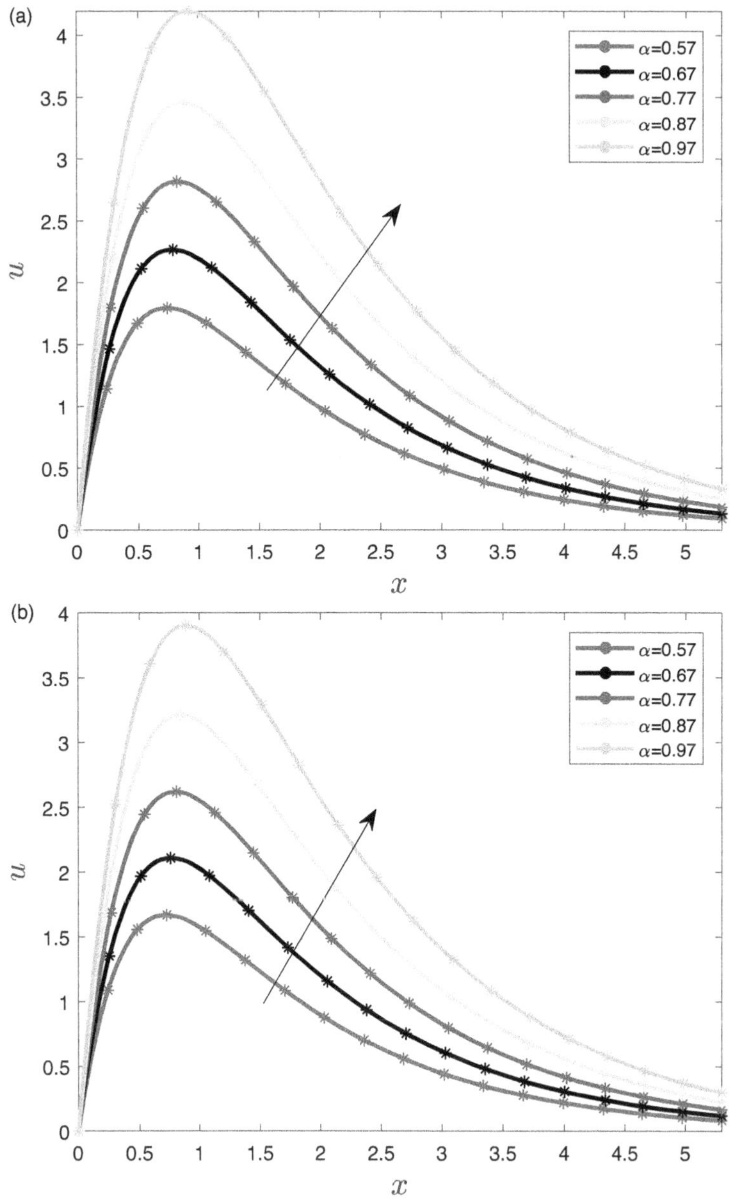

Figure 2.9 Velocities for different values of α with $Pr = 20$ (a) $Pr = 22$ (b).

gives a decrease in the values of the velocities of the considered fluid. The impact of the Prandtl number on the temperature and the velocities in the same gives a decrease in general. We continue by analyzing the influence of the thermal Grashof

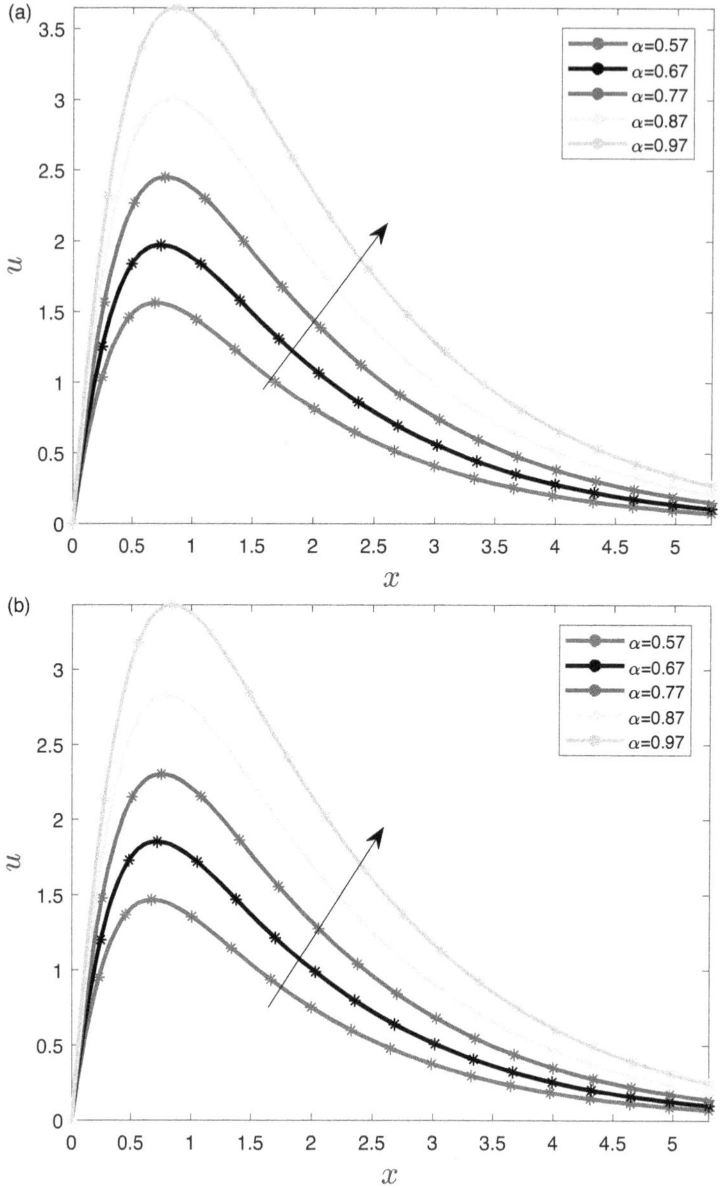

Figure 2.10 Velocities for different values of α with $Pr = 24$ (a) $Pr = 26$ (b).

number (Gr), the considered time is $t = 8$, $\beta = 0.5$, $\rho = 1.5$ and $Pr = 26$. We have Figures 2.11 and 2.12.

To see the impact of the thermal Grashof number (Gr), we compare Figures 2.11 and 2.12. We notice that the increase in the thermal Grashof number (Gr) generates

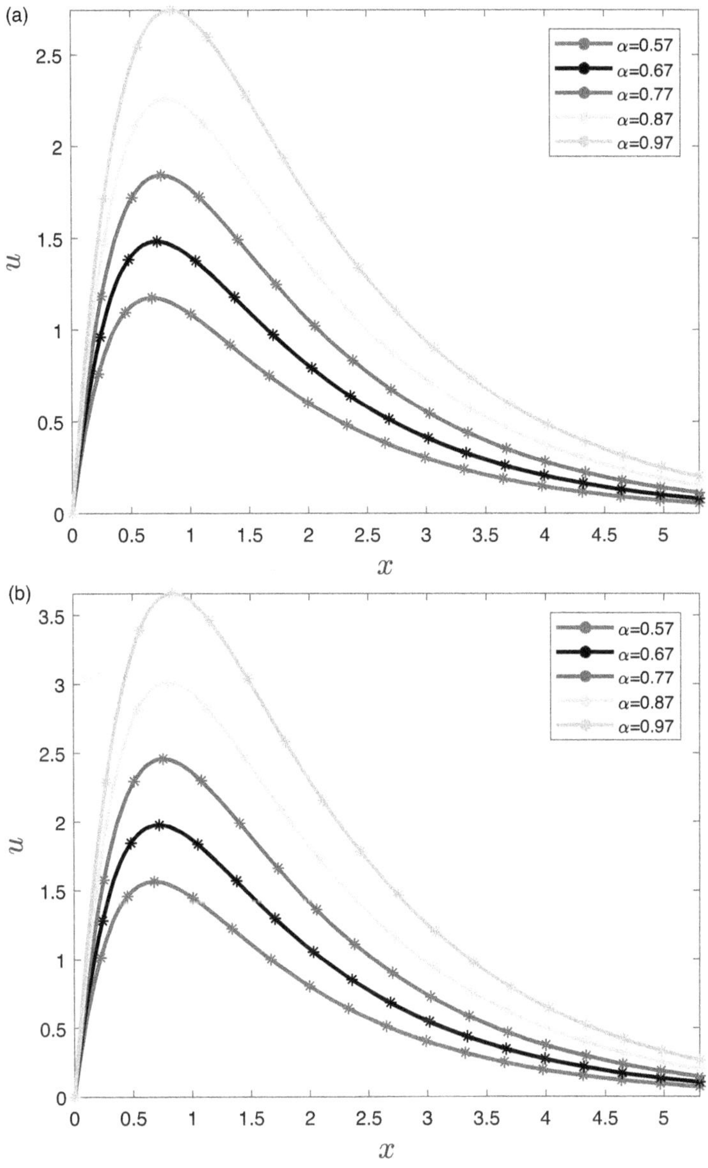

Figure 2.11 Velocities for different values of α with $Gr = 12$ (a) $Gr = 16$ (b).

an increase in the velocities of the considered fluid. This behavior can be explained by the fact that the big values of the thermal Grashof number influence positively the buoyancy forces, and then the velocity increases. We finish this section by interpreting the influence of the Jeffrey fluid parameter β, here we consider that $t = 8$, $Gr = 12$, $\rho = 1.5$ and $Pr = 26$, we have Figures 2.13 and 2.14.

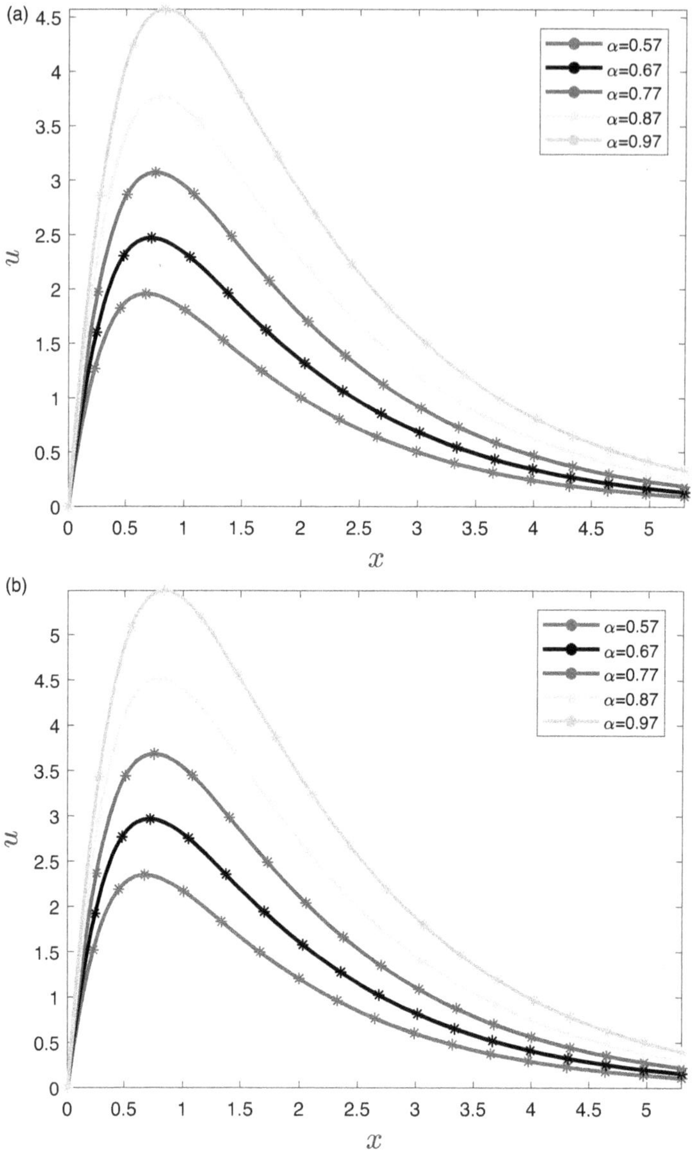

Figure 2.12 Velocities for different values of α with $Gr = 20$ (a) $Gr = 24$ (b).

In terms of comparison, we can observe that in Figures 2.13 and 2.14, the increase in the Jeffrey fluid parameter gives a decrease in the velocity of the fluid. This fact is because the increase in the Jeffrey fluid parameter gives an increase in the non-newtonian behavior, and then the thickness of the momentum boundary layer increase which in turn impacts by generating a decrease in the velocities.

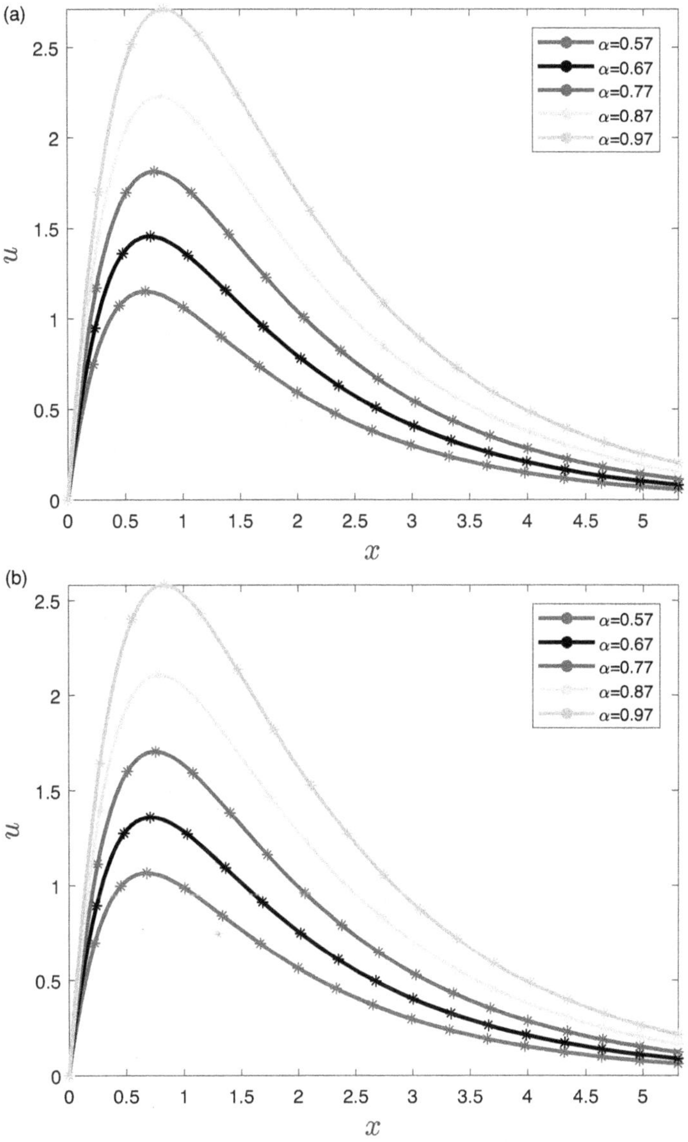

Figure 2.13 Velocities for different values of α with $\beta = 0.6$ (a) $\beta = 1$ (b).

2.6 CONCLUSION

In this chapter, we have determined the exact analytical solution of the fractional fluid model through the Laplace transform method. The particularity of this chapter is that we have used some special functions such as the Mittag-Leffler function, the Bessell function, the Gaussian error function, and the complementary Gaussian error

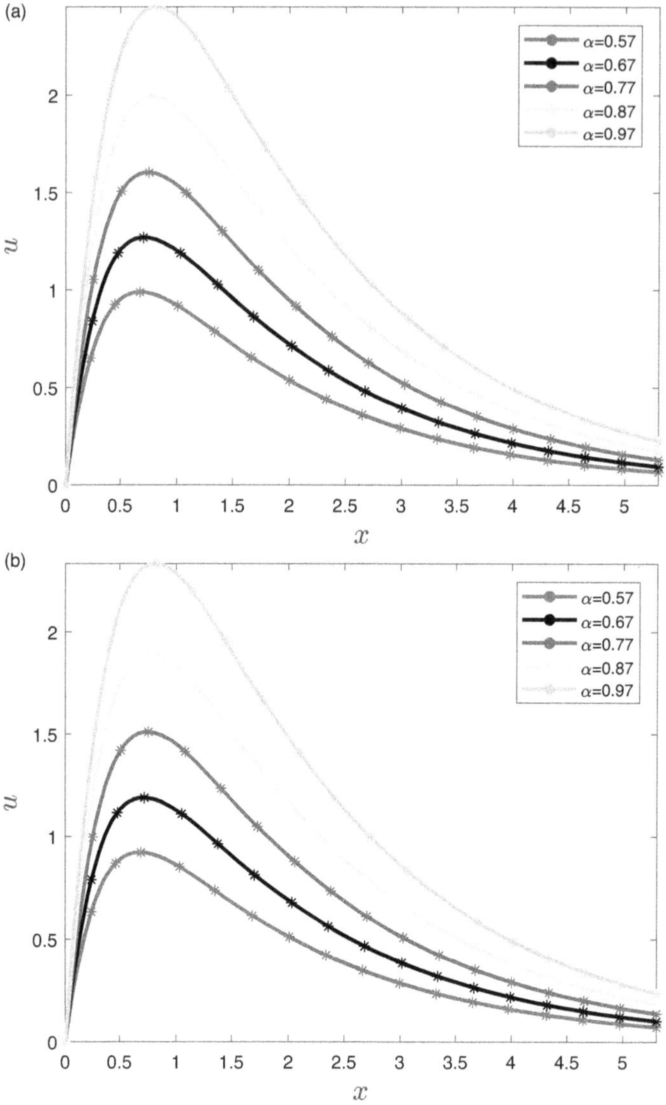

Figure 2.14 Velocities for different values of α with $\beta = 1.4$ (a) $\beta = 1.8$ (b).

function to express our obtained analytical solutions. We have analyzed the influence of the parameters of the model described by the Caputo derivative. We found that the increase of the order of the fractional operator generates an increase in the temperature and the velocities distribution. The increase in the Prandtl number generates a decrease in both the velocities and the temperature. The increase of the thermal Grashof number gives an increase in the velocities; finally, the increase of the Jeffrey fluid parameter gives a decrease in the velocity as well. For future directions of investigations, we can consider the fractional operator with the Mittag-Leffler kernel, which can generate new forms of the solutions. The changes in the boundaries and initial conditions can also be focused on in future works.

CONFLICT OF INTEREST

The authors declare that they have no conflict of interest.

References

1. M. Saqib, F. Ali, I. Khan, N. A. Sheikh, and S. A. A. Jan, Exact solutions for free convection flow of generalized Jeffrey fluid: A Caputo-Fabrizio fractional model, *Alexandria Engineering Journal* (2018), **57**, 1849–1858.
2. M. A. Imran, N. A. Shah, I. Khan, and M. Aleem, Applications of non-integer Caputo time fractional derivatives to natural convection flow subject to arbitrary velocity and Newtonian heating, *Neural Computing and Applications* (2018), **30**, 1589–1599.
3. D. Vieru, C. Fetecau, and C. Fetecau, Time-fractional free convection flow near a vertical plate with Newtonian heating and mass diffusion, *Thermal Science* (2015), **19(1)**, 85–98.
4. D. Avci, N. Ozdemir, and M. Yavuz, Fractional optimal control of diffusive transport acting on a spherical region, In: H. Singh, D. Kumar, and D. Baleanu (Eds.), *Methods of Mathematical Modelling: Fractional Differential Equations* (2019), pp. 63–82, CRC Press: Boca Raton, FL.
5. D. Kumar, J. Singh, and D. Baleanu, On the analysis of vibration equation involving a fractional derivative with Mittag-Leffler law, *Mathematical Methods in the Applied Sciences* (2019), **43(1)**, 443–457.
6. F. Ali, N. A. Sheikh, I. Khan, and M. Saqib, Magnetic field effect on blood flow of Casson fluid in axisymmetric cylindrical tube: A fractional model, *Journal of Magnetism and Magnetic Materials* (2016), **423**, 327–336.
7. H. Singh, D. Kumar, and D. Baleanu, *Methods of Mathematical Modelling Fractional Differential Equations*, (2021), CRC Press, Taylor & Francis: Boca Raton, FL.
8. A. Jajarmi, A. Yusuf, and D. Baleanu, M., A new fractional HRSV model and its optimal control: A non-singular operator approach, *Physica A: Statistical Mechanics and Its Applications* (2020), **547**, 123860.
9. X. Wang and Z. Wang, Dynamic analysis of a delayed fractional-order SIR model with saturated incidence and treatment function, *International Journal of Bifurcation and Chaos* (2018), **28(14)**, 1850180.
10. N. Sene, SIR epidemic model with Mittag–leffler fractional derivative, *Chaos, Solitons & Fractals* (2020), **137**, 109833.
11. H. Singh, H. Srivastava, and D. Baleanu, *Methods of Mathematical Modelling: Infectious Disease*, (2022), Elsevier Science: Amsterdam, The Netherlands. ISBN: 9780323998888.
12. H. Singh, Analysis for fractional dynamics of ebola virus model, *Chaos, Solitons & Fractals* (2020), **138**, 109992.
13. K. A. Abro, A fractional and analytic investigation of thermo-diffusion process on free convection flow: An application to surface modification technology, *European Physical Journal Plus* (2019), **135(1)**, 31–45.
14. H. Singh and A.-M. Wazwaz, Computational method for reaction diffusion-model arising in a spherical catalyst, *International Journal of Applied and Computational Mathematics*, (2021), **7(3)**, 65.

15. M. Caputo and M. Fabrizio, A new definition of fractional derivative without singular kernel, *Progress in Fractional Differentiation and Applications*, (2015), **1(2)**, 1–15.
16. A. Atangana, and D. Baleanu, New fractional derivatives with the nonlocal and non-singular kernel: Theory and application to heat transfer model, *Thermal Science* (2016), **20(2)**, 763–769.
17. A. A. Kilbas, H. M. Srivastava, and J. J. Trujillo, *Theory and Applications of Fractional Differential Equations*, North-Holland Mathematics Studies, vol. **204**, (2006), Elsevier: Amsterdam, The Netherlands.
18. I. Podlubny, *Fractional Differential Equations*, Mathematics in Science and Engineering, vol. **198**, (1999), Academic Press: New York.
19. J. Fahd, T. Abdeljawad, and D. Baleanu, On the generalized fractional derivatives and their Caputo modification, *Journal of Nonlinear Sciences and Applications*, (2017), **10**, 2607–2619.
20. J. Fahd and T. Abdeljawad, Generalized fractional derivatives and Laplace transform, *Discrete and Continuous Dynamical Systems-S*, (2019), **13**, 1775–1786.
21. N. Sene, A numerical algorithm applied to free convection flows of the casson fluid along with heat and mass transfer described by the Caputo Derivative, *Advances in Mathematical Physics* (2021), **2021**, 11.
22. N. A. Sheikh, F. Ali, M. Saqib, I. Khan, and S. A. A. Jan, A comparative study of Atangana-Baleanu and Caputo-Fabrizio fractional derivatives to the convective flow of a generalized Casson fluid, *The European Physical Journal Plus* (2017), **132**, 54.
23. N. A. Shah and I. Khan, Heat transfer analysis in a second-grade fluid over an oscillating vertical plate using fractional Caputo-Fabrizio derivatives, *European Physical Journal C* (2016), **76**, 362.
24. A. Khan, K. A. Abro, A. Tassaddiq, and I. Khan, Atangana-Baleanu and Caputo Fabrizio analysis of fractional derivatives for heat and mass transfer of second grade fluids over a vertical plate: A comparative study, *Entropy* (2017), **19**, 279.
25. F. Ali, M. Saqib, I. Khan, and N. A. Sheikh, Application of Caputo-Fabrizio derivatives to MHD free convection flow of generalized Walters'-B fluid model, *European Physical Journal Plus* (2016), **131**, 377.
26. B. Lohana, K. Ali Abro, and A. W. Shaikh, Thermodynamical analysis of heat transfer of gravity-driven fluid flow via fractional treatment: An analytical study, *Journal of Thermal Analysis and Calorimetry*, (2020). https://doi.org/10.1007/s10973-020-09429-w.
27. N. Sene, Analytical solutions of a class of fluids models with the Caputo fractional derivative, *Fractal and Fractional* (2022), **6(35)**. https://doi.org/10.3390/fractalfract6010035.
28. N. Sene, Fractional model and exact solutions of convection flow of an incompressible viscous fluid under the Newtonian heating and mass diffusion, *Journal of Mathematics* (2022), **2022**, 20 p. https://doi.org/10.1155/2022/8785197.
29. M. Yavuz, N. Sene, and M. Yildiz, Analysis of the influences of parameters in the fractional second-grade fluid dynamics, *Mathematics* (2022), **10**, 1125. https://doi.org/10.3390/math10071125.
30. H. M. Srivastava, A survey of some recent developments on higher transcendental functions of analytic number theory and applied mathematics, *Symmetry* (2021), **13**, 2294. https://doi.org/10.3390/sym13122294.
31. H. M. Srivastava, A. Kumar, S. Das, and K. Mehrez, Geometric properties of a certain class of Mittag-leffler-type functions, *Fractal and Fractional* (2022), **6(2)**, 54. https://doi.org/10.3390/fractalfract6020054.

3 Special Functions and Exact Solutions for Fractional Diffusion Equations with Reaction Terms

E. K. Lenzi
Universidade Estadual de Ponta Grossa
Centro Brasileiro de Pesquisas Físicas

M. K. Lenzi
Universidade Federal do Paraná

CONTENTS

3.1 INTRODUCTION

Brownian motion is one of the most important phenomena present in nature. It is present in various situations covering different fields of science, from physics to biology. It essentially started with the experiments conducted by R. Brown and was followed by the pioneer works of A. [Einstein (1905)], P. [Langevin (1908)], M. [Smoluchowski (1906)], and K. [Pearson (1905)]. The Brownian motion (or diffusion) can be connected with stochastic processes, which can be Markovian or non-Markovian. The first one is essentially characterized by a linear time dependence of the mean square displacement $\langle (x - \langle x \rangle) \rangle \propto t$ and a Gaussian behavior for the distribution of probability related to this process. The second one is characterized by a nonlinear time dependence for the mean square displacement, which

DOI: 10.1201/9781003368069-3

can be a power-law $\langle (x - \langle x \rangle) \rangle \propto t^\gamma$ ($\gamma < 1$ and $\gamma > 1$ correspond to the super and subdiffusion, respectively) or another behavior such as $\langle (x - \langle x \rangle) \rangle \propto \ln^\gamma t$ associated to a ultraslow diffusion. These situations characterize an anomalous diffusion, which may also be connected to the Lévy distributions. It is worth mentioning that the second moment is not finite when the Lévy distributions are considered. These scenarios have motivated the development of several approaches such as fractional Fokker-Planck equations [Metzler and Klafter (2000)], generalized Langevin equations [Kou and Xie (2004)], nonlinear Fokker-Planck equations [Frank (2005)], and generalized master equations [Kenkre, Montroll, and Shlesinger (1973)]. These approaches are used to investigate the systems with unusual behaviors to capture the experimental behavior and get a suitable description. Further, the diffusing elements can suffer chemical reactions that may change the concentration of species, or the reaction process may act to immobilize the diffusing elements, which after some time, switch to the motion state, so the diffusion process is altered as the process takes place [Crank (1979),Cussler and Cussler (2009)]. There are many examples of processes involving simultaneous diffusion and chemical reaction, such as in engineering [10], biological systems [Pekalski and Sznajd-Weron (1999)], chemistry [Means et al. (2006)], and physics [Lenzi et al. (2014),Méndez, Campos, and Bartumeus (2016),Ben-Avraham and Havlin (2000)], among many others. In particular, reaction-diffusion systems (see, for example, Refs. [25–27]) have attracted great attention in recent years. They have been studied in various biological and chemical systems, such as the theory of combustion [Poinsot and Veynante (2005)] and nerve impulse propagation [Manakova and Gavrilova (2018)]. They often present complex behavior and display challenging phenomena such as, for instance, pattern formation, moving fronts, and oscillations [Rinzel and Terman (1982),Szalai and De Kepper (2004),Riaz and Ray (2007)]. Solutions for diffusion-reaction systems have been explored analytically [Lenzi et al. (2014)], numerically [Iida, Ninomiya, and Yamamoto (2018)], and, more recently, through convolutional neural networks [Li et al. (2020)].

Here, we devote this chapter to the investigation of the solutions for the set of coupled fractional diffusion equations:

$$\frac{\partial^\beta}{\partial t^\beta}\rho_1(r,t) = D_1 \nabla^\mu_\eta \rho(r,t) + \mathscr{R}_1(\rho_1,\rho_2,t), \tag{3.1}$$

$$\frac{\partial^\beta}{\partial t^\beta}\rho_2(r,t) = D_2 \nabla^\mu_\eta \rho_2(r,t) + \mathscr{R}_2(\rho_1,\rho_2,t), \tag{3.2}$$

subjected to the initial conditions $\rho_1(r,t) = \varphi_1(r)$ and $\rho_2(r,t) = \varphi_2(r)$ with the boundary conditions $\lim_{r\to\infty}\rho_{1(2)}(r,t) = 0$ and $\lim_{r\to 0}\partial_r\rho_{1(2)}(r,t) = 0$. The fractional time operator is defined as follows:

$$\frac{\partial^\beta}{\partial t^\beta}\rho(r,t) = \int_0^t dt' \mathscr{K}_\beta(t-t')\frac{\partial}{\partial t'}\rho(r,t'), \tag{3.3}$$

where $\mathscr{K}_\beta(t)$ is a kernel, which can be connected to different operators such as the Caputo, i.e., $\mathscr{K}_\beta(t) = t^{-\beta}/\Gamma(1-\beta)$ [Podlubny (1999)], implies in

$$\frac{\partial^\beta}{\partial t^\beta}\rho(r,t) = \frac{1}{\Gamma(1-\beta)}\int_0^t dt' \frac{1}{(t-t')^\beta}\frac{\partial}{\partial t'}\rho(r,t'), \tag{3.4}$$

or nonsingular kernels such as ($\mathscr{K}_\beta(t) = \mathscr{N}_\beta e^{-\beta t}/(1-\beta)$

$$\frac{\partial^\beta}{\partial t^\beta}\rho(r,t) = \frac{\mathscr{N}_\beta}{1-\beta}\int_0^t dt' e^{-\frac{\beta}{1-\beta}(t-t')}\frac{\partial}{\partial t'}\rho(r,t'), \tag{3.5}$$

or $\mathscr{K}_\beta(t) = \mathscr{N}_\beta E_\beta\left(-\beta t^\beta/(1-\beta)\right)/(1-\beta)$

$$\frac{\partial^\beta}{\partial t^\beta}\rho(r,t) = \frac{\mathscr{N}_\beta}{1-\beta}\int_0^t dt' E_\beta\left(-\frac{\beta}{1-\beta}(t-t')^\beta\right)\frac{\partial}{\partial t'}\rho(r,t'), \tag{3.6}$$

where $\mathscr{N}_\beta$ is a normalization constant [Atangana and Baleanu (2016),Tateishi, Ribeiro, and Lenzi (2017),Fernandez and Baleanu (2021)]. In this manner, it is possible to cover different scenarios related to anomalous diffusion, which imply memory effects, long-range correlation, fractal structures, and among others. In particular, these fractional operators have been used in many contexts such as boundary value problems [28,31], electric circuits [29,30], and electrical impedance [Scarfone et al. (2022),Barbero, Evangelista, and Lenzi (2022)] (see also Refs. [Singh, Kumar, and Baleanu (2022),9,10]). The fractional operator in space [Lenzi and Evangelista (2021),Lenzi et al. (2022)] is defined as follows:

$$\mathscr{F}_{d,\eta}\left\{\nabla_\eta^\mu \rho(r,t);k\right\} \equiv -k^{\mu+\eta}\widehat{\widetilde{\rho}}(k,t). \tag{3.7}$$

with the integral transform given by:

$$\mathscr{F}_{d,\eta}\left\{\rho(r,t);k\right\} = \int_0^\infty dr r^{d-1}\psi_\eta(r,k)\rho(r,t)\,, \tag{3.8}$$

$$\mathscr{F}_{d,\eta}^{-1}\left\{\widetilde{\rho}(k,t);r\right\} = \int_0^\infty dk k^{d-1}\psi_\eta(r,k)\widetilde{\rho}(k,t), \tag{3.9}$$

where

$$\psi_\eta(r,k) = (kr)^{\frac{1}{2}(2-d+\eta)}\mathrm{J}_\nu\left(2(kr)^{\frac{1}{2}(2+\eta)}/(2+\eta)\right), \tag{3.10}$$

where $\nu = d/(2+\eta)-1$ and $\mathrm{J}_\nu(x)$ is the Bessel function [Evangelista and Lenzi (2018)]. It is worth stressing that Eqs. (3.8) and (3.9) may be related with a generalized Hankel transform [Ali and Kalla (1999),Garg, Rao, and Kalla (2007),Nakhi and Kalla (2003),Xie et al. (2010)]. This integral transform allows us to investigate situations often associated with solutions in terms of stretched exponential distributions. Such distributions are typical in different systems, including diffusion on fractals [O'Shaughnessy and Procaccia (1985a),O'Shaughnessy and Procaccia (1985b)], turbulence [Richardson (1926),Boffetta and Sokolov (2002)], diffusion and reaction on fractals [Daniel ben Avraham (2000)], solute transport in fractal porous media [Su et al. (2005)], and atom deposition in a porous substrate [Brault et al. (2009)]. It is worth mentioning that different approaches have been used to model the reaction of a diffusion process in the context of anomalous diffusion by using the fractional approach such as the ones presented in Refs. [Langlands, Henry, and Wearne (2007),Angstmann, Donnelly, and Henry (2013),Nec and Nepomnyashchy (2007),Henry and Wearne (2000),Seki, Wojcik, and Tachiya (2003),Lenzi et al. (2016)].

Equations (3.1) and (3.2) will be analyzed by considering the reaction term $\mathscr{R}_1(\rho_1,\rho_2,t) = k_{11}\rho_1(r,t) + k_{12}\rho_2(r,t)$ and $\mathscr{R}_2(\rho_1,\rho_2,t) = k_{21}\rho_1(r,t) + k_{22}\rho_2(r,t)$, where the constants can be related to different processes such as reversible and irreversible processes depending on the choice of the k_{11}, k_{12}, k_{21}, and k_{22}. We can also use them to simulate an intermittent motion, where the particles are trapped and switch to the motion state after some time. We express the solutions in terms of the special functions present in the fractional calculus, such as the Fox H functions and generalized Mittag-Leffler functions. These solutions make it possible to obtain a large class of behavior, which may be connected to different diffusion regimes, particularly anomalous diffusion.

3.2 DIFFUSION-REACTION

We first discuss the solutions of the previous set of equations by establishing this system's boundary and initial conditions. We consider that these equations are subjected to the boundary conditions $\lim_{r\to\infty}\rho_1(r,t) = 0$, $\lim_{r\to 0}\partial_r\rho_1(r,t) = 0$, $\lim_{r\to\infty}\rho_2(r,t) = 0$, $\lim_{r\to 0}\partial_r\rho_2(r,t) = 0$, $\rho_1(r,0) = \varphi_1(r)$, and $\rho_2(r,0) = \varphi_2(r)$. These conditions imply that we are considering free boundary conditions for the system, i.e., the system is not confined by surfaces. We start by analyzing first the case $D_1 = D_2 = D_{\mu,\eta}$ and after the scenarios obtained by considering $D_1 \neq D_2$, where the species have different diffusion coefficients. For the first case, i.e., $D_1 = D_2 = D_{\mu,\eta}$, the previous set of equations can be written as

$$\frac{\partial^\beta}{\partial t^\beta}\rho_1(r,t) = D_{\mu,\eta}\nabla^\mu_\eta\rho_1(r,t) + \mathscr{R}_1(\rho_1,\rho_2,t) \tag{3.11}$$

and

$$\frac{\partial^\beta}{\partial t^\beta}\rho_2(r,t) = D_{\mu,\eta}\nabla^\mu_\eta\rho_2(r,t) + \mathscr{R}_2(\rho_1,\rho_2,t)\,. \tag{3.12}$$

The solution for this set of equations depends on the choice of the reaction terms, which depend on the choice of the parameters k_{11}, k_{12}, k_{21}, and k_{22} and can describe different situations such as irreversible or reversible processes. In particular, we consider a reversible process, which implies in $k_{11} = -k_{21} = -k_1$ and $k_{22} = -k_{12} = -k_2$. Then, we discuss the other cases that emerge for different parameter choices related to the reaction terms. For this case, we have that

$$\frac{\partial^\beta}{\partial t^\beta}\rho_1(r,t) = D_{\mu,\eta}\nabla^\mu_\eta\rho_1(r,t) - k_1\rho_1(r,t) + k_2\rho_2(r,t) \tag{3.13}$$

and

$$\frac{\partial^\beta}{\partial t^\beta}\rho_2(r,t) = D_{\mu,\eta}\nabla^\mu_\eta \rho_2(r,t) + k_1\rho_1(r,t) - k_2\rho_2(r,t)\,. \tag{3.14}$$

To obtain the solution for Eqs. (3.13) and (3.14), we use Eqs. (3.8) and (3.9) and the Laplace transform, yielding the following expressions, remembering that $\widehat{\mathscr{K}_\beta}(s)$ is related to the kernel as presented in Eq. (4.1)

$$\begin{aligned} s\widehat{\widetilde{\rho}}_1(k,s) - \widetilde{\varphi}_1(k) = {} & \left(D_{\mu,\eta}/\widehat{\mathscr{K}_\beta}(s)\right)k^{\mu+\eta}\widehat{\widetilde{\rho}}_1(k,s) - \left(k_1/\widehat{\mathscr{K}_\beta}(s)\right)\widehat{\widetilde{\rho}}_1(k,s) \\ & + \left(k_2/\widehat{\mathscr{K}_\beta}(s)\right)\widehat{\widetilde{\rho}}_2(k,s) \end{aligned} \tag{3.15}$$

and

$$\begin{aligned} s\widehat{\widetilde{\rho}}_2(k,s) - \widetilde{\varphi}_2(k) = {} & \left(D_{\mu,\eta}/\widehat{\mathscr{K}_\beta}(s)\right)k^{\mu+\eta}\widehat{\widetilde{\rho}}_2(k,s) + \left(k_1/\widehat{\mathscr{K}_\beta}(s)\right)\widehat{\widetilde{\rho}}_1(k,s) \\ & - \left(k_2/\widehat{\mathscr{K}_\beta}(s)\right)\widehat{\widetilde{\rho}}_2(k,s)\,. \end{aligned} \tag{3.16}$$

These equations can be written as follows:

$$\begin{aligned} s\left(\widehat{\widetilde{\rho}}_1(k,s) + \widehat{\widetilde{\rho}}_2(k,s)\right) = {} & \left(D_{\mu,\eta}/\widehat{\mathscr{K}_\beta}(s)\right)k^{\mu+\eta}\left(\widehat{\widetilde{\rho}}_1(k,s) + \widehat{\widetilde{\rho}}_2(k,s)\right) \\ & + \left(\widetilde{\varphi}_1(k) + \widetilde{\varphi}_2(k)\right) \end{aligned} \tag{3.17}$$

and

$$\begin{aligned} s\left(k_1\widehat{\widetilde{\rho}}_1(k,s) - k_2\widehat{\widetilde{\rho}}_2(k,s)\right) = {} & \left(D_{\mu,\eta}/\widehat{\mathscr{K}_\beta}(s)\right)k^{\mu+\eta}\left(k_1\widehat{\widetilde{\rho}}_1(k,s) - k_2\widehat{\widetilde{\rho}}_2(k,s)\right)) \\ & + \left(k_1\widetilde{\varphi}_1(k) - k_2\widetilde{\varphi}_2(k)\right) \\ & + \left[(k_1+k_2)/\widehat{\mathscr{K}_\beta}(s)\right]\left(k_1\widehat{\widetilde{\rho}}_1(k,s) - k_2\widehat{\widetilde{\rho}}_2(k,s)\right)\,. \end{aligned} \tag{3.18}$$

The solution for Eqs. (3.17) and (3.18) is given by

$$\widehat{\widetilde{\rho}}_1(k,s) + \widehat{\widetilde{\rho}}_2(k,s) = \widehat{\widetilde{\mathscr{G}}}_1(k,s)\left(\widetilde{\varphi}_1(k) + \widetilde{\varphi}_2(k)\right)\,, \tag{3.19}$$

with

$$\widehat{\widetilde{\mathscr{G}}}_1(k,s) = \frac{1}{s + \left(D_{\mu,\eta}/\widehat{\mathscr{K}_\beta}(s)\right)k^{\mu+\eta}} \tag{3.20}$$

and

$$k_1\widehat{\widetilde{\rho}}_1(k,s) - k_2\widehat{\widetilde{\rho}}_2(k,s) = \widehat{\widetilde{\mathscr{G}}}_2(k,s)\left(k_1\widetilde{\varphi}_1(k) - k_2\widetilde{\varphi}_2(k)\right)\,, \tag{3.21}$$

with

$$\widehat{\widetilde{\mathscr{G}}}_2(k,s) = \frac{1}{s + \left(k_t/\widehat{\mathscr{K}_\beta}(s)\right) + \left(D_{\mu,\eta}/\widehat{\mathscr{K}_\beta}(s)\right)k^{\mu+\eta}}\,, \tag{3.22}$$

where $k_t = k_1 + k_2$. It is worth mentioning that Eq. (3.20) extends the standard Green function and the fractional ones to a broad class of situations. From these results, we may obtain $\widehat{\widetilde{\rho}}_1(k,s)$ and $\widehat{\widetilde{\rho}}_2(k,s)$. In particular, we have that

$$\widehat{\widetilde{\rho}}_1(k,s) = \frac{1}{k_t}\left[k_2\widehat{\widetilde{\mathscr{G}}}_1(k,s)\left(\widetilde{\varphi}_1(k)+\widetilde{\varphi}_2(k)\right)+\widehat{\widetilde{\mathscr{G}}}_2(k,s)\left(k_1\widetilde{\varphi}_1(k)-k_2\widetilde{\varphi}_2(k)\right)\right] \tag{3.23}$$

and

$$\widehat{\widetilde{\rho}}_2(k,s) = \frac{1}{k_t}\left[k_1\widehat{\widetilde{\mathscr{G}}}_1(k,s)\left(\widetilde{\varphi}_1(k)+\widetilde{\varphi}_2(k)\right)-\widehat{\widetilde{\mathscr{G}}}_2(k,s)\left(k_1\widetilde{\varphi}_1(k)-k_2\widetilde{\varphi}_2(k)\right)\right]. \tag{3.24}$$

3.2.1 CASE - $\mathscr{K}_\beta(t) = \delta(t)$, $-1 < \eta$, $\mu \neq 2$

Let us perform some choices for the kernel $\mathscr{K}_\beta(t)$, μ, and η to obtain the inverse of integral transforms. In this sense, we start by considering the case $\mathscr{K}_\beta(t) = \delta(t)$, which corresponds to the standard time derivative, with $1 < \mu \leq 2$ and $-1 < \eta$. This case has an interplay between the parameters μ and η, which correspond to the spatial integral-differential operator. The case $\mu \neq 2$ introduces the Lévy distributions, and $\eta \neq 0$ represents heterogeneity in the system. The distribution that emerges from this case mixes these two effects when $\mu \neq 2$ and $\eta \neq 0$.

For the case $\mathscr{K}_\beta(t) = \delta(t)$, after performing the inverse of Laplace transform, we have that

$$\widetilde{\rho}_1(k,t) = \frac{1}{k_1+k_2}\widetilde{\mathscr{G}}_1(k,t)\left[k_2\left(\widetilde{\varphi}_1(k)+\widetilde{\varphi}_2(k)\right)+e^{-k_t t}\left(k_1\widetilde{\varphi}_1(k)-k_2\widetilde{\varphi}_2(k)\right)\right] \tag{3.25}$$

and

$$\widetilde{\rho}_2(k,t) = \frac{1}{k_1+k_1}\widetilde{\mathscr{G}}_1(k,t)\left[k_1\left(\widetilde{\varphi}_1(k)+\widetilde{\varphi}_2(k)\right)-e^{-k_t t}\left(k_1\widetilde{\varphi}_1(k)-k_2\widetilde{\varphi}_2(k)\right)\right], \tag{3.26}$$

with $\widetilde{\mathscr{G}}_2(k,s) = e^{-k_t t}\widetilde{\mathscr{G}}_1(k,s)$ and $\widetilde{\mathscr{G}}_1(k,s) = e^{-\mathcal{D}_{\mu,\eta}k^{\mu+\eta}t}$. Applying Eqs. (3.8) and (3.9), it is possible to show that

$$\rho_1(r,t) = \frac{k_2}{k_t}\int_0^\infty dr' r'^{d-1}\bar{\mathscr{G}}_1(r,r',t) \times\left[\left(1+\frac{k_1}{k_2}e^{-k_t t}\right)\varphi_1(r')+\left(1-e^{-k_t t}\right)\varphi_2(r')\right] \tag{3.27}$$

and

$$\rho_2(r,t) = \frac{k_1}{k_t}\int_0^\infty dr' r'^{d-1}\bar{\mathscr{G}}_1(r,r',t) \times\left[\left(1+\frac{k_2}{k_1}e^{-k_t t}\right)\varphi_2(r')+\left(1-e^{-k_t t}\right)\varphi_1(r')\right], \tag{3.28}$$

where

$$\bar{\mathscr{G}}_1(r,r',t) = \int_0^\infty dkk^{d-1}\psi_\eta(r,k)\psi_\eta(r',k)\widetilde{\mathscr{G}}(k,t)\,. \tag{3.29}$$

Equation (3.29) can be written in terms of a generalized H Fox function. In this sense, by using the identity

$$E_{\alpha,\mu}(x) = \sum_{n=0}^\infty \frac{x^n}{\Gamma(\mu+\alpha n)} = \mathrm{H}^{1,1}_{1,2}\left[-x\left|{}^{(0,1)}_{(0,1),(1-\mu,\alpha)}\right.\right], \tag{3.30}$$

which connects the generalized Mittag-Leffler function and the H Fox function (or H–function), which may be defined in terms of the Mellin-Branes-type integral [Saxton (2007),Evangelista and Lenzi (2018)]

$$\mathrm{H}^{m,n}_{p,q}\left[x\left|{}^{(a_p,A_p)}_{(b_q,B_q)}\right.\right] = \mathrm{H}^{m,n}_{p,q}\left[x\left|{}^{(a_1,A_1),\ldots,(a_p,A_p)}_{(b_1,B_1),\ldots,(b_q,B_q)}\right.\right] = \frac{1}{2\pi i}\int_L \chi(\xi)x^{-\xi}d\xi$$

$$\chi(\xi) = \frac{\Pi^m_{j=1}\Gamma(b_j-B_j\xi)\Pi^n_{j=1}\Gamma(1-a_j+A_j\xi)}{\Pi^q_{j=m+1}\Gamma(1-b_j+B_j\xi)\Pi^p_{j=n+1}\Gamma(a_j-A_j\xi)} \tag{3.31}$$

where $m,n,p,$ and q are integers satisfying $0\le n\le p$ and $1\le m\le q$. Note that for $\alpha=\mu=1$ Eq. (3.30) allows us to write the Green function in terms of the H Fox function as follows:

$$\widetilde{\mathscr{G}}_1(k,t) = \mathrm{H}^{1,1}_{1,2}\left[-D_{\mu,\eta}tk^{\mu+\eta}\left|{}^{(0,1)}_{(0,1),(0,1)}\right.\right]\,. \tag{3.32}$$

Another function to be expressed in terms of the H Fox function is the Bessel function. In particular, we have that

$$J_\nu(x) = \mathrm{H}^{1,0}_{0,2}\left[\frac{x^2}{4}\left|{}^{------}_{(\frac{\nu}{2},1),(-\frac{\nu}{2},1)}\right.\right], \tag{3.33}$$

where $J_\nu(x)$ is the Bessel function of order ν. By using Eqs. (3.32) and (3.33) in Eq. (3.29), it is possible to obtain that:

$$\bar{\mathscr{G}}_1(r,r',t) = \frac{(2+\eta)^2}{2(\mu+\eta)r^{2+\eta}}\left(rr'\right)^{\frac{1}{2}(2-d+\eta)}$$
$$\times\,\mathrm{H}^{1,0,1,1,1}_{2,[0:1],0,[0:2]}\left[\begin{array}{c|c} \left(\frac{r'}{r}\right)^{2+\eta} & (\frac{2+\nu}{2},1);(\frac{2-\nu}{2},1) \\ & --;(0,\frac{\mu+\eta}{2+\eta}) \\ \frac{\overline{D}_{\mu,\eta}t^{\frac{2+\eta}{\mu+\eta}}}{r^{2+\eta}} & --;-- \\ & (\frac{\nu}{2},1);(-\frac{\nu}{2},1);(0,\frac{\mu+\eta}{2+\eta}),(0,\frac{\mu+\eta}{2+\eta}) \end{array}\right] \tag{3.34}$$

where $\overline{D}_{\mu,\eta} = (2+\eta)^2 D^{\frac{2+\eta}{\mu+\eta}}_{\mu,\eta}$ and

$$\mathrm{H}^{L,M,M_1,N,N_1}_{E,[A:C],F,[B,D]}\left[\begin{array}{c|c} & (\varepsilon_1,\omega_1),\ldots,(\varepsilon_E,\omega_E) \\ x & (a_1,\alpha_1),\ldots,(a_A,\alpha_A);(c_1,\beta_1),\ldots,(c_C,\beta_C) \\ y & (\xi_1,\varpi_1),\ldots,(\xi_F,\varpi_F) \\ & (b_1,\beta_1),\ldots,(b_B,\beta_B);(d_1,\delta_1),\ldots,(d_C,\delta_D) \end{array}\right] \tag{3.35}$$

is the generalized H−function of Fox [Mathai, Saxena, and Haubold (2009),Lenzi et al. (2010),Jiang and Xu (2010)]. For the particular case $\mu = 2$, it is possible to simplify Eq. (3.34) and obtain that

$$\bar{\mathscr{G}}_1(r,r',t) = \frac{1}{2D_{\mu,\eta}t}\left(rr'\right)^{\frac{1}{2}(2-d+\eta)} e^{-\frac{r^{2+\eta}+r'^{2+\eta}}{(2+\eta)^2 D_{\mu,\eta}t}} I_\nu\left(\frac{2\left(rr'\right)^{\frac{1}{2}(2+\eta)}}{(2+\eta)^2 D_{\mu,\eta}t}\right), \tag{3.36}$$

where $I_\nu(x)$ is the Bessel function of order ν of modified argument [Wyld and Powell (2020)]. Equation (3.36) shows that the solution is characterized in terms of a stretched exponential instead of the Gaussian one. This feature is interesting in the sense that several systems, including diffusion on fractals [O'Shaughnessy and Procaccia (1985a),O'Shaughnessy and Procaccia (1985b)], turbulence [Richardson (1926),Boffetta and Sokolov (2002)], diffusion and reaction on fractals [Daniel ben Avraham (2000)], solute transport in fractal porous media [Su et al. (2005)], atom deposition in a porous substrate [Brault et al. (2009)], are described by these distributions. Figures (11.1) and (11.2) show the behavior of Eq. (3.34) for different values of μ and η by considering $d = 3$.

Equation (3.34) for $r' = 0$ may be simplified yielding

$$\bar{\mathscr{G}}_1(r,0,t) = \frac{2+\eta}{(\mu+\eta)\Gamma\left(\frac{d}{2+\eta}\right)r^d} \mathrm{H}^{2,1}_{2,3}\left[\frac{r}{\Phi_{\eta,\mu}(t)}\middle| \begin{matrix} \left(1,\frac{1}{\mu+\eta}\right),\left(1,\frac{1}{\mu+\eta}\right) \\ \left(\frac{d}{2+\eta},\frac{1}{2+\eta}\right),\left(1,\frac{1}{\mu+\eta}\right),\left(1,\frac{1}{2+\eta}\right) \end{matrix}\right], \tag{3.37}$$

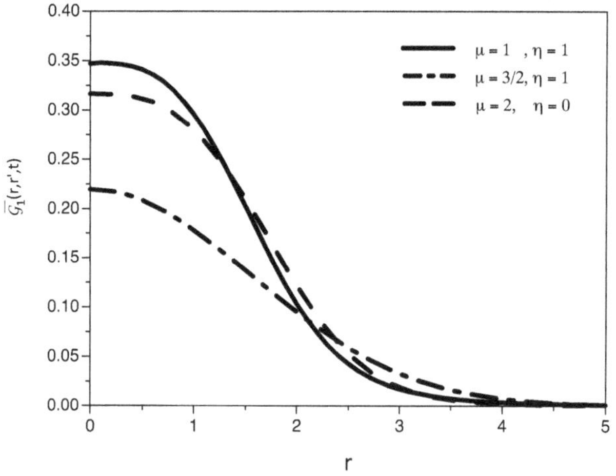

Figure 3.1 Behavior of the Green function defined by Eq. (3.34) for different values of μ with $\eta = 1$. The usual case ($\mu = 2$ and $\eta = 0$) is also shown. We consider, for simplicity, $d = 3$, $\bar{D}_{\mu,\eta}t^{\frac{2+\eta}{\mu+\eta}}/r'^{2+\eta} = 1$, and $r' = 1$.

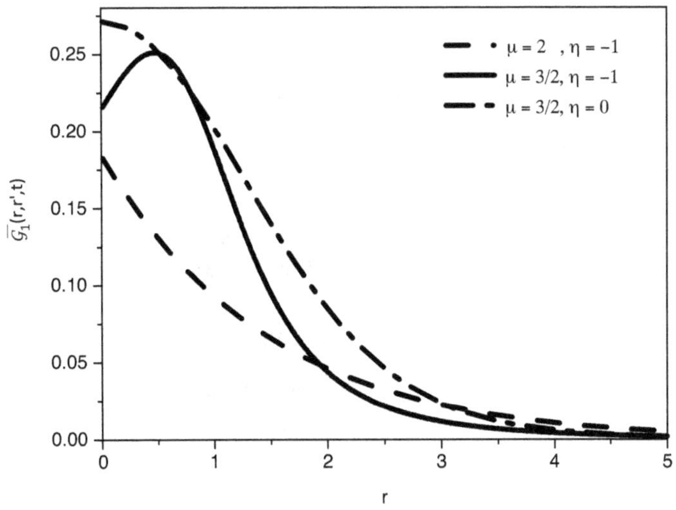

Figure 3.2 Behavior of the Green function defined by Eq. (3.34) for different values of μ with $\eta = -1$. The usual case ($\mu = 2$ and $\eta = 0$) is also shown. We consider, for simplicity, $d = 3$, $\bar{D}_{\mu,\eta} t^{\frac{2+\eta}{\mu+\eta}} / r'^{2+\eta} = 1$, and $r' = 1$.

in which

$$\Phi_{\eta,\mu}(t) = \left[(2+\eta)^{\frac{2(\mu+\eta)}{2+\eta}} D_{\mu,\eta} t\right]^{\frac{1}{\mu+\eta}}. \tag{3.38}$$

It is important to mention that for this case $\bar{\mathcal{G}}_1(r,0,t) = \mathcal{G}_1(r,t)$, as expected. Also for this case, it is possible to write Eq. (3.37) in a scaled form, i.e., $\bar{\mathcal{G}}_1(r,0,t) = \left[1/\Phi_{\eta,\mu}(t)\right]^d \bar{\mathcal{G}}_s(r/\Phi_{\eta,\mu}(t))$, where

$$\bar{\mathcal{G}}_s(z) = \frac{2+\eta}{(\mu+\eta)\Gamma\left(\frac{d}{2+\eta}\right) r^d} \mathrm{H}^{2,1}_{2,3}\left[z \left| \begin{array}{c} \left(1,\frac{1}{\mu+\eta}\right),\left(1,\frac{1}{\mu+\eta}\right) \\ \left(\frac{d}{2+\eta},\frac{1}{2+\eta}\right),\left(1,\frac{1}{\mu+\eta}\right),\left(1,\frac{1}{2+\eta}\right) \end{array}\right.\right], \tag{3.39}$$

performed for the parameters μ and η. In this sense, we have that $\langle r^2\rangle = \int_0^\infty dr r^{d-1}\left(r^2 \bar{\mathcal{G}}_1(r,0,t)\right)$, which implies in $\langle r^2\rangle = \mathcal{I}_{\mu,\eta} t^{\frac{2}{\mu+\eta}}$ for $\mathcal{I}_{\mu,\eta} = \int_0^\infty dz z^{d+1} \bar{\mathcal{G}}_s(z)$ finite. This result for the second moment implies that the free case in the absence of reaction terms is sub-, usual, and superdiffusive for $\eta+\mu > 2$, $\eta+\mu = 2$, and $\eta+\mu < 2$, respectively.

Let us consider the solution, i.e., Eqs. (3.27) and (3.28), for the initial condition $\varphi_1(r) = \delta(r)/r^{d-1}$ and $\varphi_2(r) = 0$. For this case, we have that,

$$\rho_1(r,t) = \left(1 + \frac{k_1}{k_2} e^{-k_t t}\right) \frac{k_2}{k_t} \mathcal{G}_1(r,t) \tag{3.40}$$

and

$$\rho_2(r,t) = \left(1 + e^{-k_t t}\right) \frac{k_1}{k_t} \mathscr{G}_1(r,t) \,. \tag{3.41}$$

Figure 11.3 shows the behavior of $\rho_1(r,t)$ and $\rho_2(r,t)$. In particular, we observe that the behavior for $\eta < 0$ introduces a long-tailed distribution compared to $\eta > 0$ for the same value of $\mu \neq 2$.

Figure 11.4 shows the behavior of the second moment for $\rho_1(r,t)$. In particular, we observe a crossover between two different regimes of diffusion.

3.2.2 CASE - $\mathscr{K}_\beta(t) = T^{-\beta}/\Gamma(1-\beta)$, $-1 < \eta$, $\mu \neq 2$

For the case $\mathscr{K}_\beta(t) = s^{\beta-1}$, the integral-differential operator in time is the Caputo's fractional time derivative. In this scenario, Eqs. (3.20) and (3.22) can be written as follows:

$$\widehat{\widetilde{\mathscr{G}}}_1(k,s) = \frac{s^{\beta-1}}{s^\beta + D_{\mu,\eta} k^{\mu+\eta}} \tag{3.42}$$

and

$$\widehat{\widetilde{\mathscr{G}}}_2(k,s) = \frac{s^{\beta-1}}{s^\beta + k_t + D_{\mu,\eta} k^{\mu+\eta}} \,. \tag{3.43}$$

By performing the inverse of Laplace transform, we obtain that

$$\widetilde{\mathscr{G}}_1(k,t) = E_\beta\left(-D_{\mu,\eta} t^\beta k^{\mu+\eta}\right) \tag{3.44}$$

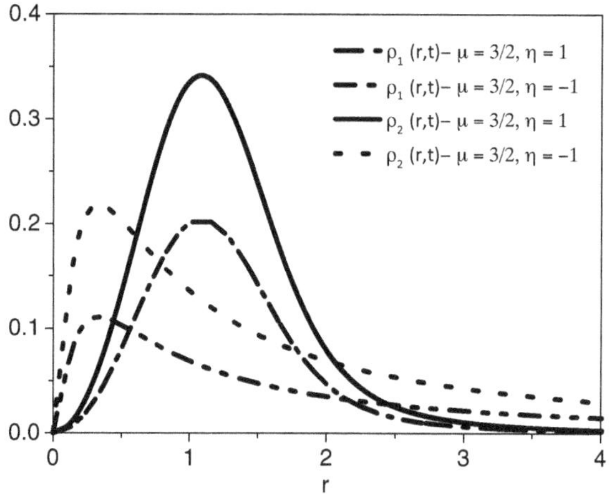

Figure 3.3 Behavior of the distributions $\rho_1(r,t)$ and $\rho_2(r,t)$ defined by Eqs. (3.40) and (3.41) for different values of μ and η. We consider, for simplicity, $d = 3$, $\bar{D}_{\mu,\eta} = 1$, $t = 1$, $k_1 = 2k_2$, $\varphi_2(r) = 0$, and $\varphi_1(r) = \delta(r)/r^{d-1}$.

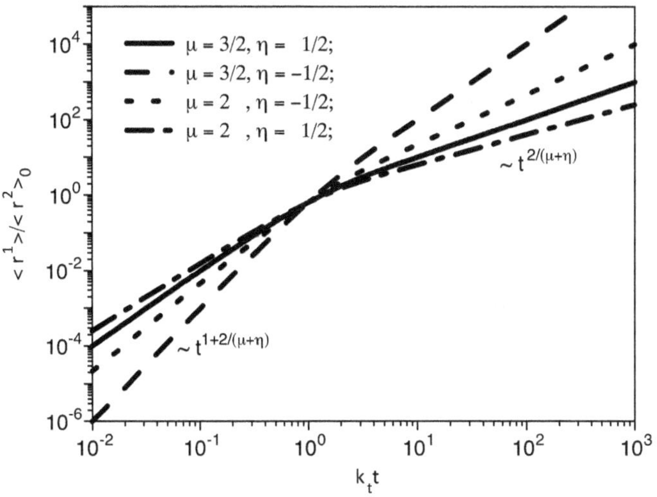

Figure 3.4 Behavior of the second moment obtained from $\rho_1(r,t)$ defined by Eqs. (3.40) for different values of μ and η. We consider, for simplicity, $d=3$, $k_1=k_2$, $\langle r^2\rangle_0=(k_2/k_t)\mathscr{I}_{\mu,\eta}$, $\varphi_2(r)=0$, $\varphi_2(r)=0$, and $\varphi_1(r)=\delta(r)/r^{d-1}$.

and

$$\widetilde{\mathscr{G}}_2(k,t)=\sum_{n=1}^{\infty}\frac{(-k_t t^{\beta})^n}{\Gamma(1+n)}E_{\beta}^{(n)}\left(-D_{\mu,\eta}k^{\mu+\eta}t^{\beta}\right), \tag{3.45}$$

where $E_{\beta}^{(n)}(x)$ is the nth derivative of the Mittag-Leffler function with respect x variable. To perform the inverse of the other integral transform and connect the solutions with the Fox H function, we use in Eq. (3.45) the following identity,

$$E_{\alpha,\mu}^{(n)}(x)=\mathrm{H}_{1,2}^{1,1}\left[-x\left|\begin{matrix}(0,1)\\(0,1),(1-(n\alpha+\mu),\alpha)\end{matrix}\right.\right], \tag{3.46}$$

which relates the nth derivative of the Mittag-Leffler function with the H Fox function. After some calculations, we have that

$$\widetilde{\mathscr{G}}_2(k,t)=\sum_{n=1}^{\infty}\frac{(-k_t t^{\beta})^n}{\Gamma(1+n)}\mathrm{H}_{1,2}^{1,1}\left[D_{\mu,\eta}k^{\mu+\eta}t^{\beta}\left|\begin{matrix}(0,1)\\(0,1),(-n\beta,\beta)\end{matrix}\right.\right]. \tag{3.47}$$

By using Eqs. (3.23) and (3.24), we obtain that

$$\begin{aligned}\rho_1(r,t)=&\frac{k_2}{k_t}\int_0^{\infty}dr' r'^{d-1}\mathscr{G}_1(r,r',t)\left(\varphi_1(r')+\varphi_2(r')\right)\\&+\frac{k_1}{k_t}\int_0^{\infty}dr' r'^{d-1}\mathscr{G}_2(r,r',t)\left(\varphi_1(r')-\frac{k_2}{k_1}\varphi_2(r')\right)\end{aligned} \tag{3.48}$$

and

$$\rho_2(k,s) = \frac{k_1}{k_t}\int_0^\infty dr' r'^{d-1}\mathscr{G}_1(r,r',t)\left(\varphi_1(r')+\varphi_2(r')\right) - \frac{k_1}{k_t}\int_0^\infty dr' r'^{d-1}\mathscr{G}_2(r,r',t)\left(\varphi_1(r')-\frac{k_2}{k_1}\varphi_2(r')\right) \tag{3.49}$$

where

$$\bar{\mathscr{G}_1}(r,r',t) = \int_0^\infty dkk^{d-1}\psi_\eta(r,k)\psi_\eta(r',k)\widetilde{\mathscr{G}_1}(k,t) = \mathscr{G}_\beta^{\bar{(0)}}(r,r',t) \tag{3.50}$$

and

$$\bar{\mathscr{G}_2}(r,r',t) = \int_0^\infty dkk^{d-1}\psi_\eta(r,k)\psi_\eta(r',k)\widetilde{\mathscr{G}_2}(k,t) = \sum_{n=1}^\infty \frac{(-k_t t^\beta)^n}{\Gamma(1+n)}\mathscr{G}_\beta^{\bar{(n)}}(r,r',t)\,, \tag{3.51}$$

with

$$\mathscr{G}_\beta^{\bar{(n)}}(r,r',t) = \frac{(2+\eta)^2}{2(\mu+\eta)r^{2+\eta}}\left(rr'\right)^{\frac{1}{2}(2-d+\eta)} \times \mathrm{H}^{1,0,1,1,1}_{2,[0:1],0,[0:2]}\left[\begin{array}{c|c} \left(\frac{r'}{r}\right)^{2+\eta} & \begin{array}{c}(\frac{2+\nu}{2},1)\,;(\frac{2-\nu}{2},1)\\ --;(0,\frac{\mu+\eta}{2+\eta})\\ --;--\end{array} \\ \frac{\overline{D}_{\mu,\eta}t^{\beta\frac{2+\eta}{\mu+\eta}}}{r^{2+\eta}} & (\frac{\nu}{2},1)\,;(-\frac{\nu}{2},1)\,;(0,\frac{\mu+\eta}{2+\eta}),(-n\beta,\beta\frac{\mu+\eta}{2+\eta}) \end{array}\right] \tag{3.52}$$

where $\mathscr{G}_\beta^{\bar{(0)}}(r,r',t) = \bar{\mathscr{G}_1}(r,r',t)$. Figure 11.5 illustrates the behavior of Eqs. (3.48) and (3.49) for different values of μ and η.

3.2.3 CASE - $\mathscr{K}_\beta(t) = \mathscr{N}'_\beta E^{-\beta' T}$, $-1 < \eta$, $\mu \neq 2$

Another possibility concerns the use of $\mathscr{K}_\beta(t) = \mathscr{N}'_\beta e^{-\beta' t}$, where $\mathscr{N}'_\beta = \mathscr{N}_\beta/(1-\beta)$ and $\beta' = \beta/(1-\beta)$ for the fractional time derivative. In this case, we have that

$$\widehat{\widetilde{\mathscr{G}}}_1(k,s) = \frac{1}{s+(s+\beta')D'_{\mu,\eta}k^{\mu+\eta}} \tag{3.53}$$

and

$$\widehat{\widetilde{\mathscr{G}}}_2(k,s) = \frac{1}{s+(s+\beta')(k'_t+D'_{\mu,\eta}k^{\mu+\eta})}\,, \tag{3.54}$$

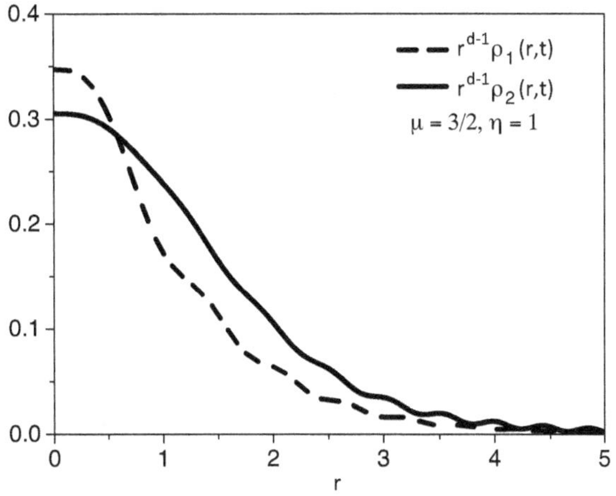

Figure 3.5 Behavior of the solutions obtained from $\rho_1(r,t)$ and $\rho_2(r,t)$ defined by Eqs. (3.48) and (3.49) for different values of μ and η. We consider, for simplicity, $D_{\mu,\eta} = 1, t = 1, d = 1$, $\beta = 1/2$, $k_1 = 2k_2 = 2$, $\varphi_2(r) = 0$, and $\varphi_1(r) = \delta(r)/r^{d-1}$.

where $D'_{\mu,\eta} = D_{\mu,\eta}/\mathcal{N}'_\beta$ and $k'_t = k_t/\mathcal{N}'_\beta$. By performing the inverse of Laplace transform, we obtain that

$$\widetilde{\mathscr{G}_1}(k,t) = \frac{1}{1+\beta' D'_{\mu,\eta} k^{\mu+\eta}} \exp\left(-\frac{\beta' D'_{\mu,\eta} t k^{\mu+\eta}}{1+\beta' D'_{\mu,\eta} k^{\mu+\eta}}\right) \tag{3.55}$$

and

$$\widetilde{\mathscr{G}_2}(k,t) = \frac{1}{1+\beta'\left(k'_t + D'_{\mu,\eta} k^{\mu+\eta}\right)} \exp\left(-\frac{\beta' t (k'_t + D'_{\mu,\eta} k^{\mu+\eta})}{1+\beta'(k'_t + D'_{\mu,\eta} k^{\mu+\eta})}\right). \tag{3.56}$$

Figure 11.6 shows the behavior of the solutions $\rho_1(r,t)$ and $\rho_2(r,t)$ for different values of β, μ, and η with the $\bar{\mathscr{G}}_1(r,r',t)$ and $\bar{\mathscr{G}}_2(r,r',t)$ obtained from Eqs. (3.55) and (3.56).

Figure 11.7 shows the behavior of $1/\rho_1^2(r,t)$ and $\rho_2^2(r,t)$ for different values of β, μ, and η with the $\bar{\mathscr{G}}_1(r,r',t)$ and $\bar{\mathscr{G}}_2(r,r',t)$ obtained from Eqs. (3.55) and (3.56). This figure illustrates the spreading of the solutions and, in particular, the anomalous spreading for a long time. In fact, the asymptotic behavior is characterized by $1/\rho_{1(2)}^2(r,t) \sim t^{\gamma_1}$ and $1/\rho_{1(2)}^2(r,t) \sim t^{\gamma_2}$, where $\gamma_1 = 0.8$ (subdiffusion, $\gamma_1 < 1$) and $\gamma_2 = 1.35$ (superdiffusion, $2 > \gamma_2 > 1$).

Now, we consider the case for which one of the species has no self-diffusion, e.g., $\mathscr{D}_2 = 0$, and the other can diffuse. This scenario is interesting and can be related to diffusion with pauses, where the diffusion process is alternated with a period of pauses (see, for example, the Refs. [Lenzi et al. (2017)] and [Lenzi et al. (2018)]). In

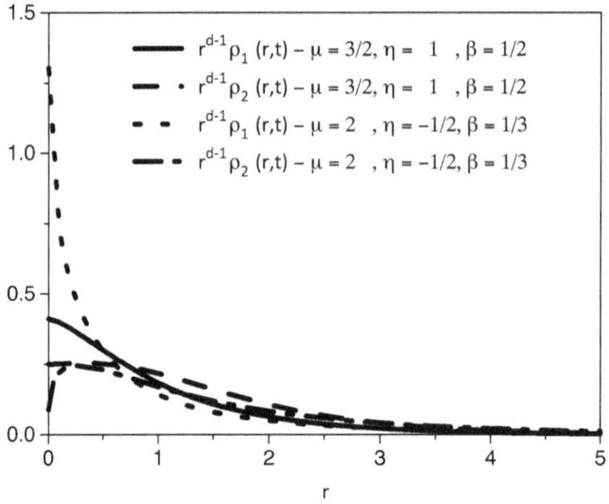

Figure 3.6 Behavior of the solutions obtained from $\rho_1(r,t)$ and $\rho_2(r,t)$ obtained from Eqs. (3.55) and (3.56) for different values of μ, β, and η. We consider, for simplicity, $\beta' D'_{\mu,\eta} t = 1$, $d = 1$, $k'_1 = 2k'_2 = 2$, $\varphi_2(r) = 0$, and $\varphi_1(r) = \delta(r)/r^{d-1}$.

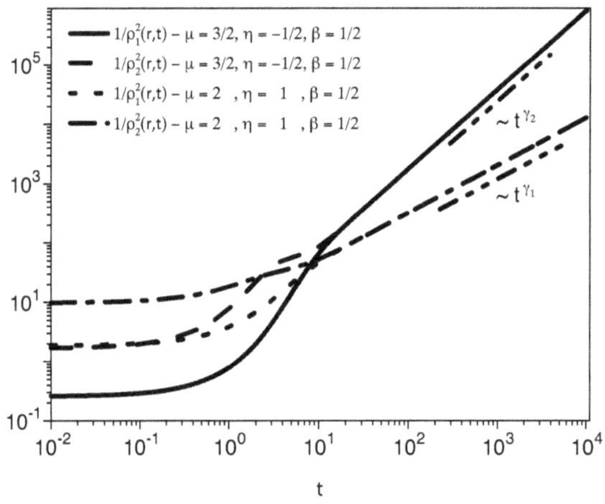

Figure 3.7 Behavior for $1/\rho_1^2(r,t)$ and $1/\rho_2^2(r,t)$ obtained from Eqs. (3.55) and (3.56) for different values of μ, β, and η. We consider, for simplicity, $\beta' D'_{\mu,\eta} t = 1$, $d = 1$, $k'_1 = 2k'_2 = 1$, $\varphi_2(r) = 0$, and $\varphi_1(r) = \delta(r)/r^{d-1}$.

this case, we have that

$$\frac{\partial^\beta}{\partial t^\beta}\rho_1(r,t) = D_{\mu,\eta}\nabla_\eta^\mu \rho_1(r,t) - k_1\rho_1(r,t) + k_2\rho_2(r,t) \tag{3.57}$$

and

$$\frac{\partial^{\beta}}{\partial t^{\beta}}\rho_2(r,t) = k_1\rho_1(r,t) - k_2\rho_2(r,t)\,. \tag{3.58}$$

The solution for this set of equations can be found by using the previous procedures. In this sense, by considering that $\varphi_2(r) = 0$, i.e., we have species in resting, it is possible to show that

$$\widehat{\widetilde{\rho}}_1(k,s) = \frac{\left(s\widehat{\mathscr{K}_{\beta}}(s)+k_2\right)\widehat{\mathscr{K}_{\beta}}(s)\widetilde{\varphi}_1(k)}{\left(s\widehat{\mathscr{K}_{\beta}}(s)+D_{\mu,\eta}k^{\mu+\eta}\right)\left(s\widehat{\mathscr{K}_{\beta}}(s)+k_2\right)+s\widehat{\mathscr{K}_{\beta}}(s)k_1} \tag{3.59}$$

and

$$\widehat{\widetilde{\rho}}_2(k,s) = \frac{k_1\widehat{\mathscr{K}_{\beta}}(s)\widetilde{\varphi}_1(k)}{\left(s\widehat{\mathscr{K}_{\beta}}(s)+D_{\mu,\eta}k^{\mu+\eta}\right)\left(s\widehat{\mathscr{K}_{\beta}}(s)+k_2\right)+s\widehat{\mathscr{K}_{\beta}}(s)k_1}\,. \tag{3.60}$$

Similarly to the case worked out above, it is necessary to choose the kernel $\widehat{\mathscr{K}_{\beta}}(s)$ to obtain the inverse of the Laplace transform. In this sense, we consider the case $\widehat{\mathscr{K}_{\beta}}(s) = s^{\beta-1}$ ($\mathscr{K}_{\beta}(t) = t^{-\beta}/\Gamma(1-\beta)$), which can be connected with the Caputo fractional time derivative. This choice yields

$$\widetilde{\rho}_1(k,t) = \widetilde{\varphi}_1(k)E_{\beta}\left(-D_{\mu,\eta}k^{\mu+\eta}t^{\beta}\right) + \sum_{n=1}^{\infty}\frac{(-k_1)^n}{\Gamma(1+n)}\int_0^t dt_n\Upsilon(t-t_n)$$
$$\times\ldots\int_0^{t_2} dt_1\Upsilon(t_2-t_1)t_1^{n\beta}E_{\beta}^{(n)}\left(-D_{\mu,\eta}k^{\mu+\eta}t_1^{\beta}\right)\widetilde{\varphi}_1(k) \tag{3.61}$$

and

$$\widetilde{\rho}_2(k,t) = k_1\int_0^t dt' t'^{\beta-1}E_{\beta,\beta}\left(\ k_2t'^{\beta}\right)\widetilde{\rho}_1(k,t-t')\,, \tag{3.62}$$

where $\Upsilon(t) = \delta(t) + k_2t^{\beta-1}E_{\beta,\beta}\left(-k_2t^{\beta}\right)$. Before performing the inverse of the other integral transform and connecting the solutions with the generalized Fox H function, we use the following identity given by Eq. (3.46), as before. By applying this result in Eq. (3.61), we have that

$$\widetilde{\rho}_1(k,t) = E_{\beta}\left(-D_{\mu,\eta}k^{\mu+\eta}t^{\beta}\right)\widetilde{\varphi}_1(k) + \sum_{n=1}^{\infty}\frac{(-k_1)^n}{\Gamma(1+n)}\int_0^t dt_n\Upsilon(t-t_n)$$
$$\times\ldots\int_0^{t_2} dt_1\Upsilon(t_2-t_1)t_1^{n\beta}\mathrm{H}_{1,2}^{1,1}\left[D_{\mu,\eta}k^{\mu+\eta}t_1^{\beta}\,\middle|\,\begin{matrix}(0,1)\\(0,1),(-n\beta,\beta)\end{matrix}\right]\widetilde{\varphi}_1(k)\,. \tag{3.63}$$

Performing the inverse of the integral transform, we obtain that

$$\rho_1(r,t) = \int_0^{\infty} dr' r'^{d-1}\varphi(r')\mathscr{G}_{\beta}^{(0)}(r,r',t) + \sum_{n=1}^{\infty}\frac{(-k_1)^n}{\Gamma(1+n)}\int_0^t dt_n\Upsilon(t-t_n)$$
$$\times\ldots\int_0^{t_2} dt_1\Upsilon(t_2-t_1)t_1^{n\beta}\int_0^{\infty} dr' r'^{d-1}\varphi_1(r')\mathscr{G}_{\beta}^{(n)}(r,r',t_1) \tag{3.64}$$

and, consequently,

$$\rho_2(r,t) = k_1 \int_0^t dt' t'^{\beta-1} E_{\beta,\beta}\left(-k_2 t'^{\beta}\right) \rho_1(r,t-t'). \tag{3.65}$$

(see, Figure 3.8). An interesting point about this case is that the diffusion of the specie 2 occurs by the reaction process and for long time, i.e., $t \to \infty$, we have that $k_1\rho_1(r,t) \approx k_2\rho_2(r,t)$ for the kernels analyzed here. This feature implies that Eqs. (3.57) and (3.58) can be approximated by the following equations:

$$\frac{\partial^{\beta}}{\partial t^{\beta}}\rho_1(r,t) = \frac{k_2}{k_1+k_2} D_{\mu,\eta} \nabla_{\eta}^{\mu} \rho_1(r,t) \tag{3.66}$$

and

$$\frac{\partial^{\beta}}{\partial t^{\beta}}\rho_2(r,t) = \frac{k_1}{k_1+k_2} D_{\mu,\eta} \nabla_{\eta}^{\mu} \rho_2(r,t)\,. \tag{3.67}$$

Equations (3.66) and (3.67) show that for long time, both species diffuse with an effective diffusion coefficient. This feature is interesting and shows that the spreading of the different species has the same spreading for long time, which will depend on the parameter β present in the fractional time derivative and the parameter η and μ connected to the fractional spatial operator. This result is also verified for the other cases analyzed above in the asymptotic limit of long times, when different kernels are considered for the fractional time derivative.

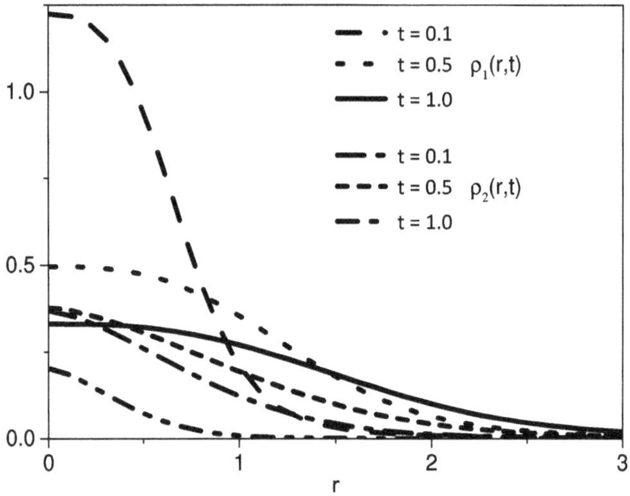

Figure 3.8 Behavior for $\rho_1(r,t)$ and $\rho_2(r,t)$ obtained from Eqs. (3.66) and (3.67), after performing the inverse of the integral transforms, for different values of μ and η. We consider, for simplicity, $D_{\mu,\eta} = 1$, $d = 1$, $k_1 = k_2 = 1$, $\beta = 1$, $\varphi_2(r) = 0$, and $\varphi_1(r) = \delta(r)/r^{d-1}$.

3.2.4 REACTION PROCESS – ARBITRARY REACTION RATES

Let us consider an arbitrary scenario for the reaction term present in Eqs. (3.13) and (3.14). By applying the previous procedure, it is possible to obtain that

$$s\widehat{\widetilde{\rho}}_1(k,s) - \widetilde{\varphi}_1(k) = -\left(D_1/\widehat{\mathscr{K}_\beta}(s)\right)k^{\mu+\eta}\widehat{\widetilde{\rho}}_1(k,s) - \left(k_{11}/\widehat{\mathscr{K}_\beta}(s)\right)\widehat{\widetilde{\rho}}_1(k,s) + \left(k_{12}/\widehat{\mathscr{K}_\beta}(s)\right)\widehat{\widetilde{\rho}}_2(k,s) \tag{3.68}$$

and

$$s\widehat{\widetilde{\rho}}_2(k,s) - \widetilde{\varphi}_2(k) = -\left(D_2/\widehat{\mathscr{K}_\beta}(s)\right)k^{\mu+\eta}\widehat{\widetilde{\rho}}_2(k,s) + \left(k_{21}/\widehat{\mathscr{K}_\beta}(s)\right)\widehat{\widetilde{\rho}}_1(k,s) - \left(k_{22}/\widehat{\mathscr{K}_\beta}(s)\right)\widehat{\widetilde{\rho}}_2(k,s)\,. \tag{3.69}$$

From this set of equations, it is possible to obtain that

$$\widehat{\widetilde{\rho}}_1(k,s) = \frac{\left(s\widehat{\mathscr{K}_\beta}(s) + D_2k^{\mu+\eta} + k_{22}\right)\widehat{\mathscr{K}_\beta}(s)\widetilde{\varphi}_1(k) + k_{12}\widehat{\mathscr{K}_\beta}(s)\widetilde{\varphi}_2(k,s)}{\left(s\widehat{\mathscr{K}_\beta}(s) + D_2k^{\mu+\eta} + k_{22}\right)\left(s\widehat{\mathscr{K}_\beta}(s) + D_1k^{\mu+\eta} + k_{11}\right) - k_{12}k_{21}} \tag{3.70}$$

and

$$\widehat{\widetilde{\rho}}_2(k,s) = \frac{\left(s\widehat{\mathscr{K}_\beta}(s) + D_1k^{\mu+\eta} + k_{11}\right)\widehat{\mathscr{K}_\beta}(s)\widetilde{\varphi}_2(k) + k_{21}\widehat{\mathscr{K}_\beta}(s)\widetilde{\varphi}_1(k,s)}{\left(s\widehat{\mathscr{K}_\beta}(s) + D_2k^{\mu+\eta} + k_{22}\right)\left(s\widehat{\mathscr{K}_\beta}(s) + D_1k^{\mu+\eta} + k_{11}\right) - k_{12}k_{21}} \tag{3.71}$$

Before applying the inverse of the integral transforms, we perform the following series expansions and obtain that

$$\widehat{\widetilde{\rho}}_1(k,s) = \frac{1}{s\widehat{\mathscr{K}_\beta}(s) + D_1k^{\mu+\eta} + k_{11}}\left[\widetilde{\varphi}_1(k) + \frac{k_{12}\widetilde{\varphi}_2(k)}{s\widehat{\mathscr{K}_\beta}(s) + D_2k^{\mu+\eta} + k_{22}}\right] + \sum_{n=1}^{\infty}\frac{(-k_{12}k_{21})^n\left[\left(s\widehat{\mathscr{K}_\beta}(s) + D_2k^{\mu+\eta} + k_{22}\right)\widetilde{\varphi}_1(k) + k_{12}\widetilde{\varphi}_2(k)\right]}{\left(s\widehat{\mathscr{K}_\beta}(s) + D_1k^{\mu+\eta} + k_{11}\right)^{n+1}\left(s\widehat{\mathscr{K}_\beta}(s) + D_2k^{\mu+\eta} + k_{22}\right)^{n+1}} \tag{3.72}$$

and

$$\widehat{\widetilde{\rho}}_2(k,s) = \frac{1}{s\widehat{\mathscr{K}_\beta}(s) + D_2k^{\mu+\eta} + k_{22}}\left[\widetilde{\varphi}_2(k) + \frac{k_{21}\widetilde{\varphi}_1(k)}{s\widehat{\mathscr{K}_\beta}(s) + D_1k^{\mu+\eta} + k_{11}}\right] + \sum_{n=1}^{\infty}\frac{(-k_{12}k_{21})^n\left[\left(s\widehat{\mathscr{K}_\beta}(s) + D_1k^{\mu+\eta} + k_{11}\right)\widetilde{\varphi}_2(k) + k_{21}\widetilde{\varphi}_1(k)\right]}{\left(s\widehat{\mathscr{K}_\beta}(s) + D_1k^{\mu+\eta} + k_{11}\right)^{n+1}\left(s\widehat{\mathscr{K}_\beta}(s) + D_2k^{\mu+\eta} + k_{22}\right)^{n+1}}\,. \tag{3.73}$$

Now applying the inverse of the integral transforms, we obtain for $\mathscr{K}_\beta = \delta(t)$ with $\eta \neq 0$ and $\mu \neq 2$ $(1 < \mu < 2)$ that

$$\begin{aligned}\rho_1(r,t) = &\int_0^t due^{-k_{11}u}\Lambda_1(r,D_1u)\\ &+k_{12}\int_0^\infty du\int_0^\infty dve^{-k_{11}u-k_{22}v}\theta(t-u-v)\Lambda_2(r,D_1u+D_2v)\\ &+\sum_{n=1}^\infty(-k_{12}k_{21})^n\left\{\int_0^\infty duu^n\int_0^\infty dvv^ne^{-k_{11}u-k_{22}v}\theta(t-u-v)\right.\\ &\left.\times\left[\frac{1}{v}\Lambda_1(r,D_1u+D_2v)+k_{12}\Lambda_2(r,D_1u+D_2v)\right]\right\}\end{aligned} \tag{3.74}$$

and

$$\begin{aligned}\rho_2(r,t) = &\int_0^t due^{-k_{22}u}\Lambda_2(r,D_2u)\\ &+k_{21}\int_0^\infty du\int_0^\infty dve^{-k_{11}u-k_{22}v}\theta(t-u-v)\Lambda_1(r,D_1u+D_2v)\\ &+\sum_{n=1}^\infty(-k_{12}k_{21})^n\left\{\int_0^\infty duu^n\int_0^\infty dvv^ne^{-k_{11}u-k_{22}v}\theta(t-u-v)\right.\\ &\left.\times\left[\frac{1}{v}\Lambda_2(r,D_1u+D_2v)+k_{21}\Lambda_1(r,D_1u+D_2v)\right]\right\}\end{aligned} \tag{3.75}$$

where $\Lambda_i(r,t) = \int_0^\infty dr'\varphi_i(r')\bar{\mathscr{G}}(r,r',t)$, $\bar{\mathscr{G}}(r,r',t) = \bar{\mathscr{G}}_1(r,r',t)$ with $D_{1(2)}t \to t$, and $\theta(t)$ is the step function. An interesting case is obtained when $k_{22} = k_{12} = 0$ and $k_{11} = k_{21} \neq 0$, i.e., one specie by reaction process produce the other specie. In this scenario, Eqs. (3.71) and (3.73) can be simplified to the following equations:

$$\widehat{\widetilde{\rho}}_1(k,s) = \frac{\widehat{\mathscr{K}_\beta}(s)\widetilde{\varphi}_1(k)}{s\widehat{\mathscr{K}_\beta}(s)+D_1k^{\mu+\eta}+k_{11}} \tag{3.76}$$

and

$$\widehat{\widetilde{\rho}}_2(k,s) = \frac{k_{11}\widehat{\mathscr{K}_\beta}(s)\widetilde{\varphi}_1(k,s)}{\left(s\widehat{\mathscr{K}_\beta}(s)+D_2k^{\mu+\eta}\right)\left(s\widehat{\mathscr{K}_\beta}(s)+D_1k^{\mu+\eta}+k_{11}\right)} \tag{3.77}$$

for the initial condition $\widetilde{\varphi}_2(k) = 0$, which for the case $\mathscr{K}_\beta = \delta(t)$ yields

$$\rho_1(r,t) = \int_0^\infty dr'r'^{d-1}\varphi_1(r')\bar{\mathscr{G}}_2'(r,r',t) \tag{3.78}$$

where $\bar{\mathscr{G}}_2(r,r',t) = \bar{\mathscr{G}}_2'(r,r',t)$ with $k_t = k_{11}$ and

$$\begin{aligned}\rho_2(r,t) = k_{11}&\int_0^\infty dr'r'^{d-1}\varphi_1(r')\\ &\times\int_0^t dt'\int_0^\infty dr''r''^{d-1}\bar{\mathscr{G}}_1'(r,r'',t-t')\bar{\mathscr{G}}_2'(r'',r',t'),\end{aligned} \tag{3.79}$$

where $\bar{\mathscr{G}}'_1(r,r',t) = \bar{\mathscr{G}}_1(r,r',t)$ with $k_1 \to k_{11}$ and $D_1 \to D_2$. For the case $\mathscr{K}_\beta(t) = \mathscr{N}'_\beta e^{-\beta' t}$, which corresponds to a nonsingular differential operator, we have that

$$\widetilde{\rho}_1(k,t) = \widetilde{\varphi}_1(k)\bar{\mathscr{G}}_2(k,t) \tag{3.80}$$

with $\widetilde{\mathscr{G}}_2(k,t)$ defined by Eq. (3.55) with $D'_{\mu,\eta} \to D'_1 = D_1/\mathscr{N}'_\beta$ and $k'_t \to k'_1 = k_{11}/\mathscr{N}'_\beta$, i.e.,

$$\widetilde{\mathscr{G}}_2(k,t) = \frac{1}{1+\beta'\left(k'_1 + D'_1 k^{\mu+\eta}\right)} \exp\left(-\frac{\beta' t(k'_1 + D'_1 k^{\mu+\eta})}{1+\beta'(k'_1 + D'_1 k^{\mu+\eta})}\right). \tag{3.81}$$

and

$$\widetilde{\rho}_2(k,t) = k_{11}\widetilde{\varphi}_1(k)\int_0^t dt' \widetilde{\mathscr{G}}_1(k,t)\widetilde{\mathscr{G}}_2(k,t-t') \tag{3.82}$$

with $\widetilde{\mathscr{G}}_1(k,t)$ defined by Eq. (3.56) with $D'_{\mu,\eta} \to D'_2 = D_2/\mathscr{N}'_\beta$, i.e.,

$$\widetilde{\mathscr{G}}_1(k,t) = \frac{1}{1+\beta' D'_2 k^{\mu+\eta}} \exp\left(-\frac{\beta' D'_2 t k^{\mu+\eta}}{1+\beta' D'_2 k^{\mu+\eta}}\right). \tag{3.83}$$

Figure 11.8 illustrates the behavior of Eqs. (3.80) and (3.82), after performing the inverse integral transform. In particular, it shows that the specie 1 is converted by the reaction process in the specie 2.

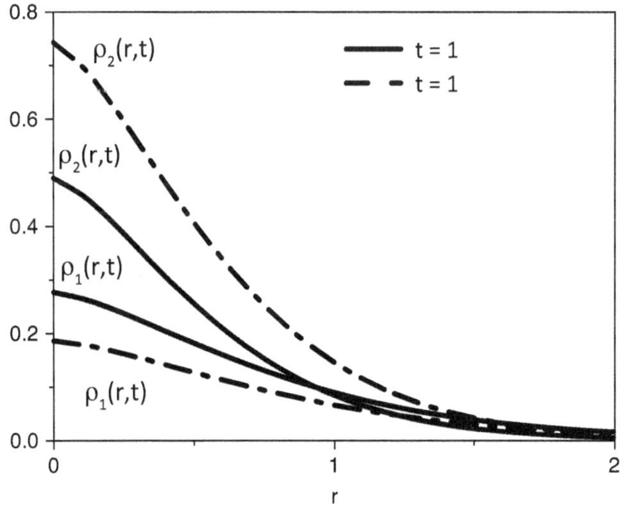

Figure 3.9 Behavior for $\rho_1(r,t)$ and $\rho_2(r,t)$ obtained from Eqs. (3.80) and (3.82), after performing the inverse integral transform. We consider, for simplicity, $D'_1 = D'_2 = 1$, $\beta = 1/2$, $d = 1$, $k'_1 = 2$, $\mu = 3/2$, $\eta = 1$, $\varphi_2(r) = 0$, and $\varphi_1(r) = \delta(r)/r^{d-1}$.

3.3 DISCUSSION AND CONCLUSION

We have investigated the solutions for fractional diffusion equations that employ fractional derivatives in space and time. This set of diffusion equations is coupled with reaction terms which may represent different processes depending on the constants related to the reaction terms. In particular, they can be reversible or irreversible, as analyzed previously. Other cases imply considering different values for the reaction constants. The boundary conditions for these equations are considered in the free space, i.e., we have no surfaces. For the fractional derivatives in the spatial variable, we have considered an extension of the Lévy distribution by incorporating a dependence of the parameter η, which implies a heterogeneous diffusion process. In fact, for the case, $\mu = 2$, the fractional derivative in space can be connected to a diffusion process with a spatial dependence on the diffusion coefficient as performed elsewhere [Lenzi and Evangelista (2021)] and, consequently, to a generalized Hankel transform. This discussion shows that this fractional operator in space considers the mixing between two different aspects which may be manifested in the system. For the fractional derivatives in time, we have considered an arbitrary situation by considering $K_\beta(t) = \delta(t)$, $\mathscr{K}_\beta(t) = \mathscr{N}_\beta' e^{-\beta' t}$, and $\mathscr{K}_\beta(t) = t^{-\beta}/\Gamma(1-\beta)$. Each one of these kernels is connected to a differential operator. In particular, we have the standard differential and fractional operators with singular and nonsingular kernels. In this scenario, we have obtained solutions for different choices of reaction terms and fractional operators. We have also analyzed the case with different diffusion coefficients. In this sense, an interesting case is obtained when one of the diffusion coefficients is absent, i.e., $D_1 = 0$ or $D_2 = 0$, implying a diffusion process with pauses. The diffusion process was analyzed and presented different behaviors depending on the choices for the set of equations. Another aspect analyzed for this case was the behavior of the solutions for long times, which allowed us to show that diffusion equations govern both species with effective diffusion coefficients. After considering these points, we have analyzed an arbitrary case scenario for the reaction rates by considering different coefficients of diffusive, i.e., $D_1 \neq D_2$. For this scenario, we have also obtained a general solution for the kernel $K_\beta(t) = \delta(t)$, with $\mu \neq 2$ and $\eta \neq 0$ in terms of the generalized Fox H functions. For the other case, we have also obtained the solutions. These results could be extended by considering the presence of drift terms to take external fields into account, dependent on space and time. Other reaction terms could be considered, in particular, nonlinear reaction terms to analyze complex processes such as impulses and physiological states in nerve membrane [FitzHugh (1961)], synthesis of materials [Shevchenko, Makogon, and Sychov (2021)], metapopulation in heterogeneous networks [Colizza, Pastor-Satorras, and Vespignani (2007)], and coagulation processes [Partohaghighi et al. (2022)]. Finally, we hope that the results found here in the H Fox functions or the generalized H Fox functions may be useful in discussing the scenarios related to anomalous diffusion.

ACKNOWLEDGMENT

E.K.L. thanks partial financial support of the CNPq under Grant No. 301715/2022-0. M.K.L. thanks partial financial support of the CNPq under Grant No. 309810/2021-3.

References

Ali and Kalla (1999). Ali, Ismail, and Shyam L Kalla. 1999. "A generalized Hankel transform and its use for solving certain partial differential equations." *The ANZIAM Journal* 41 (1): 105–117.

Angstmann, Donnelly, and Henry (2013). Angstmann, Christopher N, Isaac C Donnelly, and Bruce I Henry. 2013. "Continuous time random walks with reactions forcing and trapping." *Mathematical Modelling of Natural Phenomena* 8 (2): 17–27.

Atangana and Baleanu (2016). Atangana, Abdon, and Dumitru Baleanu. 2016. "New fractional derivatives with non-local and non-singular kernel." *Thermal Science* 20 (2): 763–769.

Barbero, Evangelista, and Lenzi (2022). Barbero, Giovanni, Luiz Roberto Evangelista, and Ervin Kaminski Lenzi. 2022. "Time-fractional approach to the electrochemical impedance: The displacement current." *Journal of Electroanalytical Chemistry* 920: 116588. https://www.sciencedirect.com/science/article/pii/S157266572200580X.

Ben-Avraham and Havlin (2000). Ben-Avraham, Daniel, and Shlomo Havlin. 2000. *Diffusion and Reactions in Fractals and Disordered Systems*. Cambridge University Press: Cambridge.

Boffetta and Sokolov (2002). Boffetta, Guido, and Igor M Sokolov. 2002. "Relative dispersion in fully developed turbulence: The Richardson's Law and intermittency corrections." *Physical Review Letters* 88: 094501. https://link.aps.org/doi/10.1103/PhysRevLett.88.094501.

Brault et al. (2009). Brault, Pascal, Christophe Josserand, Jean-Marc Bauchire, Amaël Caillard, Christine Charles, and Rod W Boswell. 2009. "Anomalous diffusion mediated by atom deposition into a porous substrate." *Physical Review Letters.* 102: 045901. https://link.aps.org/doi/10.1103/PhysRevLett.102.045901.

Colizza, Pastor-Satorras, and Vespignani (2007). Colizza, Vittoria, Romualdo Pastor-Satorras, and Alessandro Vespignani. 2007. "Reaction–diffusion processes and metapopulation models in heterogeneous networks." *Nature Physics* 3 (4): 276–282.

Crank (1979). Crank, John. 1979. *The Mathematics of Diffusion*. Oxford University Press: Oxford.

Cussler and Cussler (2009). Cussler, Edward Lansing, and Edward Lansing Cussler. 2009. *Diffusion: Mass Transfer in Fluid Systems*. Cambridge University Press: Cambridge.

Daniel ben Avraham (2000). Daniel ben Avraham, Shlomo Havlin. 2000. *Diffusion and Reactions in Fractals and Disordered Systems*. Cambridge University Press: Cambridge.

Einstein (1905). Einstein, Albert. 1905. "Über die von der molekularkinetischen Theorie der Wärme geforderte Bewegung von in ruhenden Flüssigkeiten suspendierten Teilchen." *Annals of Physics*. 322: 549–560.

Evangelista and Lenzi (2018). Evangelista, Luiz Roberto, and Ervin Kaminski Lenzi. 2018. *Fractional Diffusion Equations and Anomalous Diffusion*. Cambridge University Press: Cambridge.

Fernandez and Baleanu (2021). Fernandez, Arran, and Dumitru Baleanu. 2021. "Classes of operators in fractional calculus: A case study." *Mathematical Methods in the Applied Sciences* 44 (11): 9143–9162.

FitzHugh (1961). FitzHugh, Richard. 1961. "Impulses and physiological states in theoretical models of nerve membrane." *Biophysical Journal* 1 (6): 445–466.

Frank (2005). Frank, Till Daniel. 2005. *Nonlinear Fokker-Planck Equations: Fundamentals and Applications*. Springer Science & Business Media: Berlin, Germany.

Garg, Rao, and Kalla (2007). Garg, Mridula, Alka Rao, and Shyam L Kalla. 2007. "On a generalized finite Hankel transform." *Applied Mathematics and Computation* 190 (1): 705–711.

Henry and Wearne (2000). Henry, Bruce I, and Susan L Wearne. 2000. "Fractional reaction–diffusion." *Physica A: Statistical Mechanics and Its Applications* 276 (3–4): 448–455.

Iida, Ninomiya, and Yamamoto (2018). Iida, Masato, Hirokazu Ninomiya, and Hiroko Yamamoto. 2018. "A review on reaction–diffusion approximation." *Journal of Elliptic and Parabolic Equations* 4 (2): 565–600.

Jiang and Xu (2010). Jiang, Xiaoyun, and Mingyu Xu. 2010. "The time fractional heat conduction equation in the general orthogonal curvilinear coordinate and the cylindrical coordinate systems." *Physica A: Statistical Mechanics and Its Applications* 389 (17): 3368–3374.

Kenkre, Montroll, and Shlesinger (1973). Kenkre, Vasudev M, Elliot W Montroll, and Michael F Shlesinger. 1973. "Generalized master equations for continuous-time random walks." *Journal of Statistical Physics* 9 (1): 45–50.

Kou and Xie (2004). Kou, Samuel C, and Xiaoliang Sunney Xie. 2004. "Generalized Langevin equation with fractional Gaussian noise: Subdiffusion within a single protein molecule." *Physical Review Letters* 93 (18): 180603.

Langevin (1908). Langevin, Paul. 1908. "Sur la théorie du mouvement brownien." *Comptes-rendus de l'Académie des Sciences de Paris* 146: 530.

Langlands, Henry, and Wearne (2007). Langlands, Trevor A M, Bruce I Henry, and Susan L Wearne. 2007. "Turing pattern formation with fractional diffusion and fractional reactions." *Journal of Physics: Condensed Matter* 19 (6): 065115.

Lenzi et al. (2017). Lenzi, Ervin Kaminski, Luciano R Da Silva, Marcelo K Lenzi, Maike A F Dos Santos, Haroldo V Ribeiro, and Luiz Roberto Evangelista. 2017. "Intermittent motion, nonlinear diffusion equation and tsallis formalism." *Entropy* 19 (1): 42.

Lenzi and Evangelista (2021). Lenzi, Ervin Kaminski, and Luiz Roberto Evangelista. 2021. "Space–time fractional diffusion equations in d-dimensions." *Journal of Mathematical Physics* 62 (8): 083304.

Lenzi et al. (2010). Lenzi, Ervin Kaminski, Luiz Roberto Evangelista, Marcelo K Lenzi, Haroldo V Ribeiro, and Edmundo Capelas de Oliveira. 2010. "Solutions for a non-Markovian diffusion equation." *Physics Letters A* 374 (41): 4193–4198.

Lenzi et al. (2022). Lenzi, Ervin Kaminski, Luiz Roberto Evangelista, Rafael S Zola, and Antonio M Scarfone. 2022. "Fractional Schrödinger equation for heterogeneous media and Lévy like distributions." *Chaos, Solitons & Fractals* 163: 112564.

Lenzi et al. (2014). Lenzi, Ervin Kaminski, Marcelo K Lenzi, Rafael S Zola, Haroldo V Ribeiro, Fernanda C Zola, Luiz Roberto Evangelista, and Giane Goncalves. 2014. "Reaction on a solid surface supplied by an anomalous mass transfer source." *Physica A: Statistical Mechanics and Its Applications* 410: 399–406.

Lenzi et al. (2018). Lenzi, Ervin Kaminski, Maurício A Ribeiro, Maria E K Fuziki, Marcelo K Lenzi, and Haroldo V Ribeiro. 2018. "Nonlinear diffusion equation with reaction terms: Analytical and numerical results." *Applied Mathematics and Computation* 330: 254–265.

Lenzi et al. (2016). Lenzi, Ervin Kaminski, Rafael Menechini Neto, Angel A Tateishi, Marcelo K Lenzi, and Haroldo V Ribeiro. 2016. "Fractional diffusion equations coupled by reaction terms." *Physica A: Statistical Mechanics and Its Applications* 458: 9–16.

Li et al. (2020). Li, Angran, Ruijia Chen, Amir Barati Farimani, and Yongjie Jessica Zhang. 2020. "Reaction diffusion system prediction based on convolutional neural network." *Scientific Reports* 10 (1): 1–9.

Manakova and Gavrilova (2018). Manakova, Natalia A, and Olga V Gavrilova. 2018. "Numerical study of the process of optimizing the propagation of a nerve impulse in a membrane for a three-component model." In *2018 International Russian Automation Conference (RusAutoCon), Sochi, Russia*, 1–5. doi: 10.1109/RUSAUTOCON.2018.8501788.

Mathai, Saxena, and Haubold (2009). Mathai, Arakaparampil M, Ram Kishore Saxena, and Hans J Haubold. 2009. *The H-Function: Theory and Applications*. Springer Science & Business Media: Berlin, Germany.

Means et al. (2006). Means, Shawn, Alexander J Smith, Jason Shepherd, John Shadid, John Fowler, Richard J H Wojcikiewicz, Tomas Mazel, Gregory D Smith, and Bridget S Wilson. 2006. "Reaction diffusion modeling of calcium dynamics with realistic ER geometry." *Biophysical Journal* 91 (2): 537–557.

Méndez, Campos, and Bartumeus (2016). Méndez, Vicenc, Daniel Campos, and Frederic Bartumeus. 2016. *Stochastic Foundations in Movement Ecology*. Springer: Berlin, Germany.

Metzler and Klafter (2000). Metzler, Ralf, and Joseph Klafter. 2000. "The random walk's guide to anomalous diffusion: A fractional dynamics approach." *Physics Reports* 339 (1): 1–77.

Nakhi and Kalla (2003). Nakhi, Y Ben, and Shyam L Kalla. 2003. "Some boundary value problems of temperature fields in oil strata." *Applied Mathematics and Computation* 146 (1): 105–119.

Nec and Nepomnyashchy (2007). Nec, Yana, and Alexander A Nepomnyashchy. 2007. "Linear stability of fractional reaction-diffusion systems." *Mathematical Modelling of Natural Phenomena* 2 (2): 77–105.

O'Shaughnessy and Procaccia (1985a). O'Shaughnessy, Ben, and Itamar Procaccia. 1985a. "Analytical solutions for diffusion on fractal objects." *Physical Review Letters* 54: 455–458.

O'Shaughnessy and Procaccia (1985b). O'Shaughnessy, Ben, and Itamar Procaccia. 1985b. "Diffusion on fractals." *Physical Review A* 32: 3073–3083.

Partohaghighi et al. (2022). Partohaghighi, Mohammad, Ali Akgül, Liliana Guran, and Monica-Felicia Bota. 2022. "Novel mathematical modelling of Platelet-Poor Plasma Arising in a blood coagulation system with the fractional Caputo–Fabrizio derivative." *Symmetry* 14 (6): 1128.

Pearson (1905). Pearson, Karl. 1905. "The problem of the random walk." *Nature* 72: 294.

Pekalski and Sznajd-Weron (1999). Pekalski, Andrzej, and Katarzyna Sznajd-Weron. 1999. *Anomalous Diffusion: From Basics to Applications*. Springer: Berlin, Germany.

Podlubny (1999). Podlubny, Igor. 1999. *Fractional Differential Equations*. Academic Press: Cambridge, MA.

Poinsot and Veynante (2005). Poinsot, Thierry, and Denis Veynante. 2005. *Theoretical and Numerical Combustion*. RT Edwards, Inc.

Riaz and Ray (2007). Riaz, Syed Shahed, and Deb Shankar Ray. 2007. "Diffusion and mobility driven instabilities in a reaction-diffusion system: A review." *Indian Journal of Physics and Proceedings of the Indian Association for the Cultivation of Science - New Series* 81: 1177.

Richardson (1926). Richardson, Lewis Fry. 1926. "Atmospheric diffusion shown on a distance-neighbour graph." *Proceedings of the Royal Society A: Mathematical, Physical and Engineering Sciences*. 110 (756): 709–737.

Rinzel and Terman (1982). Rinzel, John, and David Terman. 1982. "Propagation phenomena in a bistable reaction-diffusion system." *SIAM Journal on Applied Mathematics* 42 (5): 1111–1137.

Saxton (2007). Saxton, Michael J. 2007. "A biological interpretation of transient anomalous subdiffusion. I. Qualitative model." *Biophysical Journal* 92 (4): 1178–1191.

Scarfone et al. (2022). Scarfone, Antonio M, Giovanni Barbero, Luiz Roberto Evangelista, and Ervin Kaminski Lenzi. 2022. "Anomalous diffusion and surface effects on the electric response of electrolytic cells." *Physchem* 2 (2): 163–178.

Seki, Wojcik, and Tachiya (2003). Seki, Kazuhiko, Mariusz Wojcik, and Masanori Tachiya. 2003. "Fractional reaction-diffusion equation." *The Journal of Chemical Physics* 119 (4): 2165–2170.

Shevchenko, Makogon, and Sychov (2021). Shevchenko, Vladimir Ya, Aleksei I Makogon, and Maxim M Sychov. 2021. "Modeling of reaction-diffusion processes of synthesis of materials with regular (periodic) microstructure." *Open Ceramics* 6: 100088.

Singh (2019). Singh, Harendra. 2019. "An efficient computational method for non-linear fractional Lienard equation arising in oscillating circuits." In Harendra Singh, Devendra Kumar, Dumitru Baleanu (Eds.), *Methods of Mathematical Modelling*, 39–50. CRC Press: Boca Raton, FL.

Singh (2020). Singh, Harendra. 2020. "Analysis for fractional dynamics of Ebola virus model." *Chaos, Solitons & Fractals* 138: 109992.

Singh (2021). Singh, Harendra. 2021. "Chebyshev spectral method for solving a class of local and nonlocal elliptic boundary value problems." *International Journal of Nonlinear Sciences and Numerical Simulation* 000010151520200235. doi:10.1515/ijnsns-2020-0235 .

Singh (2022). Singh, Harendra. 2022. "Solving a class of local and nonlocal elliptic boundary value problems arising in heat transfer." *Heat Transfer* 51 (2): 1524–1542.

Singh and Wazwaz (2021). Singh, Harendra, and Abdul-Majid Wazwaz. 2021. "Computational method for reaction diffusion-model arising in a spherical catalyst." *International Journal of Applied and Computational Mathematics* 7 (3): 1–11.

Singh, Kumar, and Baleanu (2019). Singh, Harendra, Devendra Kumar, and Dumitru Baleanu. 2019. *Methods of Mathematical Modelling: Fractional Differential Equations*. CRC Press: Boca Raton, FL.

Singh, Kumar, and Baleanu (2022). Singh, Harendra, Devendra Kumar, and Dumitru Baleanu. 2022. *Methods of Mathematical Modelling: Infectious Disease*. Elsevier Science: Amsterdam, Netherlands.

Singh and Srivastava (2020). Singh, Harendra, and Hari Mohan Srivastava. 2020. "Numerical investigation of the fractional-order Liénard and Duffing equations arising in oscillating circuit theory." *Frontiers in Physics* 8: 120.

Singh, Srivastava, and Nieto (2022). Singh, Harendra, Hari Mohan Srivastava, and Juan J Nieto. 2022. *Handbook of Fractional Calculus for Engineering and Science*. CRC Press: Boca Raton, FL.

Smoluchowski (1906). Smoluchowski, Marian Von. 1906. "Zur kinetischen theorie der brownschen molekularbewegung und der suspensionen." *Annals of Physics* 326: 756–780.

Su et al. (2005). Su, Ninghu, Graham C Sander, Fawang Liu, Vo Anh, and David A Barry. 2005. "Similarity solutions for solute transport in fractal porous media using a time- and scale-dependent dispersivity." *Applied Mathematical Modelling* 29 (9): 852–870.

Szalai and De Kepper (2004). Szalai, István, and Patrick De Kepper. 2004. "Turing patterns, spatial bistability, and front instabilities in a reaction- diffusion system." *The Journal of Physical Chemistry A* 108 (25): 5315–5321.

Tateishi, Ribeiro, and Lenzi (2017). Tateishi, Angel A, Haroldo V Ribeiro, and Ervin Kaminski Lenzi. 2017. "The role of fractional time-derivative operators on anomalous diffusion." *Frontiers in Physics* 5: 52.

Tripathi et al. (2021). Tripathi, Vivek Mani, Hari Mohan Srivastava, Harendra Singh, Chetan Swarup, and Sudhanshu Aggarwal. 2021. "Mathematical analysis of non-isothermal reaction–diffusion models arising in spherical catalyst and spherical biocatalyst." *Applied Sciences* 11 (21): 10423.

Wyld and Powell (2020). Wyld, Henry William, and Gary Powell. 2020. *Mathematical Methods for Physics*. CRC Press: Boca Raton, FL.

Xie et al. (2010). Xie, Kanghe, Yulin Wang, Kun Wang, and Xin Cai. 2010. "Application of Hankel transforms to boundary value problems of water flow due to a circular source." *Applied Mathematics and Computation* 216 (5): 1469–1477.

4 Computable Solution of Fractional Kinetic Equations Associated with Incomplete ℵ-Functions and 𝕄-Series

Nidhi Jolly
Maharaja Surajmal Institute

Manish Kumar Bansal
Jaypee Institute of Information Technology

CONTENTS

4.1 INTRODUCTION

Fractional Calculus (FC) is an efficacious branch of mathematical analysis that facilitates the study of derivatives and integrals of fractional order [4,21,28,31,32,40, 43–45,52,55]. Origin of FC is acknowledged by the letter written to L'Hopital by Leibniz (in 1695), inquiring about the extension of order n in $\frac{d^n f(x)}{dx^n}$ to a fraction. Distinguished mathematicians like Abel, Caputo, Liouville, Fourier, Podlubny, Riemann, etc. made phenomenal contributions to the development of FC over the years. Recently, being widely studied for its diverse applications in modeling and identification [22–27], control systems, signal processing, astrophysics, biology, robotics, heat transfer, and anomalous diffusion theory [2,6–9,11,29,30,48,50,53,54,56,58,63,64] was originally considered to be a theoretical field of pure mathematics. Fractional differential equations particularly kinetic equations for the ability to create mathematical models expressing innumerable physical phenomena have acquired noteworthy significance in control systems, dynamical systems, and mathematical physics. In

DOI: 10.1201/9781003368069-4

view of generic nature of FKE associated with special functions and fractional operators [17,49,60], the authors introduce a new generalization of FKE involving incomplete $\aleph$-functions and $\mathbb{M}$-series. The fractional differential equation established by Haubold and Mathai [17] describes explicitly the relation between the rate of change of reaction $\Phi = \Phi(v)$, production rate $p = p(\Phi)$ and destruction rate $d = d(\Phi)$ as follows:

$$\frac{d\Phi}{dv} = -d(\Phi_v) + \mathrm{p}(\Phi_v), \tag{4.1}$$

where Φ_v represents $\Phi_v(v^*) = \Phi(v - v^*), v^* > 0$.
Upon ignoring the inhomogeneities or spatial fluctuations in Φ_v, we obtain the following special case of Eq. (4.1):

$$\frac{d\Phi_i}{dv} = -\mathrm{c_i}\Phi_i(v), \tag{4.2}$$

along with the initial condition $\Phi_i(v = 0) = \Phi_0$ representing the number of density of species "i" at a given time $(v = 0), \mathrm{c_i} > 0$.
The aforementioned equation can be expressed as follows after integration (ignoring index "i"):

$$\Phi(v) - \Phi_0 = -\mathrm{c}\, {}_0D_v^{-1}\Phi(v), \tag{4.3}$$

where ${}_0D_{\mathrm{w}}^{-1}$ represents the special form of well-known Riemann-Liouville (RL) integral operator [31,32].
The fractional extension of equation (4.3) is given by:

$$\Phi(v) - \Phi_0 = -\mathrm{c}^{\beta}\, {}_0D_v^{-\beta}\Phi(v), \tag{4.4}$$

Equation (4.4) represents the familiar FKE.
Haubold and Mathai [17] gave the solution of above-mentioned FKE in the series form mentioned below:

$$\Phi(v) = \Phi_0 \sum_{\vartheta=0}^{\infty} \frac{(-1)^{\vartheta}}{\Gamma(\beta\vartheta + 1)} (\mathrm{c}v)^{\beta\vartheta}. \tag{4.5}$$

Südland et al. [46,47] in 1998 introduced the Aleph function recognizing its significance in applied mathematics, physics, engineering disciplines, etc. Taking into account the ongoing study of incomplete special functions and its applications (see, for details, Srivastava et al. [19,20], Bansal and Choi [39], Bansal et al. [35], Özarslan and Ustaoğlu [34]), the Incomplete Aleph $\aleph$-functions ${}^{(\Gamma)}\aleph^{k,l}_{m_i,n_i,\rho_i;g}(z)$ and ${}^{(\gamma)}\aleph^{k,l}_{m_i,n_i,\rho_i;g}(z)$ containing incomplete gamma functions $\gamma(u,y)$ and $\Gamma(u,y)$ (which also play an important role in engineering and science, see, [5, Chapter 9], [51]) was introduced and investigated by Bansal et al. [36] as follows:

$$\begin{aligned} {}^{(\Gamma)}\aleph^{k,l}_{m_i,n_i,\rho_i;q}(z) &= {}^{(\Gamma)}\aleph^{k,l}_{m_i,n_i,\rho_i;q}\left[z \,\middle|\, \begin{array}{c} (\mathfrak{f}_1,\mathfrak{F}_1,x),(\mathfrak{f}_j,\mathfrak{F}_j)_{2,l},[\rho_j(\mathfrak{f}_{ji},\mathfrak{F}_{ji})]_{l+1,m_i} \\ \\ (h_j,\mathfrak{H}_j)_{1,k},[\rho_j(h_{ji},\mathfrak{H}_{ji})]_{k+1,n_i} \end{array} \right] \\ &= \frac{1}{2\pi i}\int_{\mathfrak{L}} M(\xi,x) z^{-\xi} d\xi, \qquad (z \neq 0) \end{aligned} \tag{4.6}$$

where

$$M(\xi,x)=\frac{\Gamma(1-\mathfrak{f}_1-\mathfrak{F}_1\xi,x)\prod_{j=1}^{k}\Gamma(h_j+\mathfrak{H}_j\xi)\prod_{j=2}^{l}\Gamma(1-\mathfrak{f}_j-\mathfrak{F}_j\xi)}{\sum_{i=1}^{q}\rho_i\left[\prod_{j=k+1}^{n_i}\Gamma(1-h_{ji}-\mathfrak{H}_{ji}\xi)\prod_{j=l+1}^{m_i}\Gamma(\mathfrak{f}_{ji}+\mathfrak{F}_{ji}\xi)\right]}, \tag{4.7}$$

and

$$\begin{aligned}{}^{(\gamma)}\aleph_{m_i,n_i,\rho_i;q}^{k,l}(z) &= {}^{(\gamma)}\aleph_{m_i,n_i,\rho_i;q}^{k,l}\left[z\left|\begin{matrix}(\mathfrak{f}_1,\mathfrak{F}_1,x),(\mathfrak{f}_j,\mathfrak{F}_j)_{2,l},[\rho_j(\mathfrak{f}_{ji},\mathfrak{F}_{ji})]_{l+1,m_i}\\ (h_j,\mathfrak{H}_j)_{1,k},[\rho_j(h_{ji},\mathfrak{H}_{ji})]_{k+1,n_i}\end{matrix}\right.\right] \\ &= \frac{1}{2\pi i}\int_{\mathfrak{L}}N(\xi,x)z^{-\xi}d\xi, \qquad (z\neq 0)\end{aligned} \tag{4.8}$$

where

$$N(\xi,x)=\frac{\gamma(1-\mathfrak{f}_1-\mathfrak{F}_1\xi,x)\prod_{j=1}^{k}\Gamma(h_j+\mathfrak{H}_j\xi)\prod_{j=2}^{l}\Gamma(1-\mathfrak{f}_j-\mathfrak{F}_j\xi)}{\sum_{i=1}^{q}\rho_i\left[\prod_{j=k+1}^{n_i}\Gamma(1-h_{ji}-\mathfrak{H}_{ji}\xi)\prod_{j=l+1}^{m_i}\Gamma(\mathfrak{f}_{ji}+\mathfrak{F}_{ji}\xi)\right]}. \tag{4.9}$$

The functions ${}^{(\Gamma)}\aleph_{m_i,n_i,\rho_i;g}^{k,l}(z)$ and ${}^{(\gamma)}\aleph_{m_i,n_i,\rho_i;g}^{k,l}(z)$ in equations (4.6) and (4.8) respectively, exists $\forall\, x\geq 0$ under the conditions stated [36, p. 5604, Eq. (10-13)].

By setting certain parameters the Incomplete Aleph $\aleph$-functions ${}^{(\Gamma)}\aleph_{m_i,n_i,\rho_i;g}^{k,l}(z)$ and ${}^{(\gamma)}\aleph_{m_i,n_i,\rho_i;g}^{k,l}(z)$ can be transformed into several well-known functions stated as follows:

(i) Substituting $x=0$ in Eq. (4.6), we obtain the $\aleph$-function given by Sündland [46,47]:

$$\begin{aligned}&{}^{(\Gamma)}\aleph_{m_i,n_i,\rho_i;q}^{k,l}\left[z\left|\begin{matrix}(\mathfrak{f}_1,\mathfrak{F}_1,0),(\mathfrak{f}_j,\mathfrak{F}_j)_{2,l},[\rho_j(\mathfrak{f}_{ji},\mathfrak{F}_{ji})]_{l+1,m_i}\\ (h_j,\mathfrak{H}_j)_{1,k},[\rho_j(h_{ji},\mathfrak{H}_{ji})]_{k+1,n_i}\end{matrix}\right.\right] \\ &\qquad = \aleph_{m_i,n_i,\rho_i;q}^{k,l}\left[z\left|\begin{matrix}(\mathfrak{f}_j,\mathfrak{F}_j)_{1,l},[\rho_j(\mathfrak{f}_{ji},\mathfrak{F}_{ji})]_{l+1,m_i}\\ (h_j,\mathfrak{H}_j)_{1,k},[\rho_j(h_{ji},\mathfrak{H}_{ji})]_{k+1,n_i}\end{matrix}\right.\right].\end{aligned} \tag{4.10}$$

(ii) Next, setting $\rho_i=1$ in Eqs. (4.6) and (4.8), we get incomplete I-functions introduced and studied by Bansal and Kumar [37]:

$$\begin{aligned}&{}^{(\Gamma)}\aleph_{m_i,n_i,1;q}^{k,l}\left[z\left|\begin{matrix}(\mathfrak{f}_1,\mathfrak{F}_1,x),(\mathfrak{f}_j,\mathfrak{F}_j)_{2,l},[1(\mathfrak{f}_{ji},\mathfrak{F}_{ji})]_{l+1,m_i}\\ (h_j,\mathfrak{H}_j)_{1,k},[1(h_{ji},\mathfrak{H}_{ji})]_{k+1,n_i}\end{matrix}\right.\right] \\ &\qquad = {}^{(\Gamma)}I_{m_i,n_i;q}^{k,l}\left[z\left|\begin{matrix}(\mathfrak{f}_1,\mathfrak{F}_1,x),(\mathfrak{f}_j,\mathfrak{F}_j)_{2,l},(\mathfrak{f}_{ji},\mathfrak{F}_{ji})_{l+1,m_i}\\ (h_j,\mathfrak{H}_j)_{1,k},(h_{ji},\mathfrak{H}_{ji})_{k+1,n_i}\end{matrix}\right.\right].\end{aligned} \tag{4.11}$$

and

$$
{}^{(\gamma)}\aleph^{k,l}_{m_i,n_i,1;q}\left[z\,\middle|\,\begin{matrix}(\mathfrak{f}_1,\mathfrak{F}_1,x),(\mathfrak{f}_j,\mathfrak{F}_j)_{2,l},[1(\mathfrak{f}_{ji},\mathfrak{F}_{ji})]_{l+1,m_i}\\ (h_j,\mathfrak{H}_j)_{1,k},[1(h_{ji},\mathfrak{H}_{ji})]_{k+1,n_i}\end{matrix}\right]
= {}^{(\gamma)}I^{k,l}_{m_i,n_i;q}\left[z\,\middle|\,\begin{matrix}(\mathfrak{f}_1,\mathfrak{F}_1,x),(\mathfrak{f}_j,\mathfrak{F}_j)_{2,l},(\mathfrak{f}_{ji},\mathfrak{F}_{ji})_{l+1,m_i}\\ (h_j,\mathfrak{H}_j)_{1,k},(h_{ji},\mathfrak{H}_{ji})_{k+1,n_i}\end{matrix}\right]. \tag{4.12}
$$

(iii) Further, if we consider $x = 0$ and $\rho_i = 1$ in Eq. (4.6), we obtain the I-functions given by Saxena [61]:

$$
{}^{(\Gamma)}\aleph^{k,l}_{m_i,n_i,1;q}\left[z\,\middle|\,\begin{matrix}(\mathfrak{f}_1,\mathfrak{F}_1,0),(\mathfrak{f}_j,\mathfrak{F}_j)_{2,l},[1(\mathfrak{f}_{ji},\mathfrak{F}_{ji})]_{l+1,m_i}\\ (h_j,\mathfrak{H}_j)_{1,k},[1(h_{ji},\mathfrak{H}_{ji})]_{k+1,n_i}\end{matrix}\right]
= I^{k,l}_{m_i,n_i;q}\left[z\,\middle|\,\begin{matrix}(\mathfrak{f}_j,\mathfrak{F}_j)_{1,l},(\mathfrak{f}_{ji},\mathfrak{F}_{ji})_{l+1,m_i}\\ (h_j,\mathfrak{H}_j)_{1,k},(h_{ji},\mathfrak{H}_{ji})_{k+1,n_i}\end{matrix}\right]. \tag{4.13}
$$

(iv) Next, taking $q = 1$ and $\rho_i = 1$ in Eqs. (4.6) and (4.8), it reduces to recently introduced incomplete H-functions [20]:

$$
{}^{(\Gamma)}\aleph^{k,l}_{m_i,n_i,1;1}\left[z\,\middle|\,\begin{matrix}(\mathfrak{f}_1,\mathfrak{F}_1,x),(\mathfrak{f}_j,\mathfrak{F}_j)_{2,l},[1(\mathfrak{f}_{ji},\mathfrak{F}_{ji})]_{l+1,m_i}\\ (h_j,\mathfrak{H}_j)_{1,k},[1(h_{ji},\mathfrak{H}_{ji})]_{k+1,n_i}\end{matrix}\right]
= \Gamma^{k,l}_{m,n}\left[z\,\middle|\,\begin{matrix}(\mathfrak{f}_1,\mathfrak{F}_1,x),(\mathfrak{f}_j,\mathfrak{F}_j)_{2,m}\\ (h_j,\mathfrak{H}_j)_{1,n}\end{matrix}\right]. \tag{4.14}
$$

and

$$
{}^{(\gamma)}\aleph^{k,l}_{m_i,n_i,1;1}\left[z\,\middle|\,\begin{matrix}(\mathfrak{f}_1,\mathfrak{F}_1,x),(\mathfrak{f}_j,\mathfrak{F}_j)_{2,l},[1(\mathfrak{f}_{ji},\mathfrak{F}_{ji})]_{l+1,m_i}\\ (h_j,\mathfrak{H}_j)_{1,k},[1(h_{ji},\mathfrak{H}_{ji})]_{k+1,n_i}\end{matrix}\right]
= \gamma^{k,l}_{m,n}\left[z\,\middle|\,\begin{matrix}(\mathfrak{f}_1,\mathfrak{F}_1,x),(\mathfrak{f}_j,\mathfrak{F}_j)_{2,m}\\ (h_j,\mathfrak{H}_j)_{1,n}\end{matrix}\right]. \tag{4.15}
$$

(v) Substituting $q = 1$, $x = 0$ and $\rho_i = 1$ in Eq. (4.6), we get well-established and utilize Fox's H-function [18]:

$$
{}^{(\Gamma)}\aleph^{k,l}_{m_i,n_i,1;1}\left[z\,\middle|\,\begin{matrix}(\mathfrak{f}_1,\mathfrak{F}_1,0),(\mathfrak{f}_j,\mathfrak{F}_j)_{2,l},[1(\mathfrak{f}_{ji},\mathfrak{F}_{ji})]_{l+1,m_i}\\ (h_j,\mathfrak{H}_j)_{1,k},[1(h_{ji},\mathfrak{H}_{ji})]_{k+1,n_i}\end{matrix}\right] = H^{k,l}_{m,n}\left[z\,\middle|\,\begin{matrix}(\mathfrak{f}_j,\mathfrak{F}_j)_{1,m}\\ (h_j,\mathfrak{H}_j)_{1,n}\end{matrix}\right]. \tag{4.16}
$$

Some of the interesting and utilize special cases of Incomplete $\aleph$-functions are enumerated above.
Sharma et al. [41] in 2009 introduced a new modification of $\mathbb{M}$-series defined as follows:

$$ {}_{\gamma}^{\lambda,\mu}\mathbb{M}_{\sigma}\left[\begin{matrix} e_1,\ldots,e_\gamma; \\ \\ f_1,\ldots,f_\sigma; \end{matrix}\; z\right] = {}_{\gamma}^{\lambda,\mu}\mathbb{M}_{\sigma}(z) = \sum_{\kappa=0}^{\infty} \frac{(e_1)_\kappa \ldots (e_\gamma)_\kappa}{(f_1)_\kappa \ldots (f_\sigma)_\kappa} \frac{z^\kappa}{\Gamma(\lambda\kappa+\mu)}, \tag{4.17} $$

provided that, $\lambda,\mu,z \in \mathbb{C}$, $\Re(\lambda) > 0$, the thorough description of convergence conditions can be seen in Ref. [41].
We will be using the following special cases of $\mathbb{M}$-series in this manuscript:

(i) Taking $\gamma = \sigma = 0$ in Eq. (4.17), we obtain two parameter Mittag-Leffler function $E_{\lambda,\mu}(z)$ given by Wiman [62]:

$$ {}_{0}^{\lambda,\mu}\mathbb{M}_{0}(z) = E_{\lambda,\mu}(z) = \sum_{\kappa=0}^{\infty} \frac{z^\kappa}{\Gamma(\lambda\kappa+\mu)}. \tag{4.18} $$

(ii) Taking $\gamma = 0$, $\sigma = 1$ and $f_1 = 1$ in Eq. (4.17), we obtain the Wright function $\varphi(\lambda,\mu,z)$ (see, for detail, [4]):

$$ {}_{0}^{\lambda,\mu}\mathbb{M}_{1}(z) = \varphi(\lambda,\mu,z) = \sum_{\kappa=0}^{\infty} \frac{1}{\Gamma(\lambda\kappa+\mu)} \frac{z^\kappa}{\kappa!}. \tag{4.19} $$

(iii) Taking $\lambda = 1$, $\mu = 1$ and arbitrary γ,σ in Eq. (4.17), we obtain the generalized Hypergeometric function ${}_{\gamma}\mathbb{F}_{\sigma}((e_j)_1^\gamma : (f_j)_1^\sigma : z)$ (see, for detail, [4]):

$$ {}_{\gamma}^{1,1}\mathbb{M}_{\sigma}(z) = {}_{\gamma}\mathbb{F}_{\sigma}((e_j)_1^\gamma : (f_j)_1^\sigma : z) = \sum_{\kappa=0}^{\infty} \frac{(e_1)_\kappa \ldots (e_\gamma)_\kappa}{(f_1)_\kappa \ldots (f_\sigma)_\kappa} \frac{z^\kappa}{\kappa!}. \tag{4.20} $$

Using integral transforms is among the most employed method for finding solutions of integral and differential equations [1,3,10,13–15,33,42]. By modification of Natural [12] and Sumudu transform [59], a new integral transform known as J-transform was introduced by Maitama and Zhao [57, p. 1225, Eq. (2.1)] as follows:

$$ J[\lambda(t)](p,\vartheta) = V(p,\vartheta) = \vartheta \int_0^\infty \exp\left(\frac{-pt}{\vartheta}\right)\lambda(t)dt = \vartheta \lim_{\gamma\longrightarrow\infty} \int_0^\gamma \exp\left(\frac{-pt}{\vartheta}\right)\lambda(t)dt, \tag{4.21} $$

provided the limit of the above integral exists for $p,\vartheta > 0$ (p,ϑ are known as J-transform variables)

In Eq. (4.21), the function $\lambda : [0,\infty) \times \mathbb{R}$ is of exponential order defined over the following set of functions

$$ Q = \{\lambda(t) : \exists\, S, s_1, s_2 > 0, \text{ such that } |\lambda(t)| < S\exp\left(\frac{|t|}{s_i}\right) \text{ for } t \in (-1)^i \times [0,\infty)\}, \tag{4.22} $$

for $\vartheta = 1$, J-transform reduces to well-known Laplace transform [57, p. 1223, Eq. (1.1)].

The objective of this study is to find a computable solution of generalized FKE involving the product of incomplete $\aleph$-functions and $\mathbb{M}$-series established by utilizing J-transform due to its high accuracy and efficiency.

It is important to note that the condition (as defined in Eq. 4.2) on Φ_0 will remain same throughout this manuscript.

4.2 GENERALIZED FKE INVOLVING INCOMPLETE $\aleph$-FUNCTIONS AND $\mathbb{M}$-SERIES

The kinetic equations (KEs) characterize the relationship between concentrations of the materials and time. The solution of KE gives the distribution function of the dynamical states of a single particle, which often depends on the coordinates, time, and velocity. This section is dedicated to finding the solution of FKE involving the product of incomplete $\aleph$-functions and $\mathbb{M}$-series.

Theorem 4.1. *If $x > 0, \mathrm{c} > 0, \beta > 0, \alpha > 0, \rho > 0, \omega > 0, \zeta > 0, \lambda, \mu \in \mathbb{C}, \Re(\lambda) > 0$, $\mathfrak{F}_i (i = 1, \ldots, m)$ and $\mathfrak{H}_i (i = 1, \ldots, n)$, then the subsequent equation*

$$\Phi(\nu) - \Phi_0 \nu^{\alpha-1} {}^{(\Gamma)}\aleph^{k,l}_{m_i,n_i,\rho_i;q}\left[-\rho^\beta \nu^\beta \,\middle|\, \begin{matrix} A^* \\ \\ B^* \end{matrix}\right] {}^{\lambda,\mu}_{\gamma}\mathbb{M}_{\sigma}\left[\begin{matrix} e_1,\ldots,e_\gamma; \\ \\ f_1,\ldots,f_\sigma; \end{matrix} \quad \omega^\zeta \nu^\zeta\right]$$
$$= -\mathrm{c}^\beta {}_0D_\nu^{-\beta}\Phi(\nu), \tag{4.23}$$

where

$$A^* = (\mathfrak{f}_1, \mathfrak{F}_1, x), (\mathfrak{f}_j, \mathfrak{F}_j)_{2,l}, [\rho_j(\mathfrak{f}_{ji}, \mathfrak{F}_{ji})]_{l+1,m_i}$$
$$B^* = (h_j, \mathfrak{H}_j)_{1,k}, [\rho_j(h_{ji}, \mathfrak{H}_{ji})]_{k+1,n_i},$$

possess the following solution

$$\Phi(\nu) = \Phi_0 \nu^{\alpha-1} \sum_{\delta=0}^{\infty}\sum_{\kappa=0}^{\infty} (-\mathrm{c}^\beta \nu^\beta)^\delta (\omega^\zeta \nu^\zeta)^\kappa \frac{(e_1)_\kappa \ldots (e_\gamma)_\kappa}{(f_1)_\kappa \ldots (f_\sigma)_\kappa} \frac{1}{\Gamma(\lambda\kappa+\mu)}$$
$${}^{(\Gamma)}\aleph^{k,l+1}_{m_i+1,n_i+1,\rho_i;q}\left[-\rho^\beta \nu^\beta \,\middle|\, \begin{matrix} C^* \\ \\ D^* \end{matrix}\right], \tag{4.24}$$

where

$$C^* = (\mathfrak{f}_1, \mathfrak{F}_1, x), (1-\alpha-\zeta\kappa, \beta), (\mathfrak{f}_j, \mathfrak{F}_j)_{2,l}, [\rho_j(\mathfrak{f}_{ji}, \mathfrak{F}_{ji})]_{l+1,m_i}$$
$$D^* = (h_j, \mathfrak{H}_j)_{1,k}, (1-\alpha-\zeta\kappa-\beta\delta, \beta), [\rho_j(h_{ji}, \mathfrak{H}_{ji})]_{k+1,n_i}.$$

Proof. Employing J-transform (4.21) on both sides of equation (4.23) and further, using convolution theorem [57, p. 1229, Theorem 3.2] , we obtain

$$V(p,\vartheta)-\Phi_0\frac{1}{2\pi i}\sum_{\kappa=0}^{\infty}\frac{(e_1)_\kappa\dots(e_\gamma)_\kappa}{(f_1)_\kappa\dots(f_\sigma)_\kappa}\frac{(\omega^\zeta)^\kappa}{\Gamma(\lambda\kappa+\mu)}$$
$$\int_{\mathfrak{L}} M(\xi,x)(-\rho^\beta)^{-\xi}\vartheta^{\alpha-\beta\xi+\zeta\kappa+1}\frac{\Gamma(\alpha-\beta\xi+\zeta\kappa)}{p^{\alpha-\beta\xi+\zeta\kappa}}d\xi$$
$$=-c^\beta\vartheta^\beta p^{-\beta}V(p,\vartheta),$$

where $M(\xi,x)$ is defined by Eq. (4.7).

Upon simplifying the above-mentioned equation, we get

$$V(p,\vartheta)=\frac{\Phi_0}{(1+c^\beta\vartheta^\beta p^{-\beta})}\frac{1}{2\pi i}\sum_{\kappa=0}^{\infty}\frac{(e_1)_\kappa\dots(e_\gamma)_\kappa}{(f_1)_\kappa\dots(f_\sigma)_\kappa}\frac{(\omega^\zeta)^\kappa}{\Gamma(\lambda\kappa+\mu)}$$
$$\int_{\mathfrak{L}} M(\xi,x)(-\rho^\beta)^{-\xi}\vartheta^{\alpha-\beta\xi+\zeta\kappa+1} * \frac{\Gamma(\alpha-\beta\xi+\zeta\kappa)}{p^{\alpha-\beta\xi+\zeta\kappa}}d\xi,$$
$$=\Phi_0\frac{1}{2\pi i}\sum_{\delta=0}^{\infty}\sum_{\kappa=0}^{\infty}\frac{(e_1)_\kappa\dots(e_\gamma)_\kappa}{(f_1)_\kappa\dots(f_\sigma)_\kappa}\frac{(-c^\beta)^\delta(\omega^\zeta)^\kappa}{\Gamma(\lambda\kappa+\mu)}$$
$$\int_{\mathfrak{L}} M(\xi,x)(-\rho^\beta)^{-\xi}\vartheta^{\alpha-\beta\xi+\zeta\kappa+\beta\delta+1} * \frac{\Gamma(\alpha-\beta\xi+\zeta\kappa)}{p^{\alpha-\beta\xi+\beta\delta+\zeta\kappa}}d\xi.$$

The desired result (4.24) is obtained by employing inverse J-transform on both sides of the aforementioned equation. □

Theorem 4.2. *If* $x>0, c>0, \beta>0, \alpha>0, \rho>0, \omega>0, \zeta>0, \lambda,\mu\in\mathbb{C}, \Re(\lambda)>0$, $\mathfrak{F}_i(i=1,\dots,m)$ *and* $\mathfrak{H}_i(i=1,\dots,n)$, *then the subsequent equation*

$$\Phi(v)-\Phi_0 v^{\alpha-1}\,{}^{(\gamma)}\aleph^{k,l}_{m_i,n_i,\rho_i;q}\left[-\rho^\beta v^\beta\,\middle|\,\begin{matrix}A^*\\ \\ B^*\end{matrix}\right]{}^{\lambda,\mu}_{\gamma}\mathbb{M}_\sigma\left[\begin{matrix}e_1,\dots,e_\gamma;\\ \\ f_1,\dots,f_\sigma;\end{matrix}\;\omega^\zeta v^\zeta\right]$$
$$=-c^\beta{}_0D_v^{-\beta}\Phi(v), \tag{4.25}$$

where

$$A^*=(\mathfrak{f}_1,\mathfrak{F}_1,x),(\mathfrak{f}_j,\mathfrak{F}_j)_{2,l},[\rho_j(\mathfrak{f}_{ji},\mathfrak{F}_{ji})]_{l+1,m_i}$$
$$B^*=(h_j,\mathfrak{H}_j)_{1,k},[\rho_j(h_{ji},\mathfrak{H}_{ji})]_{k+1,n_i},$$

possess the following solution

$$\Phi(v)=\Phi_0 v^{\alpha-1}\sum_{\delta=0}^{\infty}\sum_{\kappa=0}^{\infty}(-c^\beta v^\beta)^\delta(\omega^\zeta v^\zeta)^\kappa\frac{(e_1)_\kappa\dots(e_\gamma)_\kappa}{(f_1)_\kappa\dots(f_\sigma)_\kappa}\frac{1}{\Gamma(\lambda\kappa+\mu)}$$
$${}^{(\gamma)}\aleph^{k,l+1}_{m_i+1,n_i+1,\rho_i;q}\left[-\rho^\beta v^\beta\,\middle|\,\begin{matrix}C^*\\ \\ D^*\end{matrix}\right], \tag{4.26}$$

where

$$C^* = (\mathfrak{f}_1, \mathfrak{F}_1, x), (1-\alpha-\zeta\kappa, \beta), (\mathfrak{f}_j, \mathfrak{F}_j)_{2,l}, [\rho_j(\mathfrak{f}_{ji}, \mathfrak{F}_{ji})]_{l+1,m_i}$$
$$D^* = (h_j, \mathfrak{H}_j)_{1,k}, (1-\alpha-\zeta\kappa-\beta\delta, \beta), [\rho_j(h_{ji}, \mathfrak{H}_{ji})]_{k+1,n_i}.$$

Proof. Theorem 4.2 can be proved by following the similar steps as stated in Theorem 4.1. □

Remark 1. *On taking* $\alpha = 1$ *and reducing incomplete* $\aleph$*-functions and* $\mathbb{M}$*-series to unity in Eqs.* (4.23) *and* (4.25)*, we can easily obtain* (4.4).

4.3 SPECIAL CASES

This section is dedicated to noteworthy corollaries and remarks obtained from Theorems 4.1 and 4.2.

Corollary 1. *If* $x > 0, \mathrm{c} > 0, \beta > 0, \alpha > 0, \rho > 0, \omega > 0, \zeta > 0, \lambda, \mu \in \mathbb{C}, \Re(\lambda) > 0$, $\mathfrak{F}_i (i = 1,\dots,m)$ *and* $\mathfrak{H}_i (i = 1,\dots,n)$*, then the subsequent equation*

$$\Phi(v) - \Phi_0 v^{\alpha-1} {}^{(\Gamma)}I^{k,l}_{m_i,n_i;q}\left[-\rho^\beta v^\beta \,\middle|\, \begin{matrix} I^* \\ J^* \end{matrix}\right] E_{\lambda,\mu}(\omega^\zeta v^\zeta) = -\mathrm{c}^\beta {}_0D_v^{-\beta}\Phi(v), \quad (4.27)$$

where

$$I^* = (\mathfrak{f}_1, \mathfrak{F}_1, x), (\mathfrak{f}_j, \mathfrak{F}_j)_{2,l}, (\mathfrak{f}_{ji}, \mathfrak{F}_{ji})_{l+1,m_i}$$
$$J^* = (h_j, \mathfrak{H}_j)_{1,k}, (h_{ji}, \mathfrak{H}_{ji})_{k+1,n_i},$$

possess the following solution

$$\Phi(v) = \Phi_0 v^{\alpha-1} \sum_{\delta=0}^{\infty}\sum_{\kappa=0}^{\infty} \frac{(-\mathrm{c}^\beta v^\beta)^\delta (\omega^\zeta v^\zeta)^\kappa}{\Gamma(\lambda\kappa+\mu)} {}^{(\Gamma)}I^{k,l+1}_{m_i+1,n_i+1;q}\left[-\rho^\beta v^\beta \,\middle|\, \begin{matrix} K^* \\ L^* \end{matrix}\right] \quad (4.28)$$

where

$$K^* = (\mathfrak{f}_1, \mathfrak{F}_1, x), (1-\alpha-\zeta\kappa, \beta), (\mathfrak{f}_j, \mathfrak{F}_j)_{2,l}, (\mathfrak{f}_{ji}, \mathfrak{F}_{ji})_{l+1,m_i}$$
$$L^* = (h_j, \mathfrak{H}_j)_{1,k}, (1-\alpha-\zeta\kappa-\beta\delta, \beta), (h_{ji}, \mathfrak{H}_{ji})_{k+1,n_i}.$$

Proof. Taking $\rho_i = 1$ and $\gamma = \sigma = 0$ (see, for details, Eqs. (4.11) and (4.18)) in Eq. (4.23), we conclude the desired result (4.28). □

Corollary 2. *If* $x > 0, \mathrm{c} > 0, \beta > 0, \alpha > 0, \rho > 0, \omega > 0, \zeta > 0, \lambda, \mu \in \mathbb{C}, \Re(\lambda) > 0$, $\mathfrak{F}_i (i = 1,\dots,m)$ *and* $\mathfrak{H}_i (i = 1,\dots,n)$*, then the subsequent equation*

$$\Phi(v) - \Phi_0 v^{\alpha-1} {}^{(\gamma)}I^{k,l}_{m_i,n_i;q}\left[-\rho^\beta v^\beta \,\middle|\, \begin{matrix} I^* \\ J^* \end{matrix}\right] \mathbb{E}_{\lambda,\mu}(\omega^\zeta v^\zeta) = -\mathrm{c}^\beta {}_0D_v^{-\beta}\Phi(v), \quad (4.29)$$

where

$$I^* = (\mathfrak{f}_1, \mathfrak{F}_1, x), (\mathfrak{f}_j, \mathfrak{F}_j)_{2,l}, (\mathfrak{f}_{ji}, \mathfrak{F}_{ji})_{l+1,m_i}$$
$$J^* = (h_j, \mathfrak{H}_j)_{1,k}, (h_{ji}, \mathfrak{H}_{ji})_{k+1,n_i},$$

possess the following solution

$$\Phi(\nu) = \Phi_0 \nu^{\alpha-1} \sum_{\delta=0}^{\infty} \sum_{\kappa=0}^{\infty} \frac{(-\mathrm{c}^{\beta}\nu^{\beta})^{\delta}(\omega^{\zeta}\nu^{\zeta})^{\kappa}}{\Gamma(\lambda\kappa+\mu)} {}^{(\gamma)}I^{k,l+1}_{m_i+1,n_i+1;q} \left[-\rho^{\beta}\nu^{\beta} \,\middle|\, \begin{matrix} K^* \\ \\ L^* \end{matrix} \right] \tag{4.30}$$

where

$$K^* = (\mathfrak{f}_1, \mathfrak{F}_1, x), (1-\alpha-\zeta\kappa, \beta), (\mathfrak{f}_j, \mathfrak{F}_j)_{2,l}, (\mathfrak{f}_{ji}, \mathfrak{F}_{ji})_{l+1,m_i}$$
$$L^* = (h_j, \mathfrak{H}_j)_{1,k}, (1-\alpha-\zeta\kappa-\beta\delta, \beta), (h_{ji}, \mathfrak{H}_{ji})_{k+1,n_i}.$$

Proof. Substituting $\rho_i = 1$ and $\gamma = \sigma = 0$ (see, for details, Eqs. (4.12) and (4.18)) in Theorem 4.2, we effectively obtain the desired result. □

Corollary 3. *If* $\mathrm{c} > 0, \beta > 0, \alpha > 0, \rho > 0, \omega > 0, \zeta > 0, \lambda, \mu \in \mathbb{C}, \Re(\lambda) > 0, \mathfrak{F}_i (i = 1, \ldots, m)$ *and* $\mathfrak{H}_i (i = 1, \ldots, n)$*, then the subsequent equation*

$$\Phi(\nu) - \Phi_0 \nu^{\alpha-1} I^{k,l}_{m_i,n_i;q} \left[-\rho^{\beta}\nu^{\beta} \,\middle|\, \begin{matrix} I^{**} \\ \\ J^{**} \end{matrix} \right] \mathbb{E}_{\lambda,\mu}(\omega^{\zeta}\nu^{\zeta}) = -\mathrm{c}^{\beta}{}_0D_{\nu}^{-\beta}\Phi(\nu), \tag{4.31}$$

where

$$I^{**} = (\mathfrak{f}_1, \mathfrak{F}_1), (\mathfrak{f}_j, \mathfrak{F}_j)_{2,l}, (\mathfrak{f}_{ji}, \mathfrak{F}_{ji})_{l+1,m_i}$$
$$J^{**} = (h_j, \mathfrak{H}_j)_{1,k}, (h_{ji}, \mathfrak{H}_{ji})_{k+1,n_i},$$

possess the following solution

$$\Phi(\nu) = \Phi_0 \nu^{\alpha-1} \sum_{\delta=0}^{\infty} \sum_{\kappa=0}^{\infty} \frac{(-\mathrm{c}^{\beta}\nu^{\beta})^{\delta}(\omega^{\zeta}\nu^{\zeta})^{\kappa}}{\Gamma(\lambda\kappa+\mu)} I^{k,l+1}_{m_i+1,n_i+1;q} \left[-\rho^{\beta}\nu^{\beta} \,\middle|\, \begin{matrix} K^{**} \\ \\ L^{**} \end{matrix} \right] \tag{4.32}$$

where

$$K^{**} = (\mathfrak{f}_1, \mathfrak{F}_1), (1-\alpha-\zeta\kappa, \beta), (\mathfrak{f}_j, \mathfrak{F}_j)_{2,l}, (\mathfrak{f}_{ji}, \mathfrak{F}_{ji})_{l+1,m_i}$$
$$L^{**} = (h_j, \mathfrak{H}_j)_{1,k}, (1-\alpha-\zeta\kappa-\beta\delta, \beta), (h_{ji}, \mathfrak{H}_{ji})_{k+1,n_i}.$$

Proof. Taking $x = 0$ in addition to Corollary 2, we conclude the desired result (4.32). □

Corollary 4. *If $x > 0, \mathrm{c} > 0, \beta > 0, \alpha > 0, \rho > 0, \omega > 0, \zeta > 0, \lambda, \mu \in \mathbb{C}, \Re(\lambda) > 0$, $\mathfrak{F}_i (i = 1, \ldots, m)$ and $\mathfrak{H}_i (i = 1, \ldots, n)$, then the subsequent equation*

$$\Phi(v) - \Phi_0 v^{\alpha-1} \Gamma^{k,l}_{m,n} \left[-\rho^\beta v^\beta \left| \begin{array}{c} (\mathfrak{f}_1, \mathfrak{F}_1, x), (\mathfrak{f}_j, \mathfrak{F}_j)_{2,m} \\ (h_j, \mathfrak{H}_j)_{1,n} \end{array} \right. \right] \varphi(\lambda, \mu, \omega^\zeta v^\zeta)$$
$$= -\mathrm{c}^\beta {}_0D_v^{-\beta} \Phi(v), \tag{4.33}$$

possess the following solution

$$\Phi(v) = \Phi_0 v^{\alpha-1} \sum_{\delta=0}^{\infty} \sum_{\kappa=0}^{\infty} \frac{(-\mathrm{c}^\beta v^\beta)^\delta (\omega^\zeta v^\zeta)^\kappa}{\Gamma(\lambda \kappa + \mu) \Gamma(\kappa + 1)}$$
$$\Gamma^{k,l+1}_{m+1,n+1} \left[-\rho^\beta v^\beta \left| \begin{array}{c} (\mathfrak{f}_1, \mathfrak{F}_1, x), (1-\alpha, \beta), (\mathfrak{f}_j, \mathfrak{F}_j)_{2,m} \\ (h_j, \mathfrak{H}_j)_{1,n}, (1-\alpha-\beta\delta, \beta) \end{array} \right. \right] \tag{4.34}$$

Proof. Considering $\rho_i = q = 1, \gamma = 0$ and $\sigma = f_1 = 1$ (see, for details, Eqs. (4.14) and (4.19)) in Eq. (4.23), we efficiently conclude the desired result (4.34). □

Corollary 5. *If $x > 0, \mathrm{c} > 0, \beta > 0, \alpha > 0, \rho > 0, \omega > 0, \zeta > 0, \lambda, \mu \in \mathbb{C}, \Re(\lambda) > 0$, $\mathfrak{F}_i (i = 1, \ldots, m)$ and $\mathfrak{H}_i (i = 1, \ldots, n)$, then the subsequent equation*

$$\Phi(v) - \Phi_0 v^{\alpha-1} \gamma^{k,l}_{m,n} \left[-\rho^\beta v^\beta \left| \begin{array}{c} (\mathfrak{f}_1, \mathfrak{F}_1, x), (\mathfrak{f}_j, \mathfrak{F}_j)_{2,m} \\ (h_j, \mathfrak{H}_j)_{1,n} \end{array} \right. \right] \varphi(\lambda, \mu, \omega^\zeta v^\zeta)$$
$$= -\mathrm{c}^\beta {}_0D_v^{-\beta} \Phi(v), \tag{4.35}$$

possess the following solution

$$\Phi(v) = \Phi_0 v^{\alpha-1} \sum_{\delta=0}^{\infty} \sum_{\kappa=0}^{\infty} \frac{(-\mathrm{c}^\beta v^\beta)^\delta (\omega^\zeta v^\zeta)^\kappa}{\Gamma(\lambda \kappa + \mu) \Gamma(\kappa + 1)}$$
$$\gamma^{k,l+1}_{m+1,n+1} \left[-\rho^\beta v^\beta \left| \begin{array}{c} (\mathfrak{f}_1, \mathfrak{F}_1, x), (1-\alpha, \beta), (\mathfrak{f}_j, \mathfrak{F}_j)_{2,m} \\ (h_j, \mathfrak{H}_j)_{1,n}, (1-\alpha-\beta\delta, \beta) \end{array} \right. \right] \tag{4.36}$$

Proof. Substituting $\rho_i = q = 1, \gamma = 0$ and $\sigma = f_1 = 1$ (see, for details, Eqs. (4.15) and (4.19)) in Theorem 4.2, we achieve the above-mentioned result. □

Corollary 6. *If $x > 0, \mathrm{c} > 0, \beta > 0, \alpha > 0, \rho > 0, \mathfrak{F}_i (i = 1, \ldots, m)$ and $\mathfrak{H}_i (i = 1, \ldots, n)$, then the subsequent equation*

$$\Phi(v) - \Phi_0 v^{\alpha-1} H^{k,l}_{m,n} \left[-\rho^\beta v^\beta \left| \begin{array}{c} (\mathfrak{f}_j, \mathfrak{F}_j)_{1,m} \\ (h_j, \mathfrak{H}_j)_{1,n} \end{array} \right. \right] {}_\gamma\mathbb{F}_\sigma((e_j)_1^\gamma : (f_j)_1^\sigma : \omega^\zeta v^\zeta)$$
$$= -\mathrm{c}^\beta {}_0D_v^{-\beta} \Phi(v), \tag{4.37}$$

possess the following solution

$$\Phi(\nu) = \Phi_0 \nu^{\alpha-1} \sum_{\delta=0}^{\infty} \sum_{\kappa=0}^{\infty} \frac{(-c^{\beta}\nu^{\beta})^{\delta}(\omega^{\zeta}\nu^{\zeta})^{\kappa}}{\Gamma(\kappa+1)} \frac{(e_1)_{\kappa}\dots(e_{\gamma})_{\kappa}}{(f_1)_{\kappa}\dots(f_{\sigma})_{\kappa}}$$

$$H^{k,l+1}_{m+1,n+1}\left[-\rho^{\beta}\nu^{\beta}\left|\begin{matrix}(1-\alpha,\beta),(\mathfrak{f}_j,\mathfrak{F}_j)_{1,m}\\ \\ (h_j,\mathfrak{H}_j)_{1,n},(1-\alpha-\beta\delta,\beta)\end{matrix}\right.\right] \tag{4.38}$$

Proof. Taking $x = 0, \rho_i = q = 1$ and $\lambda = \mu = 1$ (see, for details, Eqs. (4.16) and (4.20)) in Theorem 4.1, we deduce the result (4.38). □

Corollary 7. *If* $c > 0, \beta > 0, \alpha > 0, \rho > 0, \mathfrak{F}_i(i = 1,\dots,m)$ *and* $\mathfrak{H}_i(i = 1,\dots,n)$, *then the subsequent equation*

$$\Phi(\nu) - \Phi_0 \nu^{\alpha-1} \aleph^{k,l}_{m_i,n_i,\rho_i;q}\left[-\rho^{\beta}\nu^{\beta}\left|\begin{matrix}E^*\\ \\ F^*\end{matrix}\right.\right] = -c^{\beta}{}_0D_{\nu}^{-\beta}\Phi(\nu), \tag{4.39}$$

where

$$E^* = (\mathfrak{f}_1,\mathfrak{F}_1),(\mathfrak{f}_j,\mathfrak{F}_j)_{2,l},[\rho_j(\mathfrak{f}_{ji},\mathfrak{F}_{ji})]_{l+1,m_i}$$
$$F^* = (h_j,\mathfrak{H}_j)_{1,k},[\rho_j(h_{ji},\mathfrak{H}_{ji})]_{k+1,n_i},$$

possess the following solution

$$\Phi(\nu) = \Phi_0 \nu^{\alpha-1} \sum_{\delta=0}^{\infty} (-c^{\beta}\nu^{\beta})^{\delta} \aleph^{k,l+1}_{m_i+1,n_i+1,\rho_i;q}\left[-\rho^{\beta}\nu^{\beta}\left|\begin{matrix}G^*\\ \\ H^*\end{matrix}\right.\right] \tag{4.40}$$

where

$$G^* = (\mathfrak{f}_1,\mathfrak{F}_1),(1-\alpha-\zeta\kappa,\beta),(\mathfrak{f}_j,\mathfrak{F}_j)_{2,l},[\rho_j(\mathfrak{f}_{ji},\mathfrak{F}_{ji})]_{l+1,m_i}$$
$$H^* = (h_j,\mathfrak{H}_j)_{1,k},(1-\alpha-\zeta\kappa-\beta\delta,\beta),[\rho_j(h_{ji},\mathfrak{H}_{ji})]_{k+1,n_i}.$$

Proof. Setting $x = 0$ and reducing $\mathbb{M}$-series to unity in Theorem 4.1, we can efficiently deduce the desired result. □

Remark 2. *If we select suitable parameters to reduce incomplete H-function into Bessel function of the first kind $J_{\nu}(z)$ and Wright function to unity in Eq.* (4.33), *then the result given by Habenom et al. [16] can be easily recorded.*

Remark 3. *If we select suitable parameters to reduce incomplete $\aleph$-functions into incomplete I-functions and $\mathbb{M}$-series to unity in Eds.* (4.23) *and* (4.25), *then the result given by Bansal et al. [38, p. 5, Eq. (25) and (27)] can be easily recorded.*

4.4 CONCLUSIONS

In this chapter, we introduced a generalized FKE involving the product of incomplete $\aleph$-functions and $\mathbb{M}$-series and found its solution in the form of incomplete $\aleph$-functions using highly accurate and efficient J-transform. The significance of the present work is that we proposed a distinct computable extension of FKE in terms of incomplete $\aleph$-functions and $\mathbb{M}$-series along with effectively recording numerous corollaries and remarks of the main result. In future, we plan on accomplishing the solution of FKE involving the product of incomplete $\aleph$-functions and $\mathbb{M}$-series by using newly introduced integral transforms and studying their applications in science and engineering.

DECLARATIONS

Conflict of Interest: The authors declare that they have no known competing financial interests or personal relationships that could have appeared to influence the work reported in this paper.
Availability of Data and Material: All data are included in the paper.
Funding: None.
Authors'Contributions: Conceptualization is done by all authors. Methodology is done by MKB and investigation is done by NJ. Review and editing are done by NJ. Supervision is done by MKB. All authors read and approved the final manuscript.

References

1. A. Atangana, A note on the triple Laplace transform and its applications to some kind of third-order differential equation, *Abstr. Appl. Anal*, **2013**, 1–10, (2013).
2. A. Alsaedi, B. Ahmad, M. Alghanmi, S. K. Ntouyas, On a generalized langevin type nonlocal fractional integral multivalued problem, *Mathematics*, **7**, 1015, (2019).
3. A. Atangana, B. S. T. Alkaltani, A novel double integral transform and its applications, *J. Nonlinear Sci. Appl.*, **9**, 424–434, (2016).
4. A. A. Kilbas, H. M. Srivastava, J. J. Trujillo, *Theory and Applications of Fractional Differential Equations.* North-Holland Mathematical Studies, Vol. **204**. Elsevier (North-Holland) Science Publishers: Amsterdam, London and New York, (2006).
5. A. ErdBlyi, W. Magnus, F. Oberhettinger, F.G. Tricomi, *Higher Transcendental Functions*, Vol. **II**. McGraw-Hill: New York, (1953).
6. A. Tassaddiq, I. Khan, K. S. Nisar, Heat transfer analysis in sodium alginate based nanofluid using MoS_2 nanoparticles: Atangana-Baleanu fractional model, *Chaos Soliton Fract.*, **130**, 109445, (2020).
7. C. Copot, C. I. Muresan, K. A. Markowski, Advances in fractional order controller design and applications, *J. Appl. Nonlinear Dyn.*, **8(1)**, 1–3, (2019).
8. C. Ionescu, A. Lopes, D. Copot, J. A. T. Machado, J. H. T. Bates, The role of fractional calculus in modeling biological phenomena: A review, *Commun. Nonlinear Sci. Numer. Simul.*, **51**, 141–159, (2017).
9. D. Kumar, J. Singh, S. Kumar, A fractional model of Navier-Stokes equation arising in unsteady flow of a viscous fluid, *J. Assoc. Arab Univ. Basic Appl. Sci.*, **17**, 14–19, (2015).

10. F. B. M. Belgacem, A. A. Karaballi, Sumudu transform fundamental properties, investigations and applications, *J. Appl. Math. Stoch. Anal.*, **2006**, 1–23, (2006).
11. F. Mainardi, On the advent of fractional calculus in econophysics via continuous-time random walk, *Mathematics*, **8(4)**, 641, (2020).
12. F. B. M. Belgacem, R. Silambarasan, Theory of natural transform, *Math. Engg. Sci. Aeros.*, **3**, 99–124, (2012).
13. F. B. M. Belgacem, S. L. Kalla, A. A. Karaballi, Analytical investigations of the Sumudu transform and applications to integral production equations, *Math. Prob. Engg.*, **3**, 103–118, (2003).
14. G. Dattoli, M. R. Martinelli, P. E. Ricci, On new families of integral transforms for the solution of partial differential equations, *Integral Transforms Spec. Funct.*, **8**, 661–667, (2005).
15. H. A. Agwa, F. M. Ali, A. Kilicman, A new integral transform on time scales and its applications, *Adv. Differ. Equ.*, **60**, 1–14, (2012).
16. H. Habenom, D. L. Suthar, M. Gebeyehu, Application of Laplace transform on fractional Kinetic equation pertaining to the generalized galue type struve function, *Adv. Math. Phys.*, **2019**, 1–8, (2019).
17. H. J. Haubold, A. M. Mathai, The fractional kinetic equation and thermonuclear functions, *Astrophys. Space Sci.*, **273(1–4)**, 53–63, (2000).
18. H. M. Srivastava, K. C. Gupta, S. P. Goyal, *The H-Functions of One and Two Variables with Applications*. South Asian Publishers: New Delhi, Madras, (1982).
19. H. M. Srivastava, M. A. Chaudhry, R. P. Agarwal, The incomplete Pochhammer symbols and their applications to hypergeometric and related functions, *Integral Transforms Spec. Funct.* **23**, 659–683, (2012).
20. H. M. Srivastava, R. K. Saxena, R. K. Parmar, Some families of the incomplete H-functions and the incomplete $\bar{H}$-functions and associated integral transforms and operators of fractional calculus with applications. *Russ. J. Math. Phys.*, **25(1)**, 116–138, (2018).
21. H. M. Srivastava, N. Jolly, M. K. Bansal, R. Jain, A new integral transform associated with the λ-extended Hurwitz Lerch Zeta function, *RACSAM*, **113**, 1679–1692, (2018).
22. H. Singh, A.M. Wazwaz, Computational method for reaction diffusion-model arising in a spherical catalyst, *Int. J. Appl. Comput. Math*, **7**, 65, (2021).
23. H. Singh, Analysis for fractional dynamics of Ebola virus model, *Chaos, Solitons & Fractals*, **138**, 109992, (2020).
24. H. Singh, *An Efficient Computational Method for Non-Linear Fractional Lienard Equation Arising in Oscillating Circuits*. CRC Press: Boca Raton, FL, (2019).
25. H. Singh, H. M. Srivastava, Numerical investigation of the fractional-order liénard and duffing equations arising in oscillating circuit theory, *Front. Phys.*, **8**, 120, (2020).
26. H. Singh, H. M. Srivastava, J. J. Nieto, *Handbook of Fractional Calculus for Engineering and Science*. CRC Press: Boca Raton, FL, (2022).
27. H. Singh, D. Kumar, D. Baleanu, *Methods of Mathematical Modelling: Fractional Differential Equations*. CRC Press: Boca Raton, FL, (2019).
28. I. Podlubny, *Fractional Differential Equations*. Academic Press: California, (1999).
29. J. A. Machado, A. M. Lopes, A fractional perspective on the trajectory control of redundant and hyper-redundant robot manipulators, *Appl. Math. Model.*, **46**, 716–726, (2017).
30. J. Wang, Y. Zhou, Analysis of nonlinear fractional control systems in Banach spaces, *Nonlinear Anal. Theory Methods Appl.*, **74(17)**, 5929–5942, (2011).
31. K. Oldham, J. Spanier, *Fractional Calculus: Theory and Applications of Differentiation and Integration of Arbitrary Order*. Academic Press, New York, (1974).

32. K. S. Miller, B. Ross, *An Introduction to the Fractional Calculus and Fractional Differential Equations*. John Wiley & Sons, Inc: New York, (1993).
33. M. A. Asiru, Sumudu transform and solution of integral equations of convolution type, *Int. J. Math. Edu. Sci. Tech.*, **33**, 944–949, (2002).
34. M. A. Özarslan, C. Ustaoğlu, Some incomplete hypergeometric functions and incomplete Riemann-Liouville fractional integral operators, *Mathematics*, **7(5)**, 483, (2019).
35. M. K. Bansal, D. Kumar, K.S. Nisar, J. Singh, Application of incomplete H- functions in determination of Lambert's law, *J. Interdiscip. Math.*, **22(7)**, 1205–1212, (2019).
36. M. K. Bansal, D. Kumar, K. S. Nisar, J. Singh, Certain fractional calculus and integral transform results of incomplete ℵ-functions with applications, *Math. Meth. Appl. Sci.*, **43**, 5602–5614, (2020).
37. M. K. Bansal, D. Kumar, On the integral operators pertaining to a family of incomplete I-functions, *AIMS-Math*, **5(2)**, 1247–1259, (2020).
38. M. K. Bansal, D. Kumar, P. Harjule, J. Singh, Fractional kinetic equations associated with incomplete I-functions, *Fractal Fract.* , **4(2)**, 1–12, (2020).
39. M. K. Bansal, J. Choi, A note on pathway fractional integral formulas associated with the incomplete H-functions, *Int. J. Appl. Comput. Math.*, **5**, 133, (2019).
40. M. K. Bansal, N. Jolly, R. Jain, D. Kumar, An integral operator involving generalized Mittag-Leffler function and associated fractional calculus results, *J Anal.*, **27**, 727–740, (2018).
41. M. Sharma, R. Jain, A note on a generalized M-series as a special function of fractional calculus, *Fract. Calc. Appl. Anal.*, **12(4)**, 449–452, (2009).
42. N. Jolly, P. Harjule, R. Jain, Fractional differential equation associated with an integral operator with the H-function in the Kernel, *Glob. J. Pure Appl. Math.*, **13(7)**, 3505–3516, (2017).
43. N. Jolly, R. Jain, An investigation of composition formulae for fractional integral operators, *Palestine J. Math.*, **10(1)**, 199–208, (2021).
44. N. Jolly, M. Bansal, D. Kumar, J. Singh, Certain Mathieu-type series pertaining to incomplete H-functions, *Appl. Appl. Math.*, **15**(special issue 6), 1–14, (2020).
45. N. Jolly, M. Bansal, Several inequalities involving the generalized multi-index Mittag-Leffler functions, *Palestine J. Math.*, **11(2)**, 290–298, (2022).
46. N. Südland, B. Baumann, T. F. Nannenmacher, Open problem: Who knows about the Aleph-function? *Appl. Anal.*, **1(4)**, 401–402, (1998).
47. N. Südland, B. Baumann, T. F. Nannenmacher, Fractional driftless Fokker-Planck equation with power law diffusion coefficients, In: V. G. Gangha, E. W. Mayr, W. G. Vorozhtsov (Eds.), *Computer Algebra in Scientific Computing (CASC Konstanz 2001)*, Vol. **2001**. Springer: Berlin, pp. 513–525, (2001).
48. Q. Yang, D. Chen, T. Zhao, Y.Q. Chen, Fractional calculus in image processing: A review, *Fract. Calc. Appl. Anal.*, **19(5)**, 1222–1249, (2016).
49. R. K. Saxena, A. M. Mathai, H. J. Haubold, On fractional kinetic equations,*Astrophys. Space Sci.* , **282**, 281–287, (2002).
50. R. L. Magin, Fractional calculus models of complex dynamics in biological tissues, *Comput. Math. Appl.*, **59(5)**, 1586–1593, (2019).
51. S. Asgarani, A set of new three-parameter entropies in terms of a generalized incomplete Gamma function, *Physica A*, **392(9)**, 1972–1976, (2013).
52. S. D. Purohit, N. Jolly, M. Bansal, J. Singh, D. Kumar, Chebyshev type inequalities involving the fractional integral operator containing multi-index Mittag Leffler function in the Kernel, *Appl. Appl. Math.*, **15**(special issue 6), 29–38, (2020).

53. S. Kumar, A new analytical modelling for fractional telegraph equation via Laplace transform, *Appl. Math. Model.*, **38**, 3154–3163, (2014).
54. S. Kumar, D. Kumar, J. Singh, Fractional modelling arising in unidirectional propagation of long waves in dispersive media, *Adv. Nonlinear Anal.*, **5(4)**, 383–394, (2016).
55. S. Kumar, K. S. Nisar, R. Kumar, C. Cattani, B. Samet, A new Rabotnov fractional-exponential function-based fractional derivative for diffusion equation under external force, *Math. Meth. Appl. Sci*, **43**, 1–12, (2020).
56. S. Kumar, R. Kumar, J. Singh, K. S. Nisar, D. Kumar, An efficient numerical scheme for fractional modelof HIV-1 infection of $CD4^+$ T-cells with the effect of antiviral drug therapy, *Alex. Eng. J.*, **4**, 2053–2064, (2020).
57. S. Maitama, W. Zhao, Beyond sumudu transform and natural transform: $\mathbb{J}$-transform properties and application, *J. Appl. Anal. Comput.*, **10(4)**, 1223–1241, (2020).
58. T. Sandev, R. Metzler, A. Chechkin, From continuous time random walks to the generalized diffusion equation, *Frac. Calc. Appl. Anal.*, **21(1)**, 10–28 , (2018).
59. V. A. Ditkin, A. P. Prudnikov, *Integral Transforms and Operational Calculus*. Pergamon: Oxford, (1965).
60. V. G. Gupta, B. Sharma, F. B. M. Belgacem, On the solutions of generalized fractional kinetic equations, *Appl. Math. Sci.* , **5**, 899–910, (2011).
61. V. P. Saxena, Formal solution of certain new pair of dual integral equations involving H-functions, *Proc. Nat. Acad. Sci. India Sect. A*, **52**, 366–375, (1982).
62. A. Wiman, *Ü*ber den Fundamentalsatz in der Theorie der Funktionen Eα(x), *Acta Math.*, **29**, 191–201, (1905).
63. X. Liang, F. Gao, C. B. Zhou, Z. Wang, X. J. Yang, An anomalous diffusion model based on a new general fractional operator with the Mittag-Leffler function of Wiman type, *Adv. Differ. Equ.*, **2018(1)**, 25, (2018).
64. Y. Lui, J. Shan, N. Qi, Creep modeling and identification for piezoelectric actuators based on fractional-order system, *Mechatronics*, **23(7)**, 840–847, (2013).

5 Legendre Collocation Method for Generalized Fractional Advection Diffusion Equation

Sandeep Kumar and R. K. Pandey
Indian Institute of Technology (BHU)

Shiva Sharma
K P College Murliganj

Harendra Singh
Post Graduate College Ghazipur

CONTENTS

5.1 INTRODUCTION

In the last few decades, fractional partial differential equation (FPDE) played an important role in modelling of several physical phenomena. One form of such a FPDE is the fractional advection-diffusion equation (FADE). FADEs are used to model several natural and practical phenomena [1]– [4]. In recent years, numerical approximations for FADE were obtained by several authors using various methods. Kazem and Dehghan [5] obtained a solution of the time-fractional diffusion equation (TFDE).

DOI: 10.1201/9781003368069-5

Here, the authors used the method of lines and the solution of a system of equations obtained in form of the Mittag -Leffler matrix function. In Ref. [6], the authors solved a variable order time fractional TFDE. They first define time fractional derivative in Caputo sense and then use the operational matrix method for its solution. Safdari et al. [7] obtained a numerical solution of FADE by finite difference method and discussed its stability and convergence analysis theoretically and verify results numerically. In Ref. [8], Hendy developed a concept of TFDE for the description of transport phenomena in heterogeneous environments. For the numerical solution, author used a fully implicit discrete scheme and discussed the existence and uniqueness of solutions, convergence, and numerical stability. Tajadodi [9] studied FADE defined in terms of the Atangana–Baleanu derivative in Caputo sense by operational matrix method. In Ref. [10], the authors obtained numerical solution of the FADE using finite difference method in the temporal direction and the Chebyshev collocation method in the spatial direction. Also, they discussed the convergence and stability of the fully discrete scheme by energy method. Safdari et al. [11] solved the space-time fractional advection-diffusion equation (STFADE). Here, fractional derivative is defined in Caputo sense, and the numerical solution is found by finite difference and collocation method. The authors established the convergence and stability analysis of the proposed method and showed that the numerical convergence has the order $O(t^{2-\beta})$ in the spatial direction.

In the present work, we consider the following FADE

$$(*\mathscr{D}_t^{\gamma}u)(x,t)+a(x,t)\frac{\partial u(x,t)}{\partial x}=(\mathscr{D}_x^{\mu}u)(x,t)+s(x,t),\ \ x\in\mathscr{I}_1,\ t\in\mathscr{I}_2, \tag{5.1}$$

with initial and boundary conditions;

$$u(x,0)=\mu_1(x),\ x\in\mathscr{I}_1, \tag{5.2}$$

$$u(0,t)=\mu_2(t),\ u(1,t)=\mu_3(t),\ t\in\mathscr{I}_2, \tag{5.3}$$

where $\mathscr{I}=\mathscr{I}_1\times\mathscr{I}_2$, $\mathscr{I}_1=[0,1]$, $\mathscr{I}_2=[0,1]$ and $*\mathscr{D}_t^{\gamma}$ denotes the generalized Caputo derivative (GCD) of order γ, $0<\gamma\leq 1$ and $\mathscr{D}_x^{\mu}$ denote the Caputo derivative of order μ, $1<\mu\leq 2$. The GCD $*\mathscr{D}_t^{\gamma}$ reduces into $\mathscr{D}_x^{\mu}$ for $z(t)=t$ and $w(t)=1$, $s(x,t)$ is the source/sink function, and $z(t)$, $w(t)$ denotes the scale function and weight function, respectively. The generalized Caputo derivatives were used by several researchers in recent years. It was observed that the increasing weight function introduces the increment in the rate of diffusion and vice versa, whereas scale function has the property that increasing scale function enlarges the solution. The effects of the nonlinear scale function are shown in numerical results section of the manuscript. Numerical approximation of FADE defined in terms of GCD is obtained by several methods recently [12,13]. In Refs. [14,15], the authors obtained the solution of FADE by implicit finite difference method and discussed the stability and convergence analysis. In Ref. [11], the authors developed a method to obtain the solution of the space-time fractional advection-diffusion equation (STFADE). The authors first use a finite difference scheme on the time variable and a collocation method for space variables.

Aghdam et al. [10] used the Chebyshev collocation and finite difference method to obtain the numerical solution of STFADE and discussed the convergence in L^2 norm. In Ref. [16], Hafez et al. discussed the spectral Galerkin/Petrov–Galerkin algorithm for solving multi-dimensional STFADE with the non-smooth solution. Some more methods can be seen in Refs. [17–19] to obtain the numerical solution of FADE.

The main purpose of this chapter is to introduce the new scheme based on the finite difference and collocation method to determine the solution of FADE. The rest of the sections of this chapter are organized as follows: In Section 5.2, we discuss some basic facts about FC and Jacobi - polynomials which are needed throughout this chapter. In Section 5.3, we present function approximation and collocation method for numerical approximation. Further, we estimate the error and convergence analysis analytically in Sections 5.4 and 5.5 respectively, which defines the numerical applicability of the proposed method. In Section 5.6, we present two numerical examples to validate the proposed method. Also, we compare our results with several other methods which are presented in Section 5.6. Finally, in Section 5.7, conclusions are discussed.

5.2 MATHEMATICAL BACKGROUND OF FRACTIONAL CALCULUS

In this part, we provide the basic definitions and properties of fractional calculus. More details can be seen in Refs. [20,21].

Definition 1 ([22]). *The Caputo derivative of a function $f(t)$ of order $\gamma > 0$ defined as,*

$$(\mathscr{D}^{\gamma}_{0+;}f)(t) = \frac{1}{\Gamma(n-\gamma)}\int_0^t (t-s)^{n-\gamma-1} f^{(n)}(s)ds,\ t>0, \tag{5.4}$$

where $n-1<\gamma\leq n$, and $n \in \mathbb{Z}^+$.

Definition 2 ([22]). *Generalized fractional derivative of order $\gamma > 0$ of a function $f(t)$ with weight function $\omega(t)$ and scale function $z(t)$ is defined as,*

$$(\mathscr{D}^{\gamma}_{0+;[z;\omega;L]}f)(t) = [\omega(t)]^{-1}\left[\left(\frac{D_t}{z'(t)}\right)^{\gamma}(\omega(t)f(t))\right]. \tag{5.5}$$

Definition 3 ([22]). *The forward/left generalized fractional integral of a function $f(t)$ of fractional order $\gamma > 0$ is given as*

$$(\mathscr{I}^{\gamma}_{0+;[z;\omega]}f)(t) = \frac{[\omega(t)]^{-1}}{\Gamma(\gamma)}\int_0^t \frac{\omega(s)z'(s)f(s)}{[z(t)-z(s)]^{1-\gamma}}ds, \tag{5.6}$$

provided the integral on the right-hand side exists point wisely.

Definition 4 ([22]). *The forward/left generalized Caputo fractional derivative of order γ and of type 2 of function $f(t)$ with respect to scale function $z(t)$ and weight* *function $\omega(t)$ is given by,*

$$(\mathscr{D}^{\gamma}_{0+;[z;\omega;2]}f)(t) = (\mathscr{I}^{n-\gamma}_{0+;[z;\omega]}\mathscr{D}^{n}_{0+;[z;\omega;L]}f)(t) = \frac{[\omega(t)]^{-1}}{\Gamma(n-\gamma)}\int_0^t \frac{[\omega(s)f(s)]^{(n)}}{(z(t)-z(s))^{\gamma-n+1}}ds, \tag{5.7}$$

where $n-1 \leq \gamma < n$, $n \in \mathbb{N}$.

Definition 5 ([23]). *Shifted Legendre polynomial function is defined as,*

$$\mathscr{L}_n(t) = \sum_{r=0}^{n}(-1)^{n-r}\frac{\Gamma(n+r+1)}{\Gamma(r+1)\Gamma(n-r+1)(r)!}t^r,\ t \in [0,1], \tag{5.8}$$

satisfy the following recurrence relation in $[0,1]$,

$$\mathscr{L}_0(t) = 1,$$
$$\mathscr{L}_1(t) = 2t-1,$$

and

$$\left\{\mathscr{L}_{n+1}(t) = \frac{(2n+1)(2t-1)}{n+1}\mathscr{L}_n(t) - \frac{n}{n+1}\mathscr{L}_{n-1}(t),\ n = 1,2,3,\ldots\right\}. \tag{5.9}$$

The orthogonality relation with respect to weight function $w(t)$ is,

$$\int_0^1 \mathscr{L}_n(t)\mathscr{L}_m(t)w_1(t)dt = \tilde{h}_n\delta_{nm}, \tag{5.10}$$

where $\tilde{h}_n = \frac{1}{2n+1}$, $w_1(t) = 1$ and δ_{nm} is Kronecker delta function.

5.3 FUNCTION APPROXIMATION USING LEGENDRE POLYNOMIALS

In Ref. [24], any function $f(x) \in L^2[0,1]$ can be approximate in terms of Legendre basis as,

$$f(x) = \sum_{i=0}^{\infty} c_i\ell_i(x), \tag{5.11}$$

where c_i are the unknown coefficients, determined by,

$$c_i = \frac{1}{||\ell_i(x)||^2}\int_0^1 f(x)\ell_i(x)dx,\ \ i = 0,1,2,3..., \tag{5.12}$$

If we consider only first $(N+1)$ terms of the Eq. (5.11). Then

$$f(x) = f_N(x) = \sum_{i=0}^{N} c_i\ell_i(x), \tag{5.13}$$

where $c_i = [c_0, c_1, c_2, ...c_N]^T$, and $\ell_i(x)$ are the shifted Legendre polynomials, $\ell_i(x) = [\ell_0(x), \ell_1(x), \ell_2(x), ...\ell_N(x)]$.

5.3.1 APPROXIMATION OF A TWO-VARIABLE FUNCTION USING LEGENDRE POLYNOMIALS

As discussed in Ref. [24], for a function $f(x,t) \in L^2(\mathscr{I} = \mathscr{I}_1 \times \mathscr{I}_2)$ can be written as,

$$f(x,t) = \sum_{i=0}^{\infty}\sum_{j=0}^{\infty} a_{i,j}\mathscr{L}_i(x)\mathscr{L}_j(t), \tag{5.14}$$

where coefficients $a_{i,j}$ is determined by the formula,

$$a_{i,j} = \frac{1}{\|\mathscr{L}_i(x)\|^2}\frac{1}{\|\mathscr{L}_j(t)\|^2}\int_0^1 f(x,t)\mathscr{L}_i(x)\mathscr{L}_j(t)\,dx\,dt, i = 0,1,2,3,\ldots,$$
$$j = 0,1,2,3\ldots. \tag{5.15}$$

Again, for the seek of the good approximation of the function $f \in L^2(\mathscr{I})$, such that

$$f(x,t) \approx f_{\mathscr{N},\mathscr{K}}(x,t) = \sum_{i=0}^{\mathscr{N}}\sum_{j=0}^{\mathscr{K}} a_{i,j}\mathscr{L}_i(x)\mathscr{L}_j(t), \tag{5.16}$$

where $a_{i,j} = [a_{00}, a_{01}, a_{02}, \ldots a_{\mathscr{N}\mathscr{K}}]^T$, $\mathscr{L}_i(x)$ and $\mathscr{L}_j(t)$ are shifted Legendre polynomials with $\mathscr{L}_i(x) = [\mathscr{L}_0(x), \mathscr{L}_1(x), \mathscr{L}_2(x), \ldots, \mathscr{L}_{\mathscr{N}}(x)]$, $\mathscr{L}_j(t) = [\mathscr{L}_0(t), \mathscr{L}_1(t), \mathscr{L}_2(t), \ldots, \mathscr{L}_{\mathscr{K}}(t)]$.

5.3.2 COLLOCATION METHOD FOR GFADE

We assume the solution of Eq. (5.1) of the form,

$$u_{\mathscr{N},\mathscr{K}}(x,t) = \sum_{i=0}^{\mathscr{N}}\sum_{j=0}^{\mathscr{K}} a_{i,j}\mathscr{L}_i(x)\mathscr{L}_j(t). \tag{5.17}$$

From Eq. (5.1), we have

$$(\mathscr{D}^{\gamma}_{0+;[z;w;2]})u_{\mathscr{N},\mathscr{K}}(x,t) + v_1(x,t)\frac{\partial u_{\mathscr{N},\mathscr{K}}(x,t)}{\partial x} = \frac{\partial^2 u_{\mathscr{N},\mathscr{K}}(x,t)}{\partial x^2} + s(x,t),\ (x,t) \in \mathscr{I}, \tag{5.18}$$

initial and boundary conditions are,

$$\begin{cases} u_{\mathscr{N},\mathscr{K}}(x,0) = \sum_{i=0}^{\mathscr{N}}\sum_{j=0}^{\mathscr{K}} a_{i,j}\mathscr{L}_i(x)\mathscr{L}_j(0) = f_1(x),\ x \in \mathscr{I}_1, \\ u_{\mathscr{N},\mathscr{K}}(0,t) = \sum_{i=0}^{\mathscr{N}}\sum_{j=0}^{\mathscr{K}} a_{i,j}\mathscr{L}_i(0)\mathscr{L}_j(t) = g_1(t),\ t \in \mathscr{I}_2, \\ u_{\mathscr{N},\mathscr{K}}(1,t) = \sum_{i=0}^{\mathscr{N}}\sum_{j=0}^{\mathscr{K}} a_{i,j}\mathscr{L}_i(1)\mathscr{L}_j(t) = g_2(t),\ t \in \mathscr{I}_2. \end{cases} \tag{5.19}$$

Now, Eqs. (5.18) and (5.19) are converted into a linear system of $(\mathscr{N}+1)$ equations in $(\mathscr{K}+1)$ unknowns. For solving this system by collocation method, we consider collocation points as the root of shifted Legendre polynomial of $\mathscr{L}_i(x)$ $(0 \le i \le \mathscr{N})$ and $\mathscr{L}_j(t)$ $(0 \le j \le \mathscr{K}-1)$ respectively. So we have the points of the form (x_i, t_j) are the collocation points.

$$(\mathscr{D}^{\gamma}_{0+;[z;w;2]})u_{\mathscr{N},\mathscr{K}}(x_i,t_j) + v_1(x_i,t_j)\frac{\partial u_{\mathscr{N},\mathscr{K}}(x_i,t_j)}{\partial x} = \frac{\partial^2 u_{\mathscr{N},\mathscr{K}}(x_i,t_j)}{\partial x^2} + s(x_i,t_j), \tag{5.20}$$

and initial and boundary conditions of Eq. (5.1) given as,

$$\begin{aligned}
u_{\mathscr{N},\mathscr{K}}(x_i,0) &= \sum_{i=0}^{\mathscr{N}}\sum_{j=0}^{\mathscr{K}} a_{i,j}\mathscr{L}_i(x_i)\mathscr{L}_j(0) = f_1(x_i),\\
u_{\mathscr{N},\mathscr{K}}(0,t_j) &= \sum_{i=0}^{\mathscr{N}}\sum_{j=0}^{\mathscr{K}} a_{i,j}\mathscr{L}_i(0)\mathscr{L}_j(t_j) = g_1(t_j),\\
u_{\mathscr{N},\mathscr{K}}(1,t_j) &= \sum_{i=0}^{\mathscr{N}}\sum_{j=0}^{\mathscr{K}} a_{i,j}\mathscr{L}_i(1)\mathscr{L}_j(t_j) = g_2(t_j).
\end{aligned} \tag{5.21}$$

In this way, we have a system of $(\mathscr{N}+1)\times(\mathscr{K}+1)$ order linear algebraic equations in unknown coefficients $a_{i,j}$ from Eqs.(5.20) and (5.21). We can solve these equations by any standard method such as Gaussian elimination method or conjugate gradient method.

5.4 CONVERGENCE ANALYSIS

In this section, we study the convergence analysis of the proposed collocation method.

Theorem 5.1. *Let $u(x,t)$ be the sufficiently smooth exact solution of the Eq. (5.1) in $L^2(\mathscr{I})$ and $u_{\mathscr{N},\mathscr{K}}(x,t)$ denotes the approximate solution of Eq. (5.1) in the form of the shifted Legendre expansion of $u(x,t)$ as,*

$$u_{\mathscr{N},\mathscr{K}}(x,t) = \sum_{i=0}^{\mathscr{N}}\sum_{j=0}^{\mathscr{K}} a_{i,j}\mathscr{L}_i(x)\mathscr{L}_j(t), \tag{5.22}$$

then unknown coefficients satisfy the following condition:

$$|a_{i,j}| \le \frac{M(2i+1)}{\pi^3(i+i^2)}\frac{(2j+1)}{(j+j^2)}, \tag{5.23}$$

where M is determine by,

$$\sup|u(x,t)| \le M.$$

Proof. Any function $u(x,t)$ defined on $L^2(\mathscr{I})$ can be approximated by shifted Legendre polynomials given as,

$$u_{\mathscr{N},\mathscr{K}}(x,t) = \sum_{i=0}^{\mathscr{N}} \sum_{j=0}^{\mathscr{K}} a_{i,j} \mathscr{L}_i(x) \mathscr{L}_j(t), \tag{5.24}$$

From Eq. (5.15), we have

$$\begin{aligned} a_{i,j} &= (2i+1)(2j+1) \int_0^1 \int_0^1 u(x,t) \mathscr{L}_i(x) \mathscr{L}_j(t) dx dt, \\ &\leq (2i+1)(2j+1) \sup |(u(x,t))| \int_0^1 \int_0^1 \mathscr{L}_i(x) \mathscr{L}_j(t) dx dt, \\ &\leq (2i+1)(2j+1) M \int_0^1 \int_0^1 \mathscr{L}_i(x) \mathscr{L}_j(t) dx dt, \\ &\leq (2i+1)(2j+1) M \int_0^1 \mathscr{L}_i(x) dx \int_0^1 \mathscr{L}_j(t) dt, \end{aligned} \tag{5.25}$$

since,

$$\int_0^1 \mathscr{L}_i(x) dx = \frac{\sin(i\pi)}{(i+i^2)\pi} \leq \frac{1}{(i+i^2)\pi}.$$

Similarly, we can find for $\int_0^1 \mathscr{L}_j(t) dt$. Then from Eq. (5.25) we get

$$|a_{i,j}| \leq (2i+1)(2j+1) \frac{M}{(i+i^2)\pi} \frac{1}{(j+j^2)\pi}, \tag{5.26}$$

after simplifying Eq. (5.26),

$$|a_{i,j}| \leq \frac{M(2i+1)}{\pi^2(i+i^2)} \frac{(2j+1)}{(j+j^2)}. \tag{5.27}$$

□

Theorem 5.2. *Let $u(x,t)$ be sufficiently smooth exact solution of Eq. (5.1) and $u_{\mathscr{N},\mathscr{K}}(x,t)$ denote the numerical approximation of $u(x,t)$, and $\sup |u(x,t)| \leq M$ then $u_{\mathscr{N},\mathscr{K}}(x,t)$ converges to $u(x,t)$ uniformly. Moreover,*

$$|u(x,t) - u_{\mathscr{N},\mathscr{K}}(x,t)| \to 0 \text{ as } \mathscr{N}, \mathscr{K} \to \infty.$$

Proof. From Eqs. (5.14) and (5.24), we get

$$\begin{aligned} u(x,t) - u_{\mathscr{N},\mathscr{K}}(x,t) = &\sum_{i=\mathscr{N}+1}^{\infty} \sum_{j=0}^{\mathscr{K}} a_{i,j} \mathscr{L}_i(x) \mathscr{L}_j(t) + \sum_{i=\mathscr{N}+1}^{\infty} \sum_{j=\mathscr{K}+1}^{\infty} a_{i,j} \mathscr{L}_i(x) \mathscr{L}_j(t) \\ &+ \sum_{i=0}^{\mathscr{N}} \sum_{j=\mathscr{K}+1}^{\infty} a_{i,j} \mathscr{L}_i(x) \mathscr{L}_j(t), \end{aligned} \tag{5.28}$$

which implies

$$\left\|u(x,t)-u_{\mathscr{N},\mathscr{K}}(x,t)\right\|^2$$

$$=\left\|\sum_{i=\mathscr{N}+1}^{\infty}\sum_{j=0}^{\mathscr{K}}a_{i,j}\mathscr{L}_i(x)\mathscr{L}_j(t)+\sum_{i=\mathscr{N}+1}^{\infty}\sum_{j=\mathscr{K}+1}^{\infty}a_{i,j}\mathscr{L}_i(x)\mathscr{L}_j(t)\right.$$

$$\left.+\sum_{i=0}^{\mathscr{N}}\sum_{j=\mathscr{K}+1}^{\infty}a_{i,j}\mathscr{L}_i(x)\mathscr{L}_j(t)\right\|^2,\leq\sum_{i=\mathscr{N}+1}^{\infty}\sum_{j=0}^{\mathscr{K}}a_{i,j}^2\|\mathscr{L}_i(x)\|^2\|\mathscr{L}_j(t)\|^2$$

$$+\sum_{i=\mathscr{N}+1}^{\infty}\sum_{j=\mathscr{K}+1}^{\infty}a_{i,j}^2\|\mathscr{L}_i(x)\|^2\|\mathscr{L}_j(t)\|^2+\sum_{i=0}^{\mathscr{N}}\sum_{j=\mathscr{K}+1}^{\infty}a_{i,j}^2\|\mathscr{L}_i(x)\|^2\|\mathscr{L}_j(t)\|^2. \quad (5.29)$$

From Eq. (5.27) we have

$$|a_{i,j}|\leq\frac{M(2i+1)}{\pi^2(i+i^2)}\frac{(2j+1)}{(j+j^2)}. \quad (5.30)$$

Substituting the value of $a_{i,j}$ from Eq. (5.30) to (5.29) and taking L^2 norm, we get

$$\|u(x,t)-u_{\mathscr{N},\mathscr{K}}(x,t)\|^2\leq\sum_{i=\mathscr{N}+1}^{\infty}\sum_{j=0}^{\mathscr{K}}\left(M^2\frac{(2i+1)^2}{\pi^4(i+i^2)^2}\frac{(2j+1)^2}{(j+j^2)^2}\right)\left(\|\mathscr{L}_i(x)\|)^2(\|\mathscr{L}_j(t)\|\right)^2$$

$$+\sum_{i=\mathscr{N}+1}^{\infty}\sum_{j=\mathscr{K}+1}^{\infty}\left(M^2\frac{(2i+1)^2}{\pi^4(i+i^2)^2}\frac{(2j+1)^2}{(j+j^2)^2}\right)\left(\|\mathscr{L}_i(x)\|)^2(\|\mathscr{L}_j(t)\|\right)^2$$

$$+\sum_{i=0}^{\mathscr{N}}\sum_{j=\mathscr{K}+1}^{\infty}\left(M^2\frac{(2i+1)^2}{\pi^4(i+i^2)^2}\frac{(2j+1)^2}{(j+j^2)^2}\right)\left(\|\mathscr{L}_i(x)\|)^2(\|\mathscr{L}_j(t)\|\right)^2,$$

$$\leq M^2\left(\frac{1}{(1+\mathscr{N})^2}\left(1-\frac{1}{(1+\mathscr{K})^2}\right)+\frac{1}{(1+\mathscr{N})^2}\frac{1}{(1+\mathscr{K})^2}\right.$$

$$\left.+\frac{1}{(1+\mathscr{K})^2}\left(1-\frac{1}{(1+\mathscr{N})^2}\right)\right), \quad (5.31)$$

$$\|u(x,t)-u_{\mathscr{N},\mathscr{K}}(x,t)\|^2\leq\frac{M^2}{(1+\mathscr{N})^2(1+\mathscr{K})^2}. \quad (5.32)$$

From Eq. (5.32) it is clear that as we increase the value of $\mathscr{N},\mathscr{K}\to\infty$, $u_{\mathscr{N},\mathscr{K}}(x,t)$ converges uniformly to $u(x,t)$ under L^2 norm. □

5.5 ERROR ANALYSIS

Now we turn to discuss the error approximation of proposed collocation for GFADE.

Theorem 5.3. *Let $u(x,t)$ be sufficiently differentiable function in $L^2(\mathscr{I})$, and $(\mathscr{D}^{\gamma}_{0+;[z;\omega;2]})u_{\mathscr{N},\mathscr{K}}(x,t)$ denote the approximation of $(\mathscr{D}^{\gamma}_{0+;[z;\omega;2]})u(x,t)$. Assume that, $\sup|(\mathscr{D}^{\gamma}_{0+;[z;\omega;2]})u(x,t)|\leq\lambda$, $\forall(x,t)\in\mathscr{I}$, then $\mathscr{E}_n\to 0$ as $\mathscr{N},\mathscr{K}\to\infty$, where,*

$$\mathscr{E}_n=\|(\mathscr{D}^{\gamma}_{0+;[z;\omega;2]})u(x,t)-(\mathscr{D}^{\gamma}_{0+;[z;\omega;2]})u_{\mathscr{N},\mathscr{K}}(x,t)\|_2^2,$$

$$=\int_0^1\int_0^1\|((\mathscr{D}^{\gamma}_{0+;[z;\omega;2]})u(x,t)-(\mathscr{D}^{\gamma}_{0+;[z;\omega;2]})u_{\mathscr{N},\mathscr{K}}(x,t))\|^2\,dx\,dt. \quad (5.33)$$

Proof. Since

$$(\mathscr{D}^{\gamma}_{0+;[z;w;2]})u(x,t) = \sum_{i=0}^{\infty}\sum_{j=0}^{\infty}\sum_{j=0}^{\infty} b_{i,j,j}\mathscr{L}_i(x)\mathscr{L}_j(t), \tag{5.34}$$

where $b_{i,j}$ is given by,

$$b_{i,j,j} = \frac{1}{\tilde{h}_i\tilde{h}_j\tilde{h}_k}\int_0^1\int_0^1\int_0^1 (\mathscr{D}^{\gamma}_{0+;[z;w;2]})u(x,t)\mathscr{L}_i(x)\mathscr{L}_j(t)\,dx\,dt, \tag{5.35}$$

we approximate the derivative of function $u(x,t)$ by,

$$(\mathscr{D}^{\gamma}_{0+;[z;w;2]})u_{\mathscr{N},\mathscr{K}}(x,t) = \sum_{i=0}^{\mathscr{N}}\sum_{j=0}^{\mathscr{K}} b_{i,j}\mathscr{L}_i(x)\mathscr{L}_j(t). \tag{5.36}$$

From Eqs. (5.34) and (5.36) we get,

$$\begin{aligned}(\mathscr{D}^{\gamma}_{0+;[z;\omega;2]})&u(x,w,t) - (\mathscr{D}^{\gamma}_{0+;[z;\omega;2]})u_{\mathscr{N},\mathscr{K}}(x,t)\\ &= \sum_{i=\mathscr{N}+1}^{\infty}\sum_{j=0}^{\mathscr{K}} b_{i,j}\mathscr{L}_i(x)\mathscr{L}_j(t) + \sum_{i=N+1}^{\infty}\sum_{k=K+1}^{\infty} b_{i,j}\mathscr{L}_i(x)\mathscr{L}_j(t)\\ &\quad + \sum_{i=0}^{\mathscr{N}}\sum_{k=K+1}^{\infty} b_{i,j}\mathscr{L}_i(x)\mathscr{L}_j(t),\end{aligned} \tag{5.37}$$

by taking the L^2 norm of Eq. (5.37), we have

$$\begin{aligned}\mathscr{E}_n &= \Bigg\| \sum_{i=\mathscr{N}+1}^{\infty}\sum_{k=0}^{\mathscr{K}} b_{i,j}\mathscr{L}_i(x)\mathscr{L}_j(t) + \sum_{i=\mathscr{N}+1}^{\infty}\sum_{k=\mathscr{K}+1}^{\infty} b_{i,j}\mathscr{L}_i(x)\mathscr{L}_j(t)\\ &\quad + \sum_{i=0}^{\mathscr{N}}\sum_{k=\mathscr{K}+1}^{\infty} b_{i,j}\mathscr{L}_i(x)\mathscr{L}_j(t)\Bigg\|^2,\\ &\leq \sum_{i=\mathscr{N}+1}^{\infty}\sum_{j=0}^{\mathscr{N}} b_{i,j}^2\|\mathscr{L}_i(x)\|^2\|\mathscr{L}_j(t)\|^2 + \sum_{i=\mathscr{N}+1}^{\infty}\sum_{j=\mathscr{K}+1}^{\infty} b_{i,j}^2\|\mathscr{L}_i(x)\|^2\|\mathscr{L}_j(t)\|^2\\ &\quad + \sum_{i=0}^{\mathscr{N}}\sum_{j=\mathscr{K}+1}^{\infty} b_{i,j}^2\|\mathscr{L}_i(x)\|^2\|\mathscr{L}_j(t)\|^2,\\ &= \sum_{i=\mathscr{N}+1}^{\infty}\sum_{j=\mathscr{K}+1}^{\infty} b_{i,j}^2\tilde{h}_i\tilde{h}_j.\end{aligned} \tag{5.38}$$

Using Eq. (5.35) in Eq. (5.38), we get

$$
\begin{aligned}
|\mathscr{E}_n| &\le \sum_{i=\mathscr{N}+1}^{\infty}\sum_{j=\mathscr{K}+1}^{\infty} \frac{\sup|(\mathscr{D}^{\gamma}_{0+;[z;w;2]})u(x,t)|^2(2i+1)(2j+1)}{((i+i^2)\pi)^2((j+j^2)\pi)^2},\\
&\le \sum_{i=\mathscr{N}+1}^{\infty}\sum_{j=0}^{\mathscr{K}} \frac{\lambda^2(2i+1)(2j+1)}{((i+i^2)\pi)^2((j+j^2)\pi)^2} + \sum_{i=\mathscr{N}+1}^{\infty}\sum_{j=\mathscr{K}+1}^{\infty} \frac{\lambda^2(2i+1)(2j+1)}{((i+i^2)\pi)^2((j+j^2)\pi)^2}\\
&\quad + \sum_{i=0}^{\mathscr{N}}\sum_{j=\mathscr{K}+1}^{\infty} \frac{\lambda^2(2i+1)(2j+1)}{((i+i^2)\pi)^2((j+j^2)\pi)^2},\\
&\le \frac{\lambda^2}{\pi^4(1+\mathscr{N})^2(1+\mathscr{K})^2}.
\end{aligned}
\tag{5.39}
$$

Clearly, we obtained the following result,

$$
|\mathscr{E}_n| \le \left(\frac{\lambda^2}{\pi^4(1+\mathscr{N})^2(1+\mathscr{K})^2}\right). \tag{5.40}
$$

It is clear from the Eq. (5.40) that $\mathscr{E}_n \to 0$ as $\mathscr{N}, \mathscr{M}, \mathscr{K} \to \infty$. □

5.6 NUMERICAL RESULTS

Now, we present the numerical validation of the proposed collocation method with some examples. Numerical results of examples are presented by varying the parameters. The maximum absolute error (MAE) and absolute error (AE) are also calculated respectively by the formulae,

$$
E_N = \max|u(x,t) - u_{\mathscr{N},\mathscr{K}}(x,t)|,\ (x,t) \in \mathscr{I},
$$

$$
AE = |u(x_p,t_k) - u_{N,K}(x_p,t_k)|,\ x_p, t_k \in \mathscr{I},
$$

here, as before, $u(x,t)$ and $u_{\mathscr{N},\mathscr{K}}(x,t)$ denote the exact the approximate solution, respectively.

Example 5.1. *Consider the following problem,*

$$
(\mathscr{D}^{\gamma}_{0+;[z;\omega;2]})u(x,t) + v_1(x,t)\frac{\partial u(x,t)}{\partial x} = \frac{\partial^2 u(x,t)}{\partial x^2} + s(x,t),\ t \in \mathscr{I}_2, \tag{5.41}
$$

similar to the one studied by the authors in Ref. [11,25–28] with initial and boundary conditions given as,

$$
\begin{cases} u(x,0) = x^2(1-x),\ x \in \mathscr{I}_1,\\ u(0,t) = 0,\\ u(1,t) = 0,\end{cases} \tag{5.42}
$$

$$
s(x,t) = e^{-t}\left(-2+8x-3x^2+\frac{(-t)^{\gamma}(t)^{-\gamma}(-1+x)x^2(\gamma\Gamma(-\gamma)+\Gamma(1-\gamma,-t))}{\Gamma(1-\gamma)}\right), \gamma<1. \tag{5.43}
$$

The exact solution of the problem (5.1) is $x^2(1-x)e^{-t}$. We solved this problem by the collocation method and found the approximate solution. In Table 5.1, we have found the AE for different values of $\mathcal{N} = \mathcal{K}$ in the defined domain. It is clear from Table 5.1 that the error reduces for large values of $\mathcal{N} = \mathcal{K}$ and the numerical solution gets closure to the exact solution. We have shown the AE graphs for different values of $\mathcal{N} = \mathcal{K}$ in Figure 5.1. We see from Figure 5.1 that for a large value of the $\mathcal{N} = \mathcal{K}$ the numerical solution converges to the exact solution very rapidly than the small value of $\mathcal{N} = \mathcal{K}$. Finally, from the figures and Table 5.2, we observed that numerical results obtained by our proposed method are more accurate than the other methods presented by several authors given in Refs. [11,25–28].

Example 5.2. *Consider the Eq. (5.1) and similar problem taken in Ref. [29,30] with* $v_1(x,t) = 0$ *initial and boundary conditions,*

Table 5.1
AE for Example 5.1 with $\gamma = 0.2$

Grid Points	$\mathcal{N} = \mathcal{K} = 6$	$\mathcal{N} = \mathcal{K} = 8$	$\mathcal{N} = \mathcal{K} = 10$	$\mathcal{N} = \mathcal{K} = 12$
(0.1,0.1)	6.093×10^{-9}	4.047×10^{-12}	6.504×10^{-14}	5.211×10^{-14}
(0.2,0.2)	1.301×10^{-8}	1.701×10^{-11}	8.986×10^{-16}	7.915×10^{-14}
(0.3,0.3)	8.206×10^{-9}	1.330×10^{-11}	3.469×10^{-17}	4.761×10^{-15}
(0.4,0.4)	1.789×10^{-9}	2.427×10^{-12}	6.384×10^{-16}	7.231×10^{-15}
(0.5,0.5)	8.898×10^{-10}	9.084×10^{-12}	5.412×10^{-16}	1.361×10^{-15}
(0.6,0.6)	1.969×10^{-9}	2.096×10^{-12}	1.216×10^{-15}	2.609×10^{-15}
(0.7,0.7)	1.321×10^{-9}	7.580×10^{-13}	7.910×10^{-16}	1.444×10^{-15}
(0.8,0.8)	3.115×10^{-10}	3.495×10^{-13}	8.257×10^{-16}	5.899×10^{-16}
(0.9,0.9)	1.758×10^{-10}	1.219×10^{-13}	4.788×10^{-16}	2.165×10^{-15}

Table 5.2
Comparison of MAE for Example 5.1 with $\gamma = 0.$, $\mathcal{N} = \mathcal{K} = 8$

	MAE
Present method	1.993×10^{-13}
Method [25]	1.432×10^{-5}
Method [11]	1.693×10^{-3}
Method [26]	8.235×10^{-5}
Method [27]	2.742×10^{-5}
Method [28]	2.34×10^{-5}

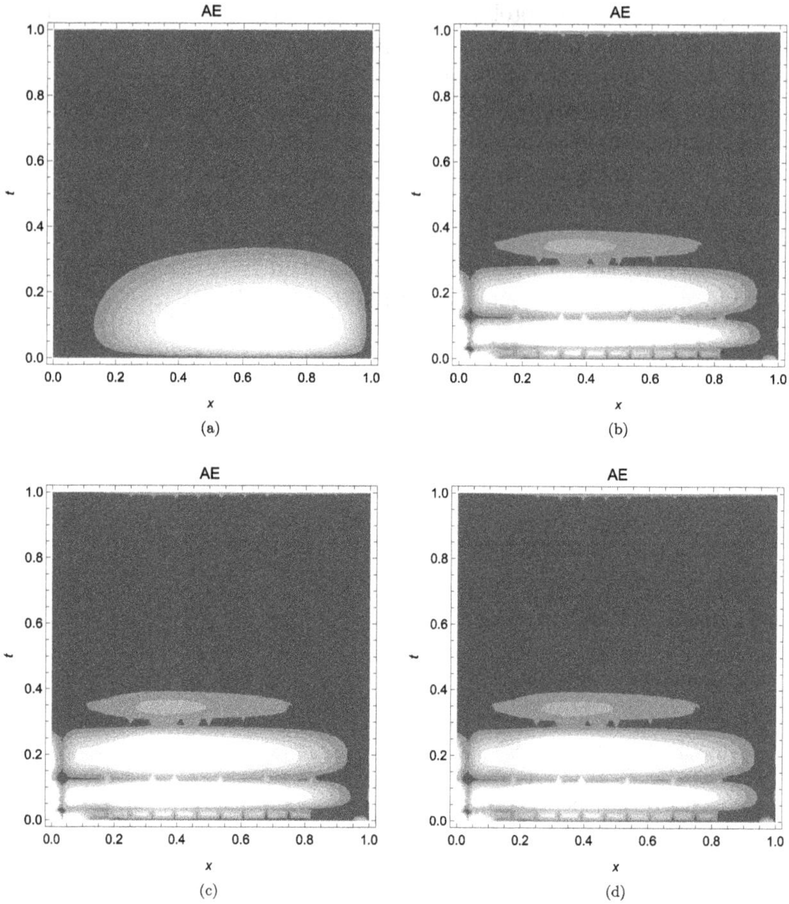

Figure 5.1 AE at different values of $\mathcal{N} = \mathcal{K} = 6, 8, 10, 12$ and $\gamma = 0.2$ for Example 5.1. (a) $\mathcal{N} = \mathcal{K} = 6$, (b) $\mathcal{N} = \mathcal{K} = 8$, (c) $\mathcal{N} = \mathcal{K} = 10$, and (d) $\mathcal{N} = \mathcal{K} = 12$.

$$\begin{aligned} &u(x,0) = 0,\ x \in \mathscr{I}_1, \\ &u(0,t) = 0,\ u(1,t) = (t)^{\beta}\sin(1),\ t \in \mathscr{I}_2, \end{aligned} \tag{5.44}$$

where

$$s(x,t) = (t)^{\beta}\sin x\left[\frac{\Gamma(\beta/\delta+1)}{\Gamma(\beta/\delta+1-\gamma)}t^{-\gamma\delta}+1\right]. \tag{5.45}$$

This problem has the exact solution $t^{\beta}\sin x$ in the unit square $\mathscr{I}_1 \times \mathscr{I}_2$. The numerical solutions of the problem are obtained for different values of $\beta = 1/2, 1, 7/2, 5$, and corresponding MAE is shown in Table 5.3. We find the approximate solution by taking $N = K = 2, 3, 4, 5, 6, 7, 8, 9, 10, 11$ and 12. We observe that numerical results obtained by our proposed method is more accurate than the methods given in

Table 5.3
MAE of Example 5.2 with $\gamma = 0.2$, $\delta = 3$ and Different Values of β

N	K	$\beta = \frac{1}{2}$	$\beta = 1$	$\beta = \frac{7}{2}$	$\beta = 5$
2	2	7.203×10^{-2}	8.573×10^{-3}	1.461×10^{-1}	3.177×10^{-1}
4	4	1.148×10^{-2}	3.879×10^{-4}	2.983×10^{-4}	4.006×10^{-3}
6	6	5.667×10^{-3}	8.798×10^{-6}	1.651×10^{-5}	2.238×10^{-6}
8	8	8.919×10^{-3}	2.452×10^{-8}	2.727×10^{-6}	6.224×10^{-9}
10	10	2.404×10^{-5}	2.140×10^{-12}	4.725×10^{-6}	2.001×10^{-7}
12	12	9.665×10^{-5}	1.139×10^{-13}	3.351×10^{-7}	8.599×10^{-13}

Refs. [29,30]. This problem is solved by taking weight $w(t) = 1$ and scale function $z(t) = t^{\delta}$. In Figure 5.2, we have shown the AE graph for the various values of β. From Table 5.3 and Figure 5.2, we conclude the applicability of the present method for the FDE whose solution contains a fractional power term.

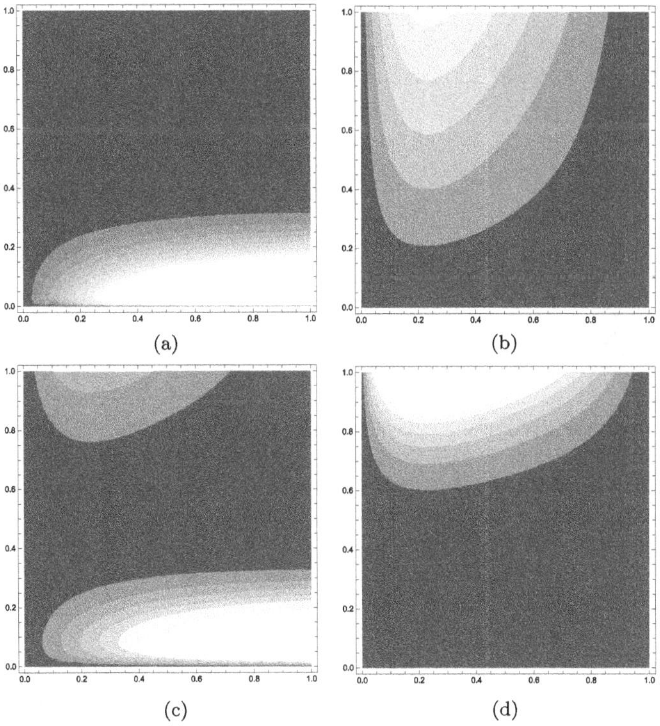

Figure 5.2 Plot of the AE at different values of β with $N = K = 6$, $w(t) = 1$, $z(t) = t^3$ and $\gamma = 0.2$ for Example 5.2. (a) $\beta = \frac{1}{2}$, (b) $\beta = 1$, (c) $\beta = \frac{7}{2}$, and (d) $\beta = 5$.

5.7 CONCLUSION

A collocation method is developed for solving GFADEs using Legendre polynomials. For the numerical simulations, we used the finite difference method on the temporal axis and the collocation method on the space variable. The convergence and error analysis of the presented method were established theoretically and verified the results numerically. The proposed method works well and produces highly accurate approximate solutions.

References

1. W. Van Beinum, J. C. Meeussen, A. C. Edwards, W. H. Van Riemsdijk, Transport of ions in physically heterogeneous systems; convection and diffusion in a column filled with alginate gel beads, predicted by a two-region model, *Water Research* 34 (7) (2000) 2043–2050.
2. T. Bocksell, E. Loth, Stochastic modeling of particle diffusion in a turbulent boundary layer, *International Journal of Multiphase Flow* 32 (10–11) (2006) 1234–1253.
3. J. S. Ferreira, M. Costa, Deterministic advection-diffusion model based on markov processes, *Journal of Hydraulic Engineering* 128 (4) (2002) 399–411.
4. N. Kumar, Unsteady flow against dispersion in finite porous media, *Journal of Hydrology* 63 (3–4) (1983) 345–358.
5. S. Kazem, M. Dehghan, Semi-analytical solution for time-fractional diffusion equation based on finite difference method of lines (mol), *Engineering with Computers* 35 (1) (2019) 229–241.
6. M. S. Dahaghin, H. Hassani, An optimization method based on the generalized polynomials for nonlinear variable-order time fractional diffusion-wave equation, *Nonlinear Dynamics* 88 (3) (2017) 1587–1598.
7. H. Safdari, H. Mesgarani, M. Javidi, Y. E. Aghdam, Convergence analysis of the space fractional-order diffusion equation based on the compact finite difference scheme, *Computational and Applied Mathematics* 39 (2) (2020) 1–15.
8. A. S. Hendy, Numerical treatment for after-effected multi-term time-space fractional advection–diffusion equations, *Engineering with Computers* 37 (2) (2020) 1–11.
9. H. Tajadodi, A numerical approach of fractional advection-diffusion equation with atangana–baleanu derivative, *Chaos, Solitons & Fractals* 130 (2020) 109527.
10. Y. E. Aghdam, H. Mesgrani, M. Javidi, O. Nikan, A computational approach for the space-time fractional advection–diffusion equation arising in contaminant transport through porous media, *Engineering with Computers* 37(4) (2020) 1–13.
11. H. Safdari, Y. E. Aghdam, J. Gómez-Aguilar, Shifted chebyshev collocation of the fourth kind with convergence analysis for the space–time fractional advection-diffusion equation, *Engineering with Computers* 38(2) (2020) 1–12.
12. S. Yadav, R. K. Pandey, A. K. Shukla, K. Kumar, High-order approximation for generalized fractional derivative and its application, *International Journal of Numerical Methods for Heat & Fluid Flow* 29 (9) (2019) 3515–3534.
13. S. Kumar, R. K. Pandey, K. Kumar, S. Kamal, T. N. Dinh, Finite difference–collocation method for the generalized fractional diffusion equation, *Fractal and Fractional* 6 (7) (2022) 387.

14. Y. Xu, O. P. Agrawal, Numerical solutions and analysis of diffusion for new generalized fractional burgers equation, *Fractional Calculus and Applied Analysis* 16 (3) (2013) 709–736.
15. Y. Xu, Z. He, Q. Xu, Numerical solutions of fractional advection–diffusion equations with a kind of new generalized fractional derivative, *International Journal of Computer Mathematics* 91 (3) (2014) 588–600.
16. R. M. Hafez, M. A. Zaky, A. S. Hendy, A novel spectral galerkin/petrov–galerkin algorithm for the multi-dimensional space–time fractional advection–diffusion–reaction equations with nonsmooth solutions, *Mathematics and Computers in Simulation* 190 (2021) 678–690.
17. H. Hassani, Z. Avazzadeh, J. A. T. Machado, Solving two-dimensional variable-order fractional optimal control problems with transcendental bernstein series, *Journal of Computational and Nonlinear Dynamics* 14 (6) (2019) 061001.
18. K. Diethelm, *The Analysis of Fractional Differential Equations: An Application-Oriented Exposition Using Differential Operators of Caputo Type*, Springer Science & Business Media: Berlin, Germany (2010).
19. A. Saadatmandi, M. Dehghan, M.-R. Azizi, The sinc–legendre collocation method for a class of fractional convection-diffusion equations with variable coefficients, *Communications in Nonlinear Science and Numerical Simulation* 17 (11) (2012) 4125–4136.
20. I. Podlubny, An introduction to fractional derivatives, fractional differential equations, methods of their solution and some of their applications, *Mathematics in Science and Engineering* 198 (1999) xxiv+–340.
21. K. Oldham, J. Spanier, *The Fractional Calculus Theory and Applications of Differentiation and Integration to Arbitrary Order*, Elsevier: Amsterdam, The Netherlands (1974).
22. O. P. Agrawal, Some generalized fractional calculus operators and their applications in integral equations, *Fractional Calculus and Applied Analysis* 15 (4) (2012) 700–711.
23. A. N. Lowan, N. Davids, A. Levenson, Table of the zeros of the Legendre polynomials of order 1-16 and the weight coefficients for Gauss' mechanical quadrature formula, *Bulletin of the American Mathematical Society* 48 (10) (1942) 739–743.
24. E. Kreyszig, *Introductory Functional Analysis with Applications*, Vol. 1. Wiley: New York (1978).
25. A. Baseri, S. Abbasbandy, E. Babolian, A collocation method for fractional diffusion equation in a long time with chebyshev functions, *Applied Mathematics and Computation* 322 (2018) 55–65.
26. S. Alavizadeh, F. M. Ghaini, Numerical solution of fractional diffusion equation over a long time domain, *Applied Mathematics and Computation* 263 (2015) 240–250.
27. M. Khader, On the numerical solutions for the fractional diffusion equation, *Communications in Nonlinear Science and Numerical Simulation* 16 (6) (2011) 2535–2542.
28. C. Tadjeran, M. M. Meerschaert, H.-P. Scheffler, A second-order accurate numerical approximation for the fractional diffusion equation, *Journal of Computational Physics* 213 (1) (2006) 205–213.
29. Q. Ding, P. J. Wong, A new approximation for the generalized fractional derivative and its application to generalized fractional diffusion equation, *Numerical Methods for Partial Differential Equations* 37 (1) (2021) 643–673.
30. C.-C. Ji, Z.-Z. Sun, A high-order compact finite difference scheme for the fractional subdiffusion equation, *Journal of Scientific Computing* 64 (3) (2015) 959–985.

6 The Incomplete Generalized Mittag-Leffler Function and Fractional Calculus Operators

Rakesh K. Parmar
Pondicherry University

Purnima Chopra
Marudhar Engineering College Bikaner

CONTENTS

6.1 INTRODUCTION, DEFINITIONS, AND PRELIMINARIES

The familiar *incomplete Gamma functions* $\gamma(s,x)$ and $\Gamma(s,x)$ defined by

$$\gamma(s,x) := \int_0^x t^{s-1}\, e^{-t}\, dt \qquad \big(\Re(s) > 0;\ x \geqq 0\big) \tag{6.1}$$

and

$$\Gamma(s,x) := \int_x^\infty t^{s-1}\, e^{-t}\, dt \qquad \big(x \geqq 0;\ \Re(s) > 0 \quad \text{when} \quad x = 0\big), \tag{6.2}$$

DOI: 10.1201/9781003368069-6

respectively, satisfy the following decomposition formula:

$$\gamma(s,x)+\Gamma(s,x):=\Gamma(s) \qquad (\Re(s)>0). \tag{6.3}$$

Throughout this chapter, $\mathbb{N}$, $\mathbb{Z}^-$, and $\mathbb{C}$ denote the sets of positive integers, negative integers, and complex numbers, respectively,

$$\mathbb{N}_0:=\mathbb{N}\cup\{0\} \qquad \text{and} \qquad \mathbb{Z}_0^-:=\mathbb{Z}^-\cup\{0\}.$$

Moreover, the parameter $x \geqq 0$ used above in Eqs. (6.1) and (6.2) *and* elsewhere in this chapter is independent of $\Re(z)$ of the complex number $z \in \mathbb{C}$.

Recently, Srivastava et al. [57] introduced and studied in a rather systematic manner the following two families of generalized incomplete hypergeometric functions:

$${}_p\gamma_q\left[\begin{array}{r}(\alpha_1,x),\alpha_2,\ldots,\alpha_p;\\ \beta_1,\ldots,\beta_q;\end{array} z\right]=\sum_{n=0}^{\infty}\frac{(\alpha_1;x)_n(\alpha_2)_n\ldots(\alpha_p)_n}{(\beta_1)_n\ldots(\beta_q)_n}\frac{z^n}{n!} \tag{6.4}$$

and

$${}_p\Gamma_q\left[\begin{array}{r}(\alpha_1,x),\alpha_2,\ldots,\alpha_p;\\ \beta_1,\ldots,\beta_q;\end{array} z\right]=\sum_{n=0}^{\infty}\frac{[\alpha_1;x]_n(\alpha_2)_n\ldots(\alpha_p)_n}{(\beta_1)_n\ldots(\beta_q)_n}\frac{z^n}{n!}, \tag{6.5}$$

where, in terms of the incomplete Gamma functions $\gamma(s,x)$ and $\Gamma(s,x)$ defined by Eqs. (6.1) and (6.2), respectively, the *incomplete* Pochhammer symbols $(\lambda;x)_\nu$ and $[\lambda;x]_\nu$ $(\lambda;\, \nu\in\mathbb{C};\, x\geqq 0)$ are defined as follows:

$$(\lambda;x)_\nu:=\frac{\gamma(\lambda+\nu,x)}{\Gamma(\lambda)} \qquad (\lambda,\nu\in\mathbb{C};\; x\geqq 0) \tag{6.6}$$

and

$$[\lambda;x]_\nu:=\frac{\Gamma(\lambda+\nu,x)}{\Gamma(\lambda)} \qquad (\lambda,\nu\in\mathbb{C};\; x\geqq 0), \tag{6.7}$$

so that, obviously, these incomplete Pochhammer symbols $(\lambda;x)_\nu$ and $[\lambda;x]_\nu$ satisfy the following decomposition relation:

$$(\lambda;x)_\nu+[\lambda;x]_\nu:=(\lambda)_\nu \qquad (\lambda;\,\nu\in\mathbb{C};\, x\geqq 0). \tag{6.8}$$

Here, and in what follows, $(\lambda)_\nu$ $(\lambda,\nu\in\mathbb{C})$ denotes the Pochhammer symbol (or the shifted factorial) which is defined (in general) by

$$(\lambda)_\nu:=\frac{\Gamma(\lambda+\nu)}{\Gamma(\lambda)}=\begin{cases}1 & (\nu=0;\;\lambda\in\mathbb{C}\setminus\{0\})\\ \lambda(\lambda+1)\ldots(\lambda+n-1) & (\nu=n\in\mathbb{N};\;\lambda\in\mathbb{C}),\end{cases} \tag{6.9}$$

it being understood *conventionally* that $(0)_0:=1$ and assumed *tacitly* that the Γ-quotient exists (see, for details, [42]).

As already observed by Srivastava et al. [57], definitions (6.4) and (6.5) readily yield the following decomposition formula:

$$
{}_p\gamma_q\left[\begin{array}{c}(\alpha_1,x),\alpha_2,\ldots,\alpha_p;\\ \beta_1,\ldots,\beta_q;\end{array} z\right]+{}_p\Gamma_q\left[\begin{array}{c}(\alpha_1,x),\alpha_2,\ldots,\alpha_p;\\ \beta_1,\ldots,\beta_q;\end{array} z\right]
= {}_pF_q\left[\begin{array}{c}\alpha_1,\alpha_2,\ldots,\alpha_p;\\ \beta_1,\ldots,\beta_q;\end{array} z\right] \tag{6.10}
$$

for the familiar generalized hypergeometric function ${}_pF_q$. Further, in the same paper, they gave a pair of Mellin-Barnes contour integral representations of incomplete generalized hypergeometric functions ${}_p\gamma_q$ and ${}_p\Gamma_q$ in terms of the incomplete Gamma functions $\gamma(s,x)$ and $\Gamma(s,x)$ defined by Eqs. (6.1) and (6.2), respectively,

$$
{}_p\gamma_q\left[\begin{array}{c}(A_p,x);\\ B_q;\end{array} z\right] = {}_p\gamma_q\left[\begin{array}{c}(a_1,x),a_2,\ldots,a_p;\\ b_1,\ldots,b_q;\end{array} z\right]
= \frac{1}{2\pi i}\frac{\prod_{j=1}^{q}\Gamma(b_j)}{\prod_{i=1}^{p}\Gamma(a_i)}\int_{\mathbb{L}}\frac{\gamma(a_1+s,x)\prod_{i=2}^{p}\Gamma(a_j+s)}{\prod_{j=1}^{q}\Gamma(b_j+s)}\Gamma(-s)(-z)^s ds
\quad (|\arg(-z)|<\pi) \tag{6.11}
$$

and

$$
{}_p\Gamma_q\left[\begin{array}{c}(A_p,x);\\ B_q;\end{array} z\right] = {}_p\Gamma_q\left[\begin{array}{c}(a_1,x),a_2,\ldots,a_p;\\ b_1,\ldots,b_q;\end{array} z\right]
= \frac{1}{2\pi i}\frac{\prod_{j=1}^{q}\Gamma(b_j)}{\prod_{i=1}^{p}\Gamma(a_i)}\int_{\mathbb{L}}\frac{\Gamma(a_1+s,x)\prod_{l=2}^{p}\Gamma(a_j+s)}{\prod_{j=1}^{q}\Gamma(b_j+s)}\Gamma(-s)(-z)^s ds
\quad (|\arg(-z)|<\pi), \tag{6.12}
$$

where $\mathscr{L}=\mathscr{L}_{(i\tau;\infty)}$ is a suitable contour of the Mellin-Barnes type starting at the point $\tau-i\infty$ and terminating at the point $\tau+i\infty\,(\tau\in\mathbb{R})$ with the usual indentations in order to separate one set of poles from the other set of poles of the integrand.

Next for $\alpha_1,\ldots,\alpha_p\in\mathbb{C}$ and, $\beta_1,\ldots,\beta_q\in\mathbb{C}\setminus\mathbb{Z}_0^-$, the incomplete Fox-Wright functions ${}_p\Psi_q^{(\gamma)}(z)$ and ${}_p\Psi_q^{(\Gamma)}(z)$ with p numerator and q denominator parameters are defined by Ref. [61](see also Refs. [25,59])

$$
{}_p\Psi_q^{(\gamma)}\left[\begin{array}{c}(\alpha_1,A_1,x),(\alpha_2,A_2),\ldots,(\alpha_p,A_p);\\ (\beta_1,B_1),\ldots,(\beta_q,B_q);\end{array} z\right]
= \sum_{n=0}^{\infty}\frac{\gamma(\alpha_1+A_1n,x)\Gamma(\alpha_2+A_2n)\ldots\Gamma(\alpha_p+A_pn)}{\Gamma(\beta_1+B_1n)\ldots\Gamma(\beta_q+B_qn)}\frac{z^n}{n!} \tag{6.13}
$$

and

$$
{}_p\Psi_q^{(\Gamma)}\left[\begin{array}{c}(\alpha_1,A_1,x),(\alpha_2,A_2),\ldots,(\alpha_p,A_p);\\(\beta_1,B_1),\ldots,(\beta_q,B_q);\end{array} z\right]
= \sum_{n=0}^{\infty}\frac{\Gamma(\alpha_1+A_1n,x)\Gamma(\alpha_2+A_2n)\ldots\Gamma(\alpha_p+A_pn)}{\Gamma(\beta_1+B_1n)\ldots\Gamma(\beta_q+B_qn)}\frac{z^n}{n!}, \tag{6.14}
$$

respectively. In view of Eq. (6.8), these families of incomplete Fox-Wright generalized hypergeometric function satisfy the following decomposition formula:

$$
{}_p\Psi_q^{(\gamma)}\left[\begin{array}{c}(\alpha_1,A_1,x),(\alpha_2,A_2),\ldots,(\alpha_p,A_p);\\(\beta_1,B_1),\ldots,(\beta_q,B_q);\end{array} z\right]
+ {}_p\Psi_q^{(\Gamma)}\left[\begin{array}{c}(\alpha_1,A_1,x),(\alpha_2,A_2),\ldots,(\alpha_p,A_p);\\(\beta_1,B_1),\ldots,(\beta_q,B_q);\end{array} z\right]
= {}_p\Psi_q\left[\begin{array}{c}(\alpha_1,A_1),\ldots,(\alpha_p,A_p);\\(\beta_1,B_1),\ldots,(\beta_q,B_q);\end{array} z\right], \tag{6.15}
$$

where ${}_p\Psi_q(z)$ is the Fox-Wright generalized hypergeometric function [21,42]. They gave a pair of Mellin-Barnes contour integral representations of incomplete Fox-Wright generalized hypergeometric functions ${}_p\Psi_q^{(\gamma)}$ and ${}_p\Psi_q^{(\Gamma)}$ defined by [61]

$$
{}_p\Psi_q^{(\gamma)}\left[\begin{array}{c}(\alpha_1,A_1,x),(\alpha_2,A_2),\ldots,(\alpha_p,A_p);\\(\beta_1,B_1),\ldots,(\beta_q,B_q);\end{array} z\right]
= \frac{1}{2\pi i}\int_{\mathbb{L}}\frac{\gamma(a_1+A_1s,x)\prod_{j=2}^{p}\Gamma(a_j+A_js)}{\prod_{j=1}^{q}\Gamma(b_j+B_js)}\Gamma(-s)(-z)^s ds
\quad (|\arg(-z)|<\pi) \tag{6.16}
$$

and

$$
{}_p\Psi_q^{(\Gamma)}\left[\begin{array}{c}(\alpha_1,A_1,x),(\alpha_2,A_2),\ldots,(\alpha_p,A_p);\\(\beta_1,B_1),\ldots,(\beta_q,B_q);\end{array} z\right]
= \frac{1}{2\pi i}\int_{\mathbb{L}}\frac{\Gamma(a_1+A_1s,x)\prod_{j=2}^{p}\Gamma(a_j+A_js)}{\prod_{j=1}^{q}\Gamma(b_j+B_js)}\Gamma(-s)(-z)^s ds
\quad (|\arg(-z)|<\pi), \tag{6.17}
$$

respectively, where $\mathscr{L}=\mathscr{L}_{(i\tau;\infty)}$ is a suitable contour of the Mellin-Barnes type starting at the point $\tau-i\infty$ and terminating at the point $\tau+i\infty\,(\tau\in\mathbb{R})$ with the usual indentations in order to separate one set of poles from the other set of poles of the integrand (see for details, [61]).

Remark 6.1: The special case when $x = 0$, both ${}_p\Psi_q^{(\gamma)}(z)$ $(p, q \in \mathbb{N}_0)$ and ${}_p\Psi_q^{(\Gamma)}(z)$ $(p, q \in \mathbb{N}_0)$ would reduce immediately to the Fox-Wright Psi function ${}_p\Psi_q(z)$ $(p, q \in \mathbb{N}_0)$.

Remark 6.2: The special case when

$$A_j = B_k = 1 \quad (j = 1, \ldots, p; \; k = 1, \ldots, q)$$

in Eqs. (6.13) and (6.14), which immediately reduce to Eqs. (6.4) and (6.5), respectively.

For various other investigations involving incomplete higher transcendental functions, the interested reader may be referred to several recent papers on the subject (see, for example, [3,7,9–11,13,18,25,57,64,65] *and* the references cited in each of these papers).

The one-parametric Mittag-Leffler function (named after the Swedish mathematician Gösta Magnus Mittag-Leffler (1846–1927)) is an entire function defined by Refs. [26,27]

$$E_\alpha(z) = \sum_{k=0}^{\infty} \frac{z^k}{\Gamma(\alpha k + 1)} \quad (\alpha \in \mathbb{C}, \; \Re(\alpha) > 0, \; z \in \mathbb{C}), \tag{6.18}$$

where $\Gamma(x)$ denotes the familiar Gamma function. A great deal of attention has been paid to the various generalizations of the Mittag-Leffler function by many researchers. The two-parametric Mittag-Leffler function was defined by (see [68]; see also [1], [19,20])

$$E_{\alpha,\beta}(z) = \sum_{n=0}^{\infty} \frac{z^n}{\Gamma(\alpha n + \beta)} \quad (\alpha, \beta \in \mathbb{C}; \; \Re(\alpha) > 0, \; \Re(\beta) > 0) \tag{6.19}$$

and the three-parametric Mittag-Leffler function introduced by Prabhaker [30, Eq. (1.3)] (see also [21]) was defined by

$$E_{\alpha,\beta}^{\rho}(z) = \sum_{n=0}^{\infty} \frac{(\rho)_n}{\Gamma(\alpha n + \beta)} \frac{z^n}{n!} \quad (\alpha, \beta, \rho \in \mathbb{C}; \; \Re(\alpha) > 0, \; \Re(\beta) > 0), \tag{6.20}$$

Further, Srivastava and Tomovaski [63, p. 200, Equation (1.13)](see also, [46]) introduced generalized Mittag-Leffler function in the following form:

$$E_{\alpha,\beta}^{\rho,\kappa}(z) = \sum_{n=0}^{\infty} \frac{(\rho)_{\kappa n}}{\Gamma(\alpha n + \beta)} \frac{z^n}{n!}, \tag{6.21}$$

$$(z, \beta, \rho \in \mathbb{C}; \; \Re(\alpha) > \max\{0, \Re(\kappa) - 1\}; \Re(\kappa) > 0, \; \Re(\beta) > 0).$$

Various authors investigated and studied several properties and applications of generalized Mittag-Leffler function in the solution of fractional order differential equation and fractional order integral equation (see, *e.g.*, [21,22,28,29,37–40,45,66]). A detailed and comprehensive account on the Mittag-Leffler function and can be found in the following review paper of Haubold et al. [16] (see also, [15]). Also, the operators of fractional integration and their applications one can refer to the following survey paper by Srivastava and Saxena [45].

Motivated essentially by the demonstrated potential for applications of these incomplete generalized hypergeometric functions, we introduce incomplete generalized Mittag-Leffler function by means of the incomplete Pochhammer symbol $(\lambda;x)_\nu$ and $[\lambda;x]_\nu$ and investigate certain basic properties including, for example, differentiation formulas, integral representations, Laplace transform, Euler-Beta transform, and Whittaker transform with their several special cases and relationship with incomplete generalized hypergeometric functions (6.5). Moreover, certain relations between the incomplete generalized Mittag-Leffler function and the Riemann-Liouville fractional integrals and derivatives are investigated. Some interesting special cases of our main results are also considered. For various other investigations involving generalizations of the hypergeometric function ${}_pF_q$ of p numerator and q denominator parameters, which were motivated essentially by the pioneering work of Srivastava et al. [57], the interested reader may be referred to several recent papers on the subject (see, for example, [3,7,9,10,18,57,64,65] *and* the references cited in each of these papers).

6.2 THE INCOMPLETE GENERALIZED MITTAG-LEFFLER FUNCTION

In this section, we introduce the incomplete generalized Mittag-Leffler Functions as follows: For $\alpha, \beta, \rho \in \mathbb{C}$, we have

$$\mathscr{E}_{\alpha,\beta}^{\rho,\kappa}(z) = \sum_{n=0}^{\infty} \frac{(\rho;x)_{\kappa n}}{\Gamma(\alpha n+\beta)} \frac{z^n}{n!} \tag{6.22}$$

$$(x \geq 0;\, z,\beta,\rho \in \mathbb{C};\ \Re(\alpha) > \max\{0, \Re(\kappa)-1\}; \Re(\kappa) > 0,\ \Re(\beta) > 0 \text{ when } x=0)$$

and

$$\Xi_{\alpha,\beta}^{\rho,\kappa}(z) = \sum_{n=0}^{\infty} \frac{[\rho;x]_{\kappa n}}{\Gamma(\alpha n+\beta)} \frac{z^n}{n!} \tag{6.23}$$

$$(x \geq 0;\, z,\beta,\rho \in \mathbb{C};\ \Re(\alpha) > \max\{0, \Re(\kappa)-1\}; \Re(\kappa) > 0,\ \Re(\beta) > 0 \text{ when } x=0)$$

Remark 6.3: The special case of Eq. (6.23) when $x = 0$ is seen to yield the known definition of the generalized Mittag-Leffler function [46,63].

In view of Eq. (6.8), these families of incomplete generalized Mittag-Leffler functions satisfy the following decomposition formula:

$$\mathscr{E}_{\alpha,\beta}^{\rho,\kappa}(z) + \Xi_{\alpha,\beta}^{\rho,\kappa}(z) = E_{\alpha,\beta}^{\rho,\kappa}(z) \tag{6.24}$$

where $E_{\alpha,\beta}^{\rho,\kappa}(z)$ is the generalized Mittag-Leffler function (6.21).

It is noted in passing that, in view of the decomposition formula (6.24), namely all Theorems and properties are valid for the functions $\mathscr{E}_{\alpha,\beta}^{\rho,\kappa}(z)$ and $\Xi_{\alpha,\beta}^{\rho,\kappa}(z)$. Here it is sufficient to discuss the properties and characteristics of the incomplete generalized Mittag-Leffler function $\Xi_{\alpha,\beta}^{\rho,\kappa}(z)$.

6.2.1 BASIC PROPERTIES OF $\Xi_{\alpha,\beta}^{\rho,\kappa}(Z)$

In this section, we present certain derivative and integral properties of the incomplete generalized Mittag-Leffler function in Eq. (6.23).

Theorem 6.1. *The following derivative formula for the incomplete generalized Mittag-Leffler function in* Eq. (6.23) *holds true*:

$$\Xi_{\alpha,\beta}^{\rho,\kappa}(z) = \beta\, \Xi_{\alpha,\beta+1}^{\rho,\kappa}(z) + \alpha z \frac{d}{dz}\, \Xi_{\alpha,\beta+1}^{\rho,\kappa}(z) \tag{6.25}$$

$$(\alpha, \beta, \rho \in \mathbb{C};\ \Re(\alpha) > 0, \Re(\beta) > 0, x \geq 0).$$

In particular, we have

$$E_{\alpha,\beta}^{\rho,\kappa}(z) = \beta E_{\alpha,\beta+1}^{\rho,\kappa}(z) + \alpha z \frac{d}{dz} E_{\alpha,\beta+1}^{\rho,\kappa}(z) \tag{6.26}$$

Proof. We find from Eq. (6.23) that

$$\begin{aligned}
\beta\, \Xi_{\alpha,\beta+1}^{\rho,\kappa}(z) + \alpha z \frac{d}{dz} \Xi_{\alpha,\beta+1}^{\rho,\kappa}(z) &= \beta\, \Xi_{\alpha,\beta+1}^{\rho,\kappa}(z) + \alpha z \frac{d}{dz} \sum_{n=0}^{\infty} \frac{[\rho;x]_{\kappa n}}{\Gamma(\alpha n + \beta + 1)} \frac{z^n}{n!} \\
&= \beta\, \Xi_{\alpha,\beta+1}^{\rho,\kappa}(z) + \sum_{n=0}^{\infty} \frac{\alpha n\, [\rho;x]_{\kappa n}}{\Gamma(\alpha n + \beta + 1)} \frac{z^n}{n!} \\
&= \beta\, \Xi_{\alpha,\beta+1}^{\rho,\kappa}(z) + \sum_{n=0}^{\infty} \frac{[\rho;x]_{\kappa n}(\alpha n + \beta - \beta)}{\Gamma(\alpha n + \beta + 1)} \frac{z^n}{n!} \\
&= \Xi_{\alpha,\beta}^{\rho,\kappa}(z).
\end{aligned}$$

The relation (6.26) follows from Eq. (6.25) when $x = 0$. □

Theorem 6.2. *Each of the following differentiation formulas for the incomplete generalized Mittag-Leffler function in* Eq. (6.23) *holds true*:

$$\left(\frac{d}{dz}\right)^m \Xi_{\alpha,\beta}^{\rho,\kappa}(z) = (\rho)_{\kappa m}\, \Xi_{\alpha,\beta+m\alpha}^{\rho+\kappa m,\kappa}(z); \tag{6.27}$$

$$\left(\frac{d}{dz}\right)^m \left[z^{\beta-1}\, \Xi_{\alpha,\beta}^{\rho,\kappa}(\omega z^{\alpha})\right] = z^{\beta-m-1}\, \Xi_{\alpha,\beta-m}^{\rho,\kappa}(\omega z^{\alpha}) \quad (\Re(\beta - m) > 0,\ m \in \mathbb{N}) \tag{6.28}$$

and

$$\left(\frac{d}{dz}\right)^m \left[z^{\beta-1}\, {}_1\Gamma_1((\rho,x);\beta;\omega z)\right] = \frac{\Gamma(\beta)}{\Gamma(\beta - m)} z^{\beta-m-1}\, {}_1\Gamma_1((\rho,x);\beta - m;\omega z), \tag{6.29}$$

where $\alpha, \beta, \rho, \omega \in \mathbb{C};\ \Re(\alpha) > 0, \Re(\beta) > 0;\ x \geq 0.$

In particular, we have

$$\left(\frac{d}{dz}\right)^m E^{\rho,\kappa}_{\alpha,\beta}(z) = (\rho)_{\kappa m} E^{\rho+\kappa m}_{\alpha,\beta+m\alpha}(z); \tag{6.30}$$

$$\left(\frac{d}{dz}\right)^m \left[z^{\beta-1} E^{\rho,\kappa}_{\alpha,\beta}(\omega z^\alpha)\right] = z^{\beta-m-1} E^{\rho,\kappa}_{\alpha,\beta-m}(\omega z^\alpha) \tag{6.31}$$

and

$$\left(\frac{d}{dz}\right)^m \left[z^{\beta-1}\ {}_1F_1(\rho;\beta;\omega z)\right] = \frac{\Gamma(\beta)}{\Gamma(\beta-m)} z^{\beta-m-1}\ {}_1F_1(\rho;\beta-m;\omega z). \tag{6.32}$$

Proof. Employing termwise differentiation m times on Eq. (6.23) gives

$$\begin{aligned}\left(\frac{d}{dz}\right)^m \Xi^\rho_{\alpha,\beta}(z) &= \sum_{n=m}^{\infty} \frac{[\rho;x]_{\kappa n}}{\Gamma(\alpha n+\beta)} \frac{z^{n-m}}{(n-m)!}\\ &= \sum_{n=0}^{\infty} \frac{[\rho;x]_{\kappa n+\kappa m}}{\Gamma(\alpha n+\beta+\alpha m)} \frac{z^n}{n!}\\ &= (\rho)_{\kappa m} \sum_{n=0}^{\infty} \frac{[\rho+\kappa m;x]_{\kappa n}}{\Gamma(\alpha n+\beta+\alpha m)} \frac{z^n}{n!}\\ &= (\rho)_{\kappa m}\ \Xi^{\rho+\kappa m,\kappa}_{\alpha,\beta+\alpha m}(z).\end{aligned}$$

Similarly as above term-by-term differentiation under the sign of summation, we have

$$\begin{aligned}\left(\frac{d}{dz}\right)^m \left[z^{\beta-1}\Xi^\rho_{\alpha,\beta}(\omega z^\alpha)\right] &= \sum_{n=0}^{\infty} \frac{[\rho;x]_{\kappa n}}{\Gamma(\alpha n+\beta)} \frac{\Gamma(\alpha n+\beta)}{\Gamma(\alpha n+\beta-m)} \omega^n \frac{z^{\alpha n+\beta-1-m}}{n!}\\ &= z^{\beta-m-1}\Xi^{\rho,\kappa}_{\alpha,\beta-m}(\omega z^\alpha).\end{aligned}$$

Furthermore, taking $\alpha = 1$ and $\kappa = 1$ in Eq. (6.28), reduces to Eq. (6.29). The special cases of Eqs. (6.27), (6.28), and (6.29) when $x = 0$ are easily seen to yield Eqs. (6.30), (6.31), and (6.32), respectively. □

6.3 INCOMPLETE FOX-H, FOX-WRIGHT REPRESENTATIONS AND MELLIN-BARNES INTEGRALS OF $\Xi^{\rho,\kappa}_{\alpha,\beta}(z)$

In this section, we obtain the relation of incomplete generalized Mittag-Leffler function $\Xi^{\rho,\kappa}_{\alpha,\beta}(z)$ in Eq. (6.23) in terms of incomplete Fox-H, Fox-Wright generalized hypergeometric functions and Mellin-Barnes integral representation.

We begin with the definitions of the incomplete H-functions $\gamma^{m,\,n}_{p,\,q}(z)$ and $\Gamma^{m,\,n}_{p,\,q}(z)$ introduced by Srivastava et al. [61] in terms of the incomplete gamma functions

$\gamma(s,x)$ and $\Gamma(s,x)$:

$$\gamma_{p,q}^{m,n}(z) = \gamma_{p,q}^{m,n}\left[z\left|\begin{matrix}(a_1,A_1,x),(a_i,A_i)_{2,p}\\(b_j,B_j)_{1,q}\end{matrix}\right.\right]$$
$$= \gamma_{p,q}^{m,n}\left[z\left|\begin{matrix}(a_1,A_1,x),(a_2,A_2),\ldots,(a_p,A_p)\\(b_1,B_1),(b_2,B_2),\ldots,(b_q,B_q)\end{matrix}\right.\right]$$
$$= \frac{1}{2\pi i}\int_{\mathscr{L}} g(s,x)\, z^{-s}\, ds \tag{6.33}$$

where

$$g(s,x) = \frac{\prod_{j=1}^{m}\Gamma(b_j+B_js)\,\gamma(1-a_1-A_1s,x)\prod_{j=2}^{n}\Gamma(1-a_j-A_js)}{\prod_{j=m+1}^{q}\Gamma(1-b_j-B_js)\prod_{j=n+1}^{p}\Gamma(a_j+A_js)} \tag{6.34}$$

and

$$\Gamma_{p,q}^{m,n}(z) = \Gamma_{p,q}^{m,n}\left[z\left|\begin{matrix}(a_1,A_1,x),(a_i,A_i)_{2,p}\\(b_j,B_j)_{1,q}\end{matrix}\right.\right]$$
$$= \Gamma_{p,q}^{m,n}\left[z\left|\begin{matrix}(a_1,A_1,x),(a_2,A_2),\ldots,(a_p,A_p)\\(b_1,B_1),(b_2,B_2),\ldots,(b_q,B_q)\end{matrix}\right.\right]$$
$$= \frac{1}{2\pi i}\int_{\mathscr{L}} G(s,x)\, z^{-s}\, ds \tag{6.35}$$

where

$$G(s,x) = \frac{\prod_{j=1}^{m}\Gamma(b_j+B_js)\,\Gamma(1-a_1-A_1s,x)\prod_{j=2}^{n}\Gamma(1-a_j-A_js)}{\prod_{j=m+1}^{q}\Gamma(1-b_j-B_js)\prod_{j=n+1}^{p}\Gamma(a_j+A_js)}. \tag{6.36}$$

The incomplete H-functions $\gamma_{p,q}^{m,n}(z)$ and $\Gamma_{p,q}^{m,n}(z)$ in Eqs. (6.33) and (6.35) exists for all $x \geq 0$ under the conditions stated in Ref. [61]. Further they gave the relation between incomplete Fox H-functions representations of Eqs. (6.11) and (6.12) of the incomplete Fox-Wright hypergeometric functions ${}_p\Psi_q^{(\gamma)}$ and ${}_p\Psi_q^{(\Gamma)}$ as

$${}_p\Psi_q^{(\gamma)}\left[\begin{matrix}(\alpha_1,A_1,x),(\alpha_2,A_2),\ldots,(\alpha_p,A_p);\\(\beta_1,B_1),\ldots,(\beta_q,B_q);\end{matrix}\; z\right]$$
$$= \gamma_{p,q+1}^{1,p}\left[-z\left|\begin{matrix}(1-\alpha_1,A_1,x),(1-\alpha_i,A_i)_{2,p}\\(0,1),(1-\beta_j,1)_{1,q}\end{matrix}\right.\right] \tag{6.37}$$

and

$$ {}_p\Psi_q^{(\Gamma)}\left[\begin{array}{c}(\alpha_1,A_1,x),(\alpha_2,A_2),\ldots,(\alpha_p,A_p);\\ (\beta_1,B_1),\ldots,(\beta_q,B_q);\end{array} z\right] $$
$$ = \Gamma_{p,q+1}^{1,p}\left[-z \,\middle|\, \begin{array}{c}(1-\alpha_1,A_1,x),(1-\alpha_i,A_i)_{2,p}\\ (0,1),(1-\beta_j,1)_{1,q}\end{array}\right]. \tag{6.38} $$

Now making use of the definitions (6.13) and (6.14) in Eqs. (6.22) and (6.23), the incomplete generalized Mittag-Leffler functions can be expressed in terms of incomplete Fox-Wright functions as follows:

$$ \mathscr{E}_{\alpha,\beta}^{\rho,\kappa}(z) = \frac{1}{\Gamma(\rho)}\,{}_1\Psi_1^{(\gamma)}\left[\begin{array}{c}(\rho,\kappa,x);\\ (\beta,\alpha);\end{array} z\right] \tag{6.39} $$

and

$$ \Xi_{\alpha,\beta}^{\rho,\kappa}(z) = \frac{1}{\Gamma(\rho)}\,{}_1\Psi_1^{(\Gamma)}\left[\begin{array}{c}(\rho,\kappa,x);\\ (\beta,\alpha);\end{array} z\right]. \tag{6.40} $$

Further if we use the relations (6.37) and (6.37) in Eqs. (6.39) and (6.40), we obtain the relations with the incomplete Fox-H-functions as follows:

$$ \mathscr{E}_{\alpha,\beta}^{\rho,\kappa}(z) = \frac{1}{\Gamma(\rho)}\,\gamma_{1,2}^{1,1}\left[-z \,\middle|\, \begin{array}{c}(1-\rho,\kappa,x)\\ (0,1),(1-\beta,\alpha)\end{array}\right] \tag{6.41} $$

and

$$ \Xi_{\alpha,\beta}^{\rho,\kappa}(z) = \frac{1}{\Gamma(\rho)}\,\Gamma_{1,2}^{1,1}\left[-z \,\middle|\, \begin{array}{c}(1-\rho,\kappa,x)\\ (0,1),(1-\beta,\alpha)\end{array}\right]. \tag{6.42} $$

Further by making use of the sum of the residues at poles, one can easily evaluate the following Mellin-Barnes contour integral representations of incomplete generalized Mittag-Leffler functions $\mathscr{E}_{\alpha,\beta}^{\rho,\kappa}(z)$ and $\Xi_{\alpha,\beta}^{\rho,\kappa}(z)$. Here we omit the proof of the theorem.

Theorem 6.3. *The following pair of Mellin-Barnes contour integral representations of incomplete generalized Mittag-Leffler functions $\mathscr{E}_{\alpha,\beta}^{\rho,\kappa}(z)$ and $\Xi_{\alpha,\beta}^{\rho,\kappa}(z)$ holds true:*

$$ \mathscr{E}_{\alpha,\beta}^{\rho,\kappa}(z) = \frac{1}{2\pi i\Gamma(\rho)}\int_{\mathscr{L}}\frac{\Gamma(-s)\gamma(\rho+\kappa s,x)}{\Gamma(\beta+\alpha s)}(-z)^s ds \tag{6.43} $$
$$ (|\arg(-z)|<\pi) $$

and

$$ \Xi_{\alpha,\beta}^{\rho,\kappa}(z) = \frac{1}{2\pi i\Gamma(\rho)}\int_{\mathscr{L}}\frac{\Gamma(-s)\Gamma(\rho+\kappa s,x)}{\Gamma(\beta+\alpha s)}(-z)^s ds \tag{6.44} $$
$$ (|\arg(-z)|<\pi), $$

respectively, where $\mathscr{L}=\mathscr{L}_{(i\tau;\infty)}$ is a suitable contour of the Mellin-Barnes type starting at the point $\tau-i\infty$ and terminating at the point $\tau+i\infty\,(\tau\in\mathbb{R})$ with the usual indentations in order to separate one set of poles from the other set of poles of the integrand.

6.4 INTEGRAL TRANSFORMS REPRESENTATIONS

In this section, we obtain Laplace transform, Whittaker transform, and Euler-Beta transform representations for the incomplete generalized Mittag-Leffler function $\Xi_{\alpha,\beta}^{\rho,\kappa}(z)$ in Eq. (6.23).

6.4.1 LAPLACE TRANSFORM

The Laplace transform (see, *e.g.*, [56]) of the function $f(z)$ is defined, as usual, by

$$L\{f(z)\} = \int_0^\infty e^{-sz} f(z)\, dz. \tag{6.45}$$

Theorem 6.4. *The following Laplace transform representation for the incomplete generalized Mittag-Leffler function* $\Xi_{\alpha,\beta}^{\rho,\kappa}(z)$ *in Eq.* (6.23) *holds true*:

$$L\{z^{a-1}\Xi_{\alpha,\beta}^{\rho,\kappa}(\omega z^\sigma)\} := \frac{s^{-a}}{\Gamma(\rho)}\, {}_2\Psi_1^{(\Gamma)}\left[\begin{array}{r} (\rho,\kappa,x),(a,\sigma);\; \frac{x}{s^\sigma} \\ (\beta,\alpha);\; \phantom{\frac{x}{s^\sigma}} \end{array}\right] \tag{6.46}$$

$$(x \geq 0; \Re(a) > 0, \Re(s) > 0, \Re(\alpha) > 0, \Re(\beta) > 0, \Re(\rho) > 0, \Re(\sigma) > 0).$$

Proof. Using the definition (6.45) of the Laplace transform, we find from Eq. (6.23)

$$\begin{aligned} L\{z^{a-1}\Xi_{\alpha,\beta}^{\rho,\kappa}(xz^\sigma)\} :&= \int_0^\infty z^{a-1} e^{-sz}\Xi_{\alpha,\beta}^{\rho,\kappa}(xz^\sigma)dz \\ &= \int_0^\infty z^{a-1} e^{-sz}\left(\sum_{n=0}^\infty \frac{[\rho;x]_{\kappa n}}{\Gamma(\alpha n+\beta)}\frac{\omega^n z^{\sigma n}}{n!}\right) dz \\ &= \sum_{n=0}^\infty \frac{[\rho;x]_{\kappa n}}{\Gamma(\alpha n+\beta)}\frac{\omega^n}{n!}\int_0^\infty z^{a+\sigma n-1} e^{-sz} dz \\ &= \sum_{n=0}^\infty \frac{[\rho;x]_{\kappa n}}{\Gamma(\alpha n+\beta)}\frac{\Gamma(a+\sigma n)}{s^{a+\sigma n}}\frac{\omega^n}{n!}, \end{aligned}$$

which, in view of the definitions (6.14), yield the desired representation (6.46). □

Corollary 6.1: *If we put* $a = \beta$, $\kappa = 1$ *and* $\sigma = \alpha$ *in Eq.* (6.46), *we get*

$$\int_0^\infty z^{\beta-1} e^{-sz}\Xi_{\alpha,\beta}^{\rho,1}(\omega z^\alpha)dz = \frac{1}{s^\beta}\, {}_1\Gamma_0\left[(\rho,x);-;\frac{\omega}{s^\alpha}\right]. \tag{6.47}$$

Remark 6.4: The special cases of Eqs. (6.46) and (6.47) when $x = 0$ are seen to yield the known Laplace transform of the generalized Mittag-Leffler function (see [36]) and (see [30, p. 8, Eq. (2.5)]; see also [21, p. 37, Eq. (2.19)]):

$$L\{z^{a-1}E_{\alpha,\beta}^{\rho,\kappa}(\omega z^\sigma)\} := \frac{s^{-a}}{\Gamma(\rho)}\, {}_2\Psi_1\left[\begin{array}{r} (\rho,\kappa),(a,\sigma);\; \frac{\omega}{s^\sigma} \\ (\beta,\alpha);\; \phantom{\frac{\omega}{s^\sigma}} \end{array}\right]$$

and

$$\int_0^\infty z^{\beta-1}e^{-sz}E_{\alpha,\beta}^{\rho,1}(\omega z^\alpha)dz = \frac{1}{s^\beta}\left(1-\frac{\omega}{s^\alpha}\right)^{-\rho}$$

respectively.

6.4.2 WHITTAKER TRANSFORMS

To obtain Whittaker Transforms, we use the following integral representation:

$$\int_0^\infty t^{\nu-1}e^{-\frac{1}{2}t}\,W_{\lambda,\mu}(t)\,dt = \frac{\Gamma(\frac{1}{2}\pm\mu+\nu)}{\Gamma(1-\lambda+\nu)} \quad (\Re(\nu\pm\mu) > -\frac{1}{2}) \tag{6.48}$$

Theorem 6.5. *The following Whittaker transform representation for the incomplete generalized Mittag-Leffler function* $\Xi_{\alpha,\beta}^{\rho,\kappa}(z)$ *in Eq.* (6.23) *holds true*:

$$\int_0^\infty t^{\rho-1}e^{-\frac{1}{2}pt}\,W_{\lambda,\mu}(pt)\,\Xi_{\alpha,\beta}^{\rho,\kappa}(\omega z^\delta)dt = \frac{s^{-a}}{\Gamma(\rho)}\;{}_3\Psi_2^{(\Gamma)}\left[\begin{array}{c}(\rho,\kappa,x),(\frac{1}{2}\pm\mu+\rho,\delta);\\ (\alpha;\beta),(1-\lambda+\rho,\delta);\end{array}\;\frac{\omega}{p^\delta}\right] \tag{6.49}$$

$(x \geq 0;\ \Re(\rho) > 0, \Re(p) > 0, \Re(\alpha) > 0, \Re(\beta) > 0, \Re(\rho) > 0, \Re(\delta) > 0).$

Proof. Using the definition (6.23) in left-hand side of Eq. (6.49), we have

$$\begin{aligned}
&\int_0^\infty t^{\rho-1}e^{-\frac{1}{2}pt}\,W_{\lambda,\mu}(pt)\,\Xi_{\alpha,\beta}^{\rho,\kappa}(\omega t^\delta)dt\\
&= \int_0^\infty t^{\rho-1}e^{-\frac{1}{2}pt}\,W_{\lambda,\mu}(pt)\left(\sum_{n=0}^{\infty}\frac{[\rho;x]_{\kappa n}}{\Gamma(\alpha n+\beta)}\frac{\omega^n t^{\delta n}}{n!}\right)dt\\
&= \sum_{n=0}^{\infty}\frac{[\rho;x]_{\kappa n}}{\Gamma(\alpha n+\beta)}\frac{\omega^n}{n!}\int_0^\infty t^{\rho+\delta n-1}e^{-\frac{1}{2}pt}\,W_{\lambda,\mu}(pt)\,dt
\end{aligned}$$

Putting $pt = \nu$ and using Eq. (6.48), we obtain

$$\int_0^\infty t^{\rho-1}e^{-\frac{1}{2}pt}\,W_{\lambda,\mu}(pt)\,\Xi_{\alpha,\beta}^{\rho,\kappa}(\omega t^\delta)dt = \frac{1}{p^\rho}\sum_{n=0}^{\infty}\frac{[\rho;x]_{\kappa n}\Gamma(\frac{1}{2}\pm\mu+\rho+\delta n)}{\Gamma(\alpha n+\beta)\Gamma(1-\lambda+\rho+\delta n)n!}\left(\frac{\omega}{p^\delta}\right)^n.$$

which, in view of the definitions (6.14), yield the desired representation (6.49). □

6.4.3 EULER-BETA TRANSFORM

The Euler-Beta transform [56] of the function $f(z)$ is defined, as usual, by

$$\mathscr{B}\{f(z);a,b\} = \int_0^1 z^{a-1}(1-z)^{b-1}f(z)\,dz. \tag{6.50}$$

Theorem 6.6. *The following Euler-Beta transform representation for the extended generalized Mittag-Leffler function* $\Xi^{\rho,\kappa}_{\alpha,\beta}(z)$ *in Eq. (6.23) holds true:*

$$\mathscr{B}\left\{\Xi^{\rho,\kappa}_{\alpha,\beta}(\omega z^{\sigma}) : a,b\right\} := \frac{\Gamma(b)}{\Gamma(\rho)}\ {}_2\Psi_2^{(\Gamma)}\left[\begin{matrix}(\rho,\kappa,x),(a,\sigma); \\ (\beta,\alpha),(a+b,\sigma);\end{matrix}\ \omega\right] \tag{6.51}$$

$$(x \geq 0;\ \Re(a) > 0,\ \Re(b) > 0,\ \Re(\alpha) > 0,\ \Re(\beta) > 0,\ \Re(\rho) > 0,\ \Re(\sigma) > 0).$$

Proof. Using the definition (6.50) of the Euler-Beta transform, we find from Eq. (6.23)

$$\begin{aligned}\mathscr{B}\left\{\Xi^{\rho,\kappa}_{\alpha,\beta}(\omega z^{\sigma}) : a,b\right\} &:= \int_0^1 z^{a-1}(1-z)^{b-1}\Xi^{\rho,\kappa}_{\alpha,\beta}(\omega z^{\sigma})dz \\ &= \int_0^1 z^{a-1}(1-z)^{b-1}\left(\sum_{n=0}^{\infty}\frac{[\rho;x]_{\kappa n}}{\Gamma(\alpha n+\beta)}\frac{\omega^n z^{\sigma n}}{n!}\right).\end{aligned} \tag{6.52}$$

Upon interchanging the order of integration and summation in Eq. (6.52), which can easily be justified by uniform convergence under the constraints stated with Eq. (6.51), we get

$$\begin{aligned}\mathscr{B}\left\{\Xi^{\rho,\kappa}_{\alpha,\beta}(\omega z^{\sigma}) : a,b\right\} &:= \sum_{n=0}^{\infty}\frac{[\rho;x]_{\kappa n}}{\Gamma(\alpha n+\beta)}\frac{\omega^n}{n!}\left(\int_0^1 z^{a+\sigma n-1}(1-z)^{b-1}dz\right) \\ &= \sum_{n=0}^{\infty}\frac{[\rho;x]_{\kappa n}}{\Gamma(\alpha n+\beta)}\frac{\Gamma(\sigma n+a)\Gamma(b)}{\Gamma(\sigma n+a+b)}\frac{\omega^n}{n!},\end{aligned}$$

which, in view of the definitions (6.14), yield the desired representation (6.51). □

Remark 6.5: Putting $a = \beta$ and $\alpha = \sigma$ in Eq. (6.51), we get

$$\int_0^{\infty} z^{\beta-1}(1-z)^{b-1}\Xi^{\rho,\kappa}_{\sigma,\beta}(\omega z^{\sigma})dz = \Gamma(b)\ \Xi^{\rho,\kappa}_{\sigma,\beta+b}(\omega).$$

Remark 6.6: The special cases of Eq. (6.51) when $x = 0$ are seen to yield the known Euler-Beta transform of the generalized Mittag-Leffler function [46, p. 8, Eq. (5.1.1) and (5.1.3)]

$$L\{z^{a-1}E^{\rho,\kappa}_{\alpha,\beta}(\omega z^{\sigma})\} := \frac{s^{-a}}{\Gamma(\rho)}\ {}_2\Psi_2\left[\begin{matrix}(\rho,\kappa),(a,\sigma); & \omega \\ (\beta,\alpha);(a+b,\sigma) & s^{\sigma}\end{matrix}\right]$$

and

$$\int_0^{\infty} z^{\beta-1}(1-z)^{b-1}E^{\rho,\kappa}_{\sigma,\beta}(\omega z^{\sigma})dz = \Gamma(b)\ E^{\rho,\kappa}_{\sigma,\beta+b}(\omega)$$

respectively.

6.5 FRACTIONAL CALCULUS OPERATORS

In this section, we derive certain interesting properties of the incomplete generalized Mittag-Leffler function $\Xi^{\rho,\kappa}_{\alpha,\beta}(z)$ in Eq. (6.23) associated with *right-sided* Riemann-Liouville fractional integral operator I^{μ}_{a+} and the *right-sided* Riemann-Liouville fractional derivative operator D^{μ}_{a+}, which are defined as (see, *e.g.*, [17,19,21]):

$$\left(I^{\mu}_{a+}\varphi\right)(y) = \frac{1}{\Gamma(\mu)}\int_a^y \frac{\varphi(t)}{(y-t)^{1-\mu}}dt \quad (\mu \in \mathbb{C}, \Re(\mu) > 0) \tag{6.53}$$

and

$$\left(D^{\mu}_{a+}\varphi\right)(y) = \left(\frac{d}{dy}\right)^n \left(I^{n-\mu}_{a+}\varphi\right)(y) \quad \left(\mu \in \mathbb{C}, \Re(\mu) > 0; n = [\Re(\mu)] + 1\right). \tag{6.54}$$

where $[y]$ means the greatest integer not exceeding real y.

A generalization of Riemann-Liouville fractional derivative operator $D^{\mu}_{\alpha+}$ in Eq. (6.54) by introducing a *right-sided* Riemann-Liouville fractional derivative operator $D^{\mu,\nu}_{a+}$ of order $0 < \mu < 1$ and type $0 \leqq \nu \leqq 1$ with respect to y by Hilfer (see, *e.g.*, [18])as follows:

$$\left(D^{\mu,\nu}_{a+}\varphi\right)(y) = \left(I^{\nu(1-\mu)}_{a+}\frac{d}{dy}\right)\left(I^{(1-\nu)(1-\mu)}_{a+}\varphi\right)(y) \quad \left(\mu \in \mathbb{C}, \Re(\mu) > 0; n = [\Re(\mu)] + 1\right). \tag{6.55}$$

The generalization (6.55) yields the classical Riemann-Liouville fractional derivative operator D^{μ}_{a+} when $\nu = 0$.

$$\left(I^{a}_{0+}[t^{\rho-1}E^{\delta}_{\beta,\rho}(\omega t^{\beta})]\right)(y) = y^{a+\rho-1}E^{\delta}_{\beta,a+\rho}(\omega y^{\beta}). \tag{6.56}$$

Theorem 6.7. *Let* $a \in \mathbb{R}_+ = [0,\infty)$, $\alpha, \beta, \rho, \mu, \omega \in \mathbb{C}$ *and* $\Re(\alpha) > 0, \Re(\beta) > 0, \Re(\mu) > 0$. *Then for* $x \geqq 0$ *and* $y > a$, *there holds the relations*:

$$\left(I^{\mu}_{a+}\left[(t-a)^{\beta-1}\,\Xi^{\rho,\kappa}_{\alpha,\beta}(\omega(t-a)^{\rho})\right]\right)(y)$$
$$= (y-a)^{\beta+\mu-1}\,\Xi^{\rho,\kappa}_{\alpha,\beta+\mu}(\omega(y-a)^{\rho}) \tag{6.57}$$

$$\left(D^{\mu}_{a+}\left[(t-a)^{\beta-1}\,\Xi^{\rho,\kappa}_{\alpha,\beta}(\omega(t-a)^{\rho})\right]\right)(y)$$
$$= (y-a)^{\beta-\mu-1}\,\Xi^{\rho,\kappa}_{\alpha,\beta-\mu}(\omega(y-a)^{\rho}) \tag{6.58}$$

and

$$\left(D^{\mu,\nu}_{a+}\left[(t-a)^{c-1}\,\Xi^{\rho,\kappa}_{\alpha,\beta}(\omega(t-a)^{\rho})\right]\right)(y)$$
$$= (y-a)^{\beta-\mu-1}\,\Xi^{\rho,\kappa}_{\alpha,\beta-\mu}(\omega(y-a)^{\rho}) \tag{6.59}$$

Proof. By virtue of the formulas (6.53) and (6.23), the term-by-term fractional integration and the application of the relation [19]:

$$\left(I_{a+}^{\alpha}[(t-a)^{\beta-1}]\right)(y)=\frac{\Gamma(\beta)}{\Gamma(\alpha+\beta)}(y-a)^{\alpha+\beta-1}\quad(\alpha,\beta\in\mathbb{C},\,\Re(\alpha)>0,\,\Re(\beta)>0)\tag{6.60}$$

yields for $y>a$

$$\begin{aligned}&\left(I_{a+}^{\mu}[(t-a)^{\beta-1}\,\Xi_{\alpha,\beta}^{\rho,\kappa}(\omega(t-a)^{\rho})]\right)(y)\\&\qquad=\left(I_{a+}^{\mu}\left[\sum_{n=0}^{\infty}\frac{[\rho;x]_{\kappa n}}{\Gamma(\alpha n+\beta)n!}\omega^{n}\,(t-a)^{\alpha n+\beta-1}\right]\right)(y)\\&\qquad=(y-a)^{\beta+\mu-1}\,\Xi_{\alpha,\beta+\mu}^{\rho,\kappa}(\omega(y-a)^{\rho})\end{aligned}\tag{6.61}$$

Next, by Eqs. (6.54) and (6.23), we find that

$$\begin{aligned}&\left(D_{a+}^{\mu}[(t-a)^{\beta-1}\,\Xi_{\alpha,\beta}^{\rho,\kappa}(\omega(t-a)^{\rho})]\right)(y))\\&\quad=\left(\frac{d}{dy}\right)^{n}\left(I_{a+}^{n-\mu}[(t-a)^{\beta-1}\,\Xi_{\alpha,\beta}^{\rho,\kappa}(\omega(t-a)^{\rho})]\right)(y)\\&\quad=\left(\frac{d}{dy}\right)^{n}\left[(y-a)^{\beta+n-\mu-1}\,\Xi_{\alpha,\beta+n-\mu}^{\rho+1}(\omega(y-a)^{\rho})\right]\end{aligned}\tag{6.62}$$

Applying Eq. (6.28), we are led to the desired result (6.58).

Finally, by Eqs. (6.55) and (6.23), we have

$$\begin{aligned}&\left(D_{a+}^{\mu,\nu}\left[(t-a)^{\beta-1}\,\Xi_{\alpha,\beta}^{\rho,\kappa}(\omega(t-a)^{\rho})\right]\right)(y))\\&\quad=\left(D_{a+}^{\mu,\nu}\left[\sum_{n=0}^{\infty}\frac{[\rho;x]_{\kappa n}}{\Gamma(\alpha n+\beta)n!}\omega^{n}\,(t-a)^{\alpha n+\beta-1}\right]\right)(y)\\&\quad=\sum_{n=0}^{\infty}\frac{[\rho;x]_{\kappa n}}{\Gamma(\alpha n+\beta)n!}\omega^{n}\left(D_{a+}^{\mu,\nu}\left[(t-a)^{\alpha n+\beta-1}\right]\right)(y)\end{aligned}\tag{6.63}$$

Using the known relation of Srivastava and Tomovski [63, p. 203, Equation (2.18)]

$$\begin{aligned}&\left(D_{\alpha+}^{\mu,\nu}\left[(t-\alpha)^{\lambda-1}\right]\right)(y)\\&\quad=\frac{\Gamma(\lambda)}{\Gamma(\lambda-\mu)}(y-\alpha)^{\lambda-\mu-1}\quad(y>\alpha;\,0<\mu<1;\,0\leqq\nu\leqq1;\,\Re(\lambda)>0)\end{aligned}\tag{6.64}$$

in Eq. (6.63), we are led to the desired result (6.59). □

6.6 APPLICATION TO THE SOLUTION OF FRACTIONAL KINETIC EQUATION

During the last few decades, fractional kinetic equations of different forms have been widely used in describing and solving several important problems of physics and astrophysics. If an arbitrary reaction is characterized by a time-dependent quantity $N = N(t)$, then it is possible to calculate the rate $\frac{dN}{dt}$ by the mathematical equation

$$\frac{dN}{dt} = -d + p \tag{6.65}$$

where d is the destruction rate and p is the production rate of N. In general, the destruction rate d and the production rate p depend on the quantity $N(t)$ itself: $d = d(N)$ or $p = p(N)$, which is a complicated dependence because the destruction or production at time t depends not only on $N(t)$ but also on the past history $N(\eta)$, $\eta < t$, of variable N. Formally, this can be described by the following equation:

$$\frac{dN}{dt} = -d(N_t) + s(N_t), \tag{6.66}$$

where $N(t)$ is the function defined by $N_t(t^\star) = N(t - t^\star), t^\star > 0$.

Haubold and Mathai [17](see also, [39–41]) studied a special case of this equation (6.66), namely standard kinetic equation

$$\frac{dN}{dt} = -\varepsilon_i N_i(t) \tag{6.67}$$

with the initial condition $N_i(t = 0) = N_0$ representing the number density of the species i at time $t = 0$ and the constant $\varepsilon_i > 0$. If we remove the index i and integrate the standard kinetic equation (6.67), then

$$N(t) - N_0 = -\varepsilon_0 D_t^{-1} N(t), \tag{6.68}$$

where D_t^{-1} is the particular case of the Riemann-Liouville integral operator ${}_0D_t^{-1}$ defined as [21]

$${}_0D_t^{-1} f(t) = \frac{1}{\Gamma(\nu)} \int_0^t (t-x)^{\nu-1} f(x) dx \quad (t > 0, \Re(\nu) > 0). \tag{6.69}$$

A fractional generalization of the standard kinetic equation (6.68) is given in Refs. [17,39–41] as follows:

$$N(t) - N_0 = -\varepsilon^\nu {}_0D_t^{-1} N(t). \tag{6.70}$$

They obtained the solution of (6.70) as

$$N(\tau) = N_0 \sum_{k=0}^{\infty} \frac{(-1)^k}{\Gamma(\nu k + 1)} (\varepsilon \tau)^{\nu k}. \tag{6.71}$$

Definition 1. *[21] Let $f(t)$ be a real(or complex)-valued function of the (time) variable $t > 0$ and s is a real or complex parameter. Then the Laplace transform of the function $f(t)$ is defined by*

$$L\{f(t):s\} = \int_0^\infty e^{-st} f(t)\, dt = F(s) \ (\Re(s) > 0)). \tag{6.72}$$

The Laplace transform of the power function t^ν is given by the following relation []

$$L\{t^\nu : s\} = \frac{\Gamma(\nu+1)}{s^{\nu+1}} \ (\Re(\nu) > -1; (\Re(s) > 0)). \tag{6.73}$$

We know that the Laplace transform of the Riemann-Liouville fractional integral is given by [21]

$$L\left\{{}_0D_t^{-\nu} f(t) : s\right\} = \frac{1}{s^\nu} L\{f(t):s\} \tag{6.74}$$

In this section, we obtain the solution of generalized fractional kinetic equation (6.68) involving generalized incomplete Mittag-Leffler function (6.23) by applying the Laplace transform technique.

Theorem 6.8. *For all $c, \nu, h > 0; c \neq h$, if $x \geq 0$; $t, \beta, \rho \in \mathbb{C}$; $\Re(\alpha) > max\{0, \Re(\kappa) - 1\}; \Re(\kappa) > 0$, $\Re(\beta) > 0$ when $x = 0, |ht| \leq 1$, then equation*

$$N(t) - N_0\, \Xi_{\alpha,\beta}^{\rho,\kappa}(ht) = -c^\nu\, {}_0D_t^{-\nu} N(t), \tag{6.75}$$

has the solution

$$N(t) = N_0 \sum_{n-0}^{\infty} \frac{[\rho;x]_{\kappa n} (ht)^n}{\Gamma(\alpha n+\beta)} \times E_{\nu,k+1}(-c^\nu t^\nu).$$

Proof. Employing the Laplace transform (6.72) to the equation (6.75) and using the definition of generalized incomplete Mittag-Leffler function (6.23), we have

$$N^*(s) = N_0 \left(\int_0^\infty e^{-st} \frac{[\rho;x]_{\kappa n}}{\Gamma(\alpha n+\beta) n!} (ht)^n\, dt \right) - c^\nu s^{-\nu} N^*(s) \tag{6.76}$$

where $N^*(s) = L\{N(t);s\}$. Now under the given conditions in Theorem 6.8 and computing the integral in Eq. (6.76) term by term and using the relation (6.73), we have

$$\begin{aligned}
\left(1 + \left(\frac{c}{s}\right)^\nu\right) N^*(s) &= N_0 \sum_{n=0}^{\infty} \frac{[\rho;x]_{\kappa n}}{\Gamma(\alpha n+\beta) n!} \frac{\Gamma(n+1) h^n}{s^{n+1}} \\
\Rightarrow N^*(s) &= N_0 \sum_{n=0}^{\infty} \frac{[\rho;x]_{\kappa n}}{\Gamma(\alpha n+\beta)} \frac{h^n}{s^{n+1}} \left(1 + \left(\frac{c}{s}\right)^\nu\right)^{-1} \\
&= N_0 \sum_{n=0}^{\infty} \frac{[\rho;x]_{\kappa n}}{\Gamma(\alpha n+\beta)} \frac{h^n}{s^{n+1}} \times \sum_{\ell=0}^{\infty} (-1)^\ell \left(\frac{c}{s}\right)^{\nu\ell},
\end{aligned}$$

where we have applied the binomial series expansion

$$\left(1+\left(\frac{c}{s}\right)^{\nu}\right)^{-1}=\sum_{\ell=0}^{\infty}(-1)^{\ell}\left(\frac{c}{s}\right)^{\nu\ell}\quad\left(\left|\frac{c}{s}\right|<1\right).$$

Taking the inverse Laplace transform and applying the relation $L^{-1}\{s^{-\nu};t\}=\frac{t^{\nu-1}}{\Gamma(\nu)}$ $(\Re(\nu)>0)$, we get

$$\begin{aligned}N(t)=L^{-1}\{N^{*}(s);t\}&=N_0\sum_{n=0}^{\infty}\frac{[\rho;x]_{\kappa n}(ht)^{n}}{\Gamma(\alpha n+\beta)}\times\sum_{\ell=0}^{\infty}\frac{(-1)^{\ell}(ct)^{\nu\ell}}{\Gamma(\nu\ell+k+1)},\\&=N_0\sum_{n=0}^{\infty}\frac{[\rho;x]_{\kappa n}(ht)^{n}}{\Gamma(\alpha n+\beta)}\times E_{\nu,n+1}(-c^{\nu}t^{\nu}),\end{aligned}$$

where we have used the celebrated series expansion of Mittag-Leffler function [19]

$$E_{\xi,\zeta}(z)=\sum_{\ell=0}^{\infty}\frac{z^{\ell}}{\Gamma(\xi\ell+\zeta)}.$$

This completes the proof of Theorem 6.8. □

6.7 FURTHER REMARKS AND OBSERVATIONS

Special functions have been studied widely in recent years. Many researchers in this field are actively doing various works related to extensions of some higher transcendental function, for more details see the recent papers [1,5–9,20,24–27,29]. In this chapter, motivated essentially by the demonstrated potential for applications of several families of incomplete generalized higher transcendental functions, for example, incomplete generalized hypergeometric functions ${}_p\gamma_q$ and ${}_p\Gamma_q$, incomplete Fox-Wright generalized hypergeometric functions ${}_p\Psi_q^{(\gamma)}(z)$ and ${}_p\Psi_q^{(\Gamma)}(z)$, incomplete Fox H-functions $\gamma_{p,q}^{m,n}(z)$ and $\Gamma_{p,q}^{m,n}(z)$, *etc.* studied by various authors, we introduce family of the incomplete generalized Mittag-Leffler function $\mathscr{E}_{\alpha,\beta}^{\rho,\kappa}(z)$ and $\Xi_{\alpha,\beta}^{\rho,\kappa}(z)$ by means of the incomplete Pochhammer symbol $(\lambda;x)_\nu$ and $[\lambda;x]_\nu$ and investigate certain basic properties including, for example, differentiation formulas, integral representations, Laplace transform, Euler-Beta transform and Whittaker transform with their several special cases and relationship with incomplete Fox-Wright generalized hypergeometric functions (6.14). Moreover, certain relations between the incomplete generalized Mittag-Leffler function and the Riemann-Liouville fractional integrals and derivatives are investigated. Some interesting special cases of our main results are also considered. Finally, as an application, we established the solution to the kinetic equations. Mathematical modeling of real-world problems usually leads to fractional differential equations and various other problems involving special functions in mathematical physics and their extensions and generalizations in one or more variables [29–37,48]. The results obtained in this chapter are also useful in such diverse and widely used fields of engineering and sciences such as electromagnetism, viscoelasticity, fluid dynamics, electrochemistry, biological population modeling, optics, and signal processing.

Bibliography

1. R. P. Agarwal, A propos d'une Note M. Pierre Humbert, *C. R. Acad. Sci. Paris* **236** (1953), 2031–2032.
2. R. P. Agarwal, A. Kılıcman, R. K. Parmar and A. K. Rathie, Certain generalized fractional calculus formulas and integral transforms involving (p,q)–Mathieu-type series, *Adv. Differ. Equ.* **221**, (2019), 1–11 . https://doi.org/10.1186/s13662-019-2142-0.
3. A. Cetinkaya, The incomplete second Appell hypergeometric functions, *Appl. Math. Comput.* **219** (2013), 8332–8337.
4. J. Choi and R. K. Parmar, An extension of the generalized Hurwitz-Lerch zeta function of two variables, *Filomat*, **31**(1) (2017), 91–96.
5. J. Choi and R. K. Parmar, Fractional integration and differentiation of the (p,q)–extended Bessel function, *Bull. Korean Math. Soc.*, **55**(2) (2018), 599–610.
6. J. Choi and R. K. Parmar, Fractional calculus of the (p,q)–extended Struve function, *Far. East J. Math. Sci.*, **103**(2) (2018), 541–559.
7. J. Choi and R. K. Parmar, The incomplete Lauricella and Fourth Appell functions, *Far. East J. Math. Sci.*, **96**, (2015), 315–328.
8. J. Choi, R. K. Parmar, and P. Chopra, Extended Mittag-Leffler function and associated fractional calculus operators, *Georgian Math. J.*, **27(2)** (2020), 199–209.
9. J. Choi, R. K. Parmar and P. Chopra, The incomplete Srivastava's triple hypergeometric functions γ_B^H and Γ_B^H, *Filomat*, **30** (2016), 1779–1787.
10. J. Choi, R. K. Parmar and P. Chopra, The incomplete Lauricella and First Appell functions and associated properties, *Honam Math. J.* **36** (2014), 531–542.
11. J. Choi and R. K. Parmar, The incomplete Srivastava's triple hypergeometric functions γ_A^H and Γ_A^H, *Miskolc Math. Notes*, **19** (2018), 191–200.
12. J. Choi, R. K. Parmar and R. K. Raina, Extension of generalized Hurwitz-Lerch Zeta function and associated properties, *Kyungpook Math. J.*, **57**(3) (2017), 393–400.
13. J. Choi, R. K. Parmar and H. M. Srivastava, The incomplete Lauricella Functions of several variables and associated properties and formulas, *Kyungpook Math. J.*,**58** (2018), 19–35.
14. A. Erdélyi, W. Magnus, F. Oberhettinger and F. G. Tricomi, *Higher Transcendental Functions*, Vol. **1**, McGraw-Hill Book Company: New York, Toronto and London (1953).
15. R. Gorenflo, A. Kilbas, F. Mainardi and S. Rogosin, *Mittag-Lffler Function, Related Topics and Applications*, Springer-Verlag: Berlin Heidelberg (2014).
16. H. J. Haubold, A. M. Mathai and R. K. Saxena, Mittag-Leffler function and their applications, *J. Appl. Math.*, **2011** (2011), 51 p. doi: 10.1135/ 2011/298628.
17. H. J. Haubold and A. M. Mathai, The fractional kinetic equation and thermonuclear functions, *Astrophys. Space Sci.*, **327** (2000), 53–63.
18. R. Hilfer (Ed.), *Applications of Fractional Calculus in Physics*, World Scientific Publishing Company: Singapore, New Jersey, London and Hong Kong (2000).
19. P. Humbert, Quelques resultats d'le fonction de Mittag-Leffler, *C. R. Acad. Sci. Paris* **236** (1953), 1467–1468.
20. P. Humbert and R. P. Agarwal, Sur la fonction de Mittag-Leffler et quelques unes de ses generalizations, *Bull. Sci. Math.* **77**(2) (1953), 180–185.
21. A. A. Kilbas, M. Saigo and R. K. Saxena, Generalized Mittag-Leffler function and generalized fractional calculus operators, *Integr. Transforms Spec. Funct.* **15** (2004), 31–49.
22. A. A. Kilbas, M. Saigo and R. K. Saxena, Solution of Volterra integro-differential equations with generalized Mittag-Leffler function in the Kernels, *J. Integr. Equations Appl.* **14** (2002), 377–396.

23. A. A. Kilbas, H. M. Srivastava and J.J. Trujillo, *Theory and Applications of Fractional Differential Equations*, North-Holland Mathematical Studies, Vol. **204**, Elsevier (North-Holland) Science Publishers: Amsterdam, London and New York (2006).
24. M. J. Luo, R. K. Parmar and R. K. Raina, On extended Hurwitz–Lerch zeta function, *J. Math. Anal. Appl.* **448** (2017), 1281–1304.
25. A. M. Mathai, R. K. Saxena and H. J. Haubold, *The H-Functions: Theory and Applications*, Springer: New York, 2010.
26. G. M. Mittag-Leffler, Sur la nouvelle fonction $E_\alpha(x)$, *C. R. Acad. Sci. Paris* **137** (1903), 554–558.
27. G. M. Mittag-Leffler, Sur la representation analytiqui d'une fonction monogene (cinquieme note), *Acta Math.* **29** (1905), 101–181.
28. J. Paneva-Konovska, The convergence of series in multi-index Mittag-Leffler functions, *Integr. Transforms Spec. Funct.* **23** (2012), 207–221.
29. R. K. Parmar, A class of extended Mittag–Leffler functions and their properties related to integral transforms and fractional calculus, *Mathematics* **3(4)** (2015), 1069–1082.
30. T. R. Prabhakar, A singular integral equation with a generalized Mittag-Leffler function in the kernel, *Yokohama Math. J.* **19** (1971), 7–15.
31. E. D. Rainville, *Special Functions*, Macmillan Company: New York (1960); Reprinted by Chelsea Publishing Company: Bronx, New York (1971).
32. R. K Parmar, J. Choi and S. D. Purohit, Further generalization of the extended Hurwitz-Lerch Zeta functions, *Bol. Soc. Paranaense Mat.* **37**(1) (2019), 177–190.
33. R. K. Parmar and R. K. Raina, On a certain extension of the Hurwitz-Lerch Zeta function, *Ann. West Univ. Timisoara-Math.*, **52(2)** (2014), 157–170.
34. T. K. Pogány and R. K. Parmar, On p–extended Mathieu series, *Rad Hrvat. Akad. Znan. Umjet. Mat. Znan.* **22** (2018), 107–117.
35. S. G. Samko, A. A. Kilbas and O. I. Marichev, *Fractional Integrals and Derivatives: Theory and Applications*, Gordon and Breach: Yverdon (1993).
36. R. K. Saxena, Certain properties of generalized Mittag-Leffler function, *Proceedings of the Third Annual Conference of the Society for Special Functions and Their Applications*, Chennai, India (2002), pp. 75–81.
37. R. K. Saxena, S. L. Kalla and V. S. Kiryakova, Relations connecting multi-index Mittag–Leffler functions and Riemann-Liouville fractional calculus, *Algebras Groups Geom.* **20** (2003), 363–386.
38. R. K. Saxena, A. M. Mathai and H. J. Haubold, On fractional kinetic equations, *Astrophys. Space Sci.*, **282** (2002), 281–287.
39. R. K. Saxena, A. M. Mathai and H. J. Haubold, On generalized fractional kinetic equations, *Physica A* **344** (2004), 657–664.
40. R. K.Saxena, A. M. Mathai and H. J. Haubold, Unified fractional kinetic equations and a fractional diffusion equation,*Astrophys. Space Sci.* **290** (2004), 241–245.
41. R. K. Saxena, A. M. Mathai and H. J. Haubold, On generalized fractional kinetic equations, *Physica A* **344** (2004), 657–664.
42. R. K. Saxena and S. L. Kalla, On the solutions of certain fractional kinetic equations, *Appl. Math. Comput.* **199** (2008), 504–511.
43. R. K. Parmar and R. K. Saxena, The incomplete generalized τ-hypergeometric and second τ-appell functions, *J. Korean Math. Soc.* , **53** (2016), 363–379.
44. R. K. Parmar and R. K. Saxena, Incomplete extended Hurwitz-Lerch Zeta Functions and associated properties, *Commun. Korean Math. Soc.*, **32** (2017), 287–304.

45. R. K. Saxena and M. Saigo, Certain properties of fractional calculus operators associated with generalized Mittag-Leffler function, *Fract. Calc. Appl. Anal.* **8** (2005), 141–154.
46. A. K. Shukla, J. C. Prajapati, On a generalization of Mittag–Leffler function and its properties, *J. Math. Anal. Appl.*, **336** (2007), 797–811.
47. H. Singh and A. M. Wazwaz, Computational method for reaction diffusion-model arising in a spherical catalyst. *Int. J. Appl. Comput. Math.* **7(3)**, (2021), 1–11.
48. H. Singh, Analysis for fractional dynamics of Ebola virus model, *Chaos, Solitons Fractals*, **138** (2020), 109992,
49. H. Singh, Solving a class of local and nonlocal elliptic boundary value problems arising in heat transfer, *Heat Transfer* **51** (2021), 1524–1542.
50. H. Singh, An efficient computational method for non-linear fractional Lienard equation arising in oscillating circuits, In: H. Singh, D. Kumar, and D. Baleanu (Eds.), *Methods of Mathematical Modelling: Fractional Differential Equations*, CRC Press, Taylor & Francis Group: Boca Raton, FL (2019), pp. 39–50.
51. H. Singh, Chebyshev spectral method for solving a class of local and nonlocal elliptic boundary value problems, *Int. J. Nonlinear Sci. Numer. Simul.* (2021), 000010151520200235.
52. H. Singh and H. M. Srivastava, Numerical investigation of the fractional-order Liénard and Duffing equations arising in oscillating circuit theory, *Front. Phys.* **8** (2020) , 120.
53. H. Singh, H. M. Srivastava and J. J. Nieto, *Handbook of Fractional Calculus for Engineering and Science*. CRC Press, Taylor & Francis Group: Boca Raton, FL (2022).
54. H. Singh, D. Kumar and D. Baleanu, *Methods of Mathematical Modelling: Fractional Differential Equations*, CRC Press, Taylor & Francis: Boca Raton, FL (2019).
55. H. Singh, H. M. Srivastava and D. Baleanu, *Methods of Mathematical Modelling: Infectious Disease*, Elsevier Science: Amsterdam, Netherlands (2022).
56. I. N. Sneddon, *The Use of the Integral Transforms*. Tata McGraw-Hill: New Delhi (1979).
57. H. M. Srivastava, M. A. Chaudhry and R. P. Agarwal, The incomplete Pochhammer symbols and their applications to hypergeometric and related functions, *Integr. Transforms Spec. Funct.* **23** (2012), 659–683.
58. H. M. Srivastava and P. W. Karlsson, *Multiple Gaussian Hypergeometric Series*, Halsted Press (Ellis Horwood Limited, Chichester), John Wiley & Sons: New York, Chichester, Brisbane and Toronto (1985).
59. H. M. Srivastava and T. K. Pogány, Inequalities for a unified Voigt functions in several variables, *Russ. J. Math. Phys.* **14**(2) (2007), 194–200.
60. H. M. Srivastava, R. K. Parmar and P. Chopra, Some families of generalized complete and incomplete elliptic-type integrals, *J. Nonlinear Sci. Appl.* **10** (2017), 1162–1182 .
61. H. M. Srivastava, R. K. Saxena and R. K. Parmar. Some families of the incomplete H-functions and the incomplete $\bar{H}$-functions and associated integral transforms and operators of fractional calculus with applications, *Russ. J. Math. Phys.*, **25(1)**, (2018), 116–138.
62. H. M. Srivastava and R. K. Saxena, Operators of fractional integration and their applications, *Appl. Math. Comput.*, **118**, (2001), 1–52.
63. H. M. Srivastava and Ẑ. Tomovski, Fractional calculus with an integral operator containing a generalized Mittag-Leffler function in the kernal, *Appl. Math. Comput.* **211** (2009), 198–210.
64. R. Srivastava, Some properties of a family of incomplete hypergeometric functions, *Russ. J. Math. Phys.* **20** (2013), 121–128.

65. R. Srivastava and N. E. Cho, Generating functions for a certain class of incomplete hypergeometric polynomials, *Appl. Math. Comput.* **219** (2012), 3219–3225.
66. D. L. Suthar, R. K. Parmar, and S. D. Purohit, Fractional calculus with complex order and generalized hypergeometric functions, *J. Nonlinear Sci. Letters A*, **8**(2) (2017), 156–161.
67. V. M. Tripathi, H. M. Srivastava, H. Singh, C. Swarup and S. Aggarwal, Mathematical analysis of non-isothermal reaction-diffusion models arising in spherical catalyst and spherical biocatalyst, *Appl. Sci.* 11(21) (2021), 10423.
68. A. Wiman, Über den fundamental satz in der theorie der functionen $E_\alpha(x)$, *Acta Math.* **29** (1905), 191–201.

7 Numerical Solution of Fractional Order Diffusion Equation Using Fibonacci Neural Network

Kushal Dhar Dwivedi
S. N. Government Post-Graduate College

CONTENTS

7.1 INTRODUCTION

The fractional diffusion equation plays a very important role in many fields such as physics, statistics, neuroscience, economy, etc. Many phenomena have been modeled by the fractional diffusion equation. But, finding the exact solution is not always possible. Things become more complex when the diffusion model is fractional and has two dimensions. So researchers keen their interests in finding numerical solutions and developed many efficient and accurate methods to find the numerical solutions of different types of diffusion models viz., Abbasbandy and Shirzadi [1] have solved two-dimensional diffusion equation with the integral condition, Mohebbi et al. [10] have developed meshless method to solve the two-dimensional sub-diffusion equation, Abbaszadeh and Mohebbi [2] have developed a numerical

DOI: 10.1201/9781003368069-7

method to find the solution of the two-dimensional modified anomalous fractional sub-diffusion equation with a nonlinear source term, Fan and Liu [6] have solved the two-dimensional distributed order space-fractional diffusion equation on an irregular convex domain, Tuan et al. [25] solved a two-dimensional diffusion equation modeled in transport phenomenon, Vivek et al. [26] have solved non-isothermal reaction-diffusion models of spherical catalysts and bio-catalyst. The Ebola virus, the spherical catalyst model, the Lienard equation originating in oscillating circuits, and other diffusion models have all been addressed by Singh et al. [25,27–31] who have also authored a book on fractional calculus and mathematical modeling [9,10].

In this chapter, the author discusses a method to solve the following two-dimensional fractional order diffusion equation (FDE)

$$\frac{\partial^{\gamma} u(x,y,t)}{\partial t^{\gamma}} = f_1(x,y,t)\frac{\partial^{\alpha} u(x,y,t)}{\partial x^{\alpha}} + f_2(x,y,t)\frac{\partial^{\beta} u(x,y,t)}{\partial y^{\beta}} + f_3(x,y,t), \tag{7.1}$$

with initial and boundary conditions

$$\begin{aligned} u(x,y,0) &= g_1(x,y), \\ u(0,y,t) &= g_2(y,t), \\ u(1,y,t) &= g_3(y,t), \\ u(x,0,t) &= g_4(x,t), \\ u(x,1,t) &= g_5(x,t). \end{aligned} \tag{7.2}$$

where $\frac{\partial^{\gamma}}{\partial t^{\gamma}}, \frac{\partial^{\alpha}}{\partial x^{\alpha}}, \frac{\partial^{\beta}}{\partial x^{\beta}}$ are the fractional order derivatives of ordered γ, α and β respectively in Caputo Sense.

As the world is moving toward automation, artificial intelligence (AI) is playing a crucial role in it. AI is gaining popularity due to its ability to do unbelievable tasks such as finding patterns in data which is nearly impossible for humans, speech recognition, virtual assistant, self-driving cars, and many more. An artificial neural network (ANN) helps us build an AI agent for such tasks. ANN is nothing but a mathematical mimic of the human brain. At this stage, we have enough numbers of numerical techniques to solve highly complex diffusion models, but the method becomes complex whenever researchers try to develop it for the complex models. The author believes that this problem can be solved with AI. More precisely, the author believes that ANN can help build highly accurate and mathematically simple numerical methods to solve complex diffusion models. Many researchers have solved different types of differential models using neural networks viz., Fang et al. [7] have used ANN to solve single-delay differential equations, Berg and Nyström [3] have to solve partial differential equation using deep ANN, Panghal and Kumar [11] have developed a method to solve an ordinary and partial differential equation by using ANN, Li et al. [9] have developed a numerical technique using convolutional-type ANN to solve reaction-diffusion equation.

In this chapter, the author has extended the numerical method developed by himself [5] to solve a two-dimensional fractional diffusion equation. The author first constructed a neural network of three layers, input, output, and one hidden layer. Different degrees of the Fibonacci polynomial in the hidden layers are used as an activation function to add non-linearity to the input. After that considered model is transformed into a non-constraints maximization problem.

This chapter is arranged in the following order. In Section 7.2, the author has given some necessary definitions. In Section 7.3, the author discussed the architecture of the Fibonacci neural network (FNN) and discussed how to use it to solve the considered model (7.1) with initial and boundary condition (7.2). In Section 7.4, the method is used to solve some examples to verify its accuracy. In Section 7.5, the particular case of the considered model (7.1) is solved with the developed numerical technique, and the behavior of the solution is discussed.

7.2 DEFINITIONS

7.2.1 CAPUTO FRACTIONAL ORDER DERIVATIVE

In Caputo sense the fractional differential operator is defined as [36]

$$\frac{\partial^{\alpha} f(t)}{\partial t^{\alpha}} = \frac{1}{\Gamma(n-\alpha)} \int_0^t (t-\tau)^{(n-\alpha-1)} f^{(n)}(\tau) d\tau, \quad \alpha > 0, \tau > 0,$$

where $n-1 < \alpha < n, \quad n \in N.$

The operator D^{α} satisfies the following properties for $n-1 < \alpha < n:$

$$\frac{\partial^{\alpha} t^k}{\partial t^{\alpha}} = \begin{cases} 0 & k \in 0,1,2,...,\lceil \alpha \rceil, \\ \frac{\Gamma(k+1)}{\Gamma(k+1-\alpha)} t^{k-\alpha}, & k \in N, \ k \geq \lceil \alpha \rceil, \end{cases} \tag{7.3}$$

where the notation $\lceil \cdot \rceil$ denotes the ceiling function, which is defined by the set of equation $\lceil \alpha \rceil = \min\{n \in Z | n \geq \alpha\}$, Z is the set of integers. I^{α} is the Riemann-Liouville integral operator of order α.

7.2.2 PROPERTIES OF FIBONACCI POLYNOMIAL

It is known that Fibonacci Polynomial can be constructed by the following recurrence relation

$$F_{m+2}(x) = xF_{m+1}(x) + F_m(x), \quad m \geq 0,$$

with initial conditions as

$$F_0(x) = 0, \quad F_1(x) = 1.$$

From above relation, the explicit form of the series is obtained as

$$F_m(x) = \sum_{r=0}^{\lfloor \frac{m-1}{2} \rfloor} \binom{m-r-1}{r} x^{m-2r-1}, \tag{7.4}$$

where $\lfloor \cdot \rfloor$ denotes the floor function, which is defined by $\lfloor \alpha \rfloor = \max\{n \in Z | n \leq \alpha\}$, Z is the set of integers.

The above equation can be rewritten as

$$F_i(x) = \sum_{j=0}^{i} \frac{(\frac{i+j-1}{2})!}{j!(\frac{i-j-1}{2})!} x^j, \quad (j+i) = \text{odd}, \quad i \geq 0. \tag{7.5}$$

The Caputo order derivative of Fibonacci polynomial of degree i can be obtained by using Eq. (7.3) in Eq. (7.5) as

$$D^\alpha F_i(x) = \begin{cases} 0 & i \in 0,1,2,\ldots,\lceil \alpha \rceil, \\ \sum_{\substack{j=\lceil \alpha \rceil \\ (j+i)=odd}}^{i} \frac{(\frac{i+j-1}{2})!}{(\frac{i-j-1}{2})!(j-\alpha)!} x^{j-\alpha}, & i \in N, \quad i \geq \lceil \alpha \rceil, \end{cases} \tag{7.6}$$

7.3 FNN AND METHOD TO APPLY TO SOLVE CONSIDERED MODEL

In this section, the authors discuss about the FNN and how to apply it to solve the fractional order partial differential equation like model (7.1).

Figure 7.1 depicts the considered Fibonacci neural network (FNN). The Fibonacci neural network is drawn by taking $l \times m \times n$ numbers of nodes in the hidden layer. Various degrees of the Fibonacci polynomial have been used as an activation function in the nodes of the hidden layer. The output layer contains linear combinations of the hidden layers with the weights w_{ijk}. During the training process of the FNN, these weights w's will be updated by an appropriate optimization technique. The number of nodes in the hidden layer will depend on the considered problem.

7.3.1 METHOD TO USE FNN TO SOLVE TWO-DIMENSIONAL FDE

In this subsection, the author illustrates the numerical method to solve the considered model (7.1). Let us consider the trial solution of the considered problem

$$N(x,y,t) = \sum_{i=1}^{l} \sum_{j=1}^{m} \sum_{k=1}^{n} w_{ijk} F_i(x) F_j(y) F_k(t). \tag{7.7}$$

Then considered model (7.1) transformed into

$$\frac{\partial^\gamma N(x,y,t)}{\partial t^\gamma} = f_1(x,y,t)\frac{\partial^\alpha N(x,y,t)}{\partial x^\alpha} + f_2(x,y,t)\frac{\partial^\beta N(x,y,t)}{\partial y^\beta} + f_3(x,y,t), \tag{7.8}$$

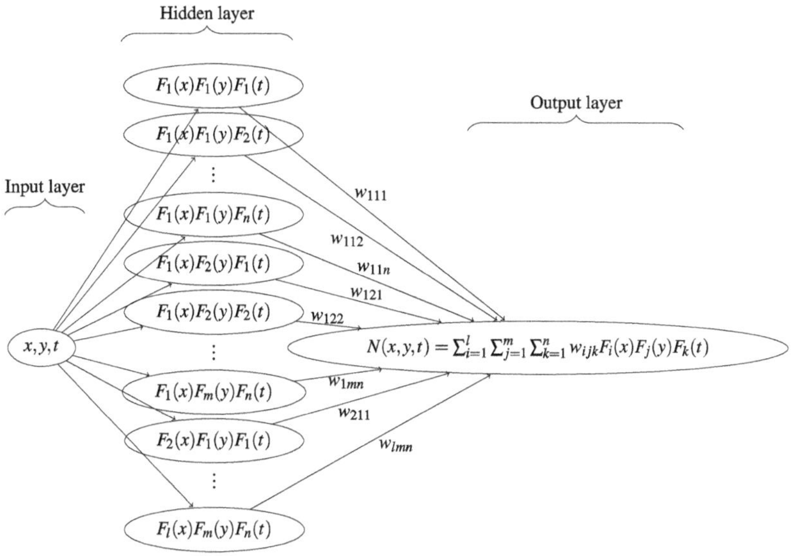

Figure 7.1 Structure of Fibonacci neural network.

with initial and boundary conditions

$$\begin{aligned} N(x,y,0) &= g_1(x,y), \\ N(0,y,t) &= g_2(y,t), \\ N(1,y,t) &= g_3(y,t), \\ N(x,0,t) &= g_4(x,t), \\ N(x,1,t) &= g_5(x,t). \end{aligned} \tag{7.9}$$

where

$$\begin{aligned} \frac{\partial N(x,y,t)}{\partial t} &= \sum_{i=1}^{l}\sum_{j=1}^{m}\sum_{k=1}^{n} w_{ijk}F_i(x)F_j(y)\frac{\partial F_k(t)}{\partial t}, \\ \frac{\partial^\alpha N(x,y,t)}{\partial x^\alpha} &= \sum_{i=1}^{l}\sum_{j=1}^{m}\sum_{k=1}^{n} w_{ijk}\frac{\partial^\alpha F_i(x)}{\partial x^\alpha}F_j(y)F_k(t), \\ \frac{\partial^\beta N(x,y,t)}{\partial y^\beta} &= \sum_{i=1}^{l}\sum_{j=1}^{m}\sum_{k=1}^{n} w_{ijk}F_i(x)\frac{\partial^\beta F_j(y)}{\partial y^\beta}F_k(t). \end{aligned} \tag{7.10}$$

The values of equation (7.10) can be obtained from Eq. (7.6). Now the above model (7.8) with the condition (7.9) can be transformed into the following non-constrained minimization problem

$$\min_{w}\Bigg\{\sum_{x_p,y_p,t_q}\Bigg(\Big(\frac{\partial N(x_p,y_p,t_q)}{\partial t}-f_1(x_p,y_p,t_q)\frac{\partial^{\alpha}N(x_p,y_p,t_q)}{\partial x^{\alpha}}$$
$$-f_2(x_p,y_p,t_p)\frac{\partial^{\beta}u(x_p,y_p,t_p)}{\partial y^{\beta}}-f_3(x_p,y_p,t_q)\Big)^2+(N(x_p,y_p,0)-g_1(x_p,y_p))^2$$
$$+(N(0,y_p,t_q)-g_2(y_p,t_q))^2+(N(1,y_p,t_q)-g_3(y_p,t_q))^2+(N(x_p,0,t_p)$$
$$-g_4(y_p,t_p))^2+(N(x_p,1,t_p)-g_5(y_p,t_p))^2\Bigg)\Bigg\}, \tag{7.11}$$

where x_p, y_p, t_p are the training points. In the next section, the author discusses the method to train the FNN to minimize the weights w's of the neural network.

7.3.2 LEARNING ALGORITHM FOR FNN

Marquardt's [13] method is actually a combination of Cauchy's and Newton's optimization method. This is also an gradient based method. The process of using Marquardt's method to minimize Eq. (7.3.2) as follows

Step 1. Define $w^{(0)}$ (Randomly chose but not identical)
M =maximum number of iteration
ε = convergence criteria

Step 2. Set $k = 0,\ \lambda^{(0)} = 10^4$.

Step 3. Calculate $\nabla E(w^{(k)})$

Step 4. Is $E(w^{(k)}) < \varepsilon$?
Yes: Go to **Step 11**
No: Continue

Step 5. Is $k \geq M$? Yes: Go to **Step 11**
No: Continue

Step 6. Calculate $s(w^{(k)}) = \left[H^{(k)} + \lambda^{(k)} I\right]^{-1} \nabla E(w^{(k)})$

Step 7. $w^{(k+1)} = w^{(k)} - s(w^{(k)})$

Step 8. Is $E(w^{(k+1)}) < E(w^{(k)})$?
Yes: Go to **Step 9**
No: Go to **Step 10**

Step 9. Set $\lambda^{(k+1)} = \lambda^{(k)}/4$ and $k = k+1$. Go to **Step 3**

Step 10. $\lambda^{(k)} = 2\lambda^{(k)}$. Go to **Step 6**

Step 11. Print the result and stop.

Where

$$E(w^k) = \min_{w}\Bigg\{\sum_{x_p,y_p,t_q}\Bigg(\Big(\frac{\partial N(x_p,y_p,t_q)}{\partial t} - f_1(x_p,y_p,t_q)\frac{\partial^\alpha N(x_p,y_p,t_q)}{\partial x^\alpha}$$

$$- f_2(x_p,y_p,t_p)\frac{\partial^\beta u(x_p,y_p,t_p)}{\partial y^\beta} - f_3(x_p,y_p,t_q)\Big)^2 + (N(x_p,y_p,0) - g_1(x_p,y_p))^2$$

$$+ (N(0,y_p,t_q) - g_2(y_p,t_q))^2 + (N(1,y_p,t_q) - g_3(y_p,t_q))^2 + (N(x_p,0,t_p)$$

$$- g_4(y_p,t_p))^2 + (N(x_p,1,t_p) - g_5(y_p,t_p))^2\Bigg)\Bigg\},$$

$$\nabla E(w^{(k)}) = \begin{bmatrix} \frac{\partial E(w^{(k)})}{\partial w_{111}} \\ \frac{\partial E(w^{(k)})}{\partial w_{112}} \\ \vdots \\ \frac{\partial E(w^{(k)})}{\partial w_{11n}} \\ \frac{\partial E(w^{(k)})}{\partial w_{121}} \\ \vdots \\ \frac{\partial E(w^{(k)})}{\partial w_{1mn}} \\ \frac{\partial E(w^{(k)})}{\partial w_{2mn}} \\ \vdots \\ \frac{\partial E(w^{(k)})}{\partial w_{lmn}} \end{bmatrix},$$

$$H = \begin{bmatrix} \frac{\partial^2 E(w^{(k)})}{\partial w_{111}^2} & \frac{\partial^2 E(w^{(k)})}{\partial w_{111}\partial w_{112}} & \cdots & \frac{\partial^2 E(w^{(k)})}{\partial w_{111}\partial w_{11n}} & \cdots & \frac{\partial^2 E(w^{(k)})}{\partial w_{111}\partial w_{1mn}} & \cdots & \frac{\partial^2 E(w^{(k)})}{\partial w_{111}\partial w_{lmn}} \\ \vdots & \vdots & \cdots & \vdots & \cdots & \vdots & \cdots & \vdots \\ \frac{\partial^2 E(w^{(k)})}{\partial w_{11n}\partial w_{111}} & \frac{\partial^2 E(w^{(k)})}{\partial w_{11n}\partial w_{112}} & \cdots & \frac{\partial^2 E(w^{(k)})}{\partial w_{11n}\partial w_{11n}} & \cdots & \frac{\partial^2 E(w^{(k)})}{\partial w_{11n}\partial w_{1mn}} & \cdots & \frac{\partial^2 E(w^{(k)})}{\partial w_{11n}\partial w_{lmn}} \\ \vdots & \vdots & \cdots & \vdots & \cdots & \vdots & \cdots & \vdots \\ \frac{\partial^2 E(w^{(k)})}{\partial w_{1mn}\partial w_{111}} & \frac{\partial^2 E(w^{(k)})}{\partial w_{1mn}\partial w_{112}} & \cdots & \frac{\partial^2 E(w^{(k)})}{\partial w_{1mn}\partial w_{11n}} & \cdots & \frac{\partial^2 E(w^{(k)})}{\partial w_{1mn}\partial w_{1mn}} & \cdots & \frac{\partial^2 E(w^{(k)})}{\partial w_{1mn}\partial w_{lmn}} \\ \vdots & \vdots & \cdots & \vdots & \cdots & \vdots & \cdots & \vdots \\ \frac{\partial^2 E(w^{(k)})}{\partial w_{lmn}\partial w_{111}} & \frac{\partial^2 E(w^{(k)})}{\partial w_{lmn}\partial w_{112}} & \cdots & \frac{\partial^2 E(w^{(k)})}{\partial w_{lmn}\partial w_{11n}} & \cdots & \frac{\partial^2 E(w^{(k)})}{\partial w_{lmn}\partial w_{1mn}} & \cdots & \frac{\partial^2 E(w^{(k)})}{\partial w_{lmn}^2} \end{bmatrix}.$$

In the next section, the authors apply the discussed method in this section to some of the examples that have the exact solutions to verify the accuracy of the method.

7.4 NUMERICAL EXAMPLE

In the section, the author has used the above discussed method on certain examples to verify it efficiency. The author has taken maximum absolute error and is defined as

$$E_{\max}(t) = \operatorname*{Max}_{a\le x,y\le b} |U_{\text{Exact}}(x,y,t) - U_{\text{Numerical}}(x,y,t)|, \tag{7.12}$$

where $U_{\text{Exact}}(x,y,t)$ and $U_{\text{Numerical}}(x,y,t)$ are exact and numerical solution of the considered problem.

Example 7.1. Consider the following two-dimensional heat equation

$$\frac{\partial u(x,y,t)}{\partial t} = \frac{\partial^2 u(x,y,t)}{\partial x^2} + \frac{\partial^2 u(x,y,t)}{\partial y^2}, \tag{7.13}$$

with initial and boundary conditions

$$\begin{aligned} u(0,y,t) &= (1-y)e^t, \\ u(1,y,t) &= (1-y)e^{1+t}, \\ u(x,0,t) &= e^{x+t}, \\ u(x,1,t) &= 0, \end{aligned} \tag{7.14}$$

where $x \in [0,1]$, $y \in [0,1]$ and $tt \in [0,1]$ have the exact solution $u(x,y,t) = (1-y)e^{x+t}$ [8]. By training the FNN for $l = m = n = 6$ over 46 iterations on 125 input points, this example has been solved. The input points were obtained by dividing the x, y and t spaces by five equal points. Table 7.1 provides the maximum absolute error for various time levels. Figure 7.2 has been plotted to compare the exact and numerical solution. It is evident from Figure 7.2 and Table 7.1 that the proposed technique operates pretty effectively even with a very small number of hidden nodes and iterations.

Example 7.2. Consider the following fractional order partial differential equation as

$$\frac{\partial u(x,y,t)}{\partial t} = f_1(x,y,t)\frac{\partial^{1.8} u(x,y,t)}{\partial x^{1.8}} + f_2(x,y,t)\frac{\partial^{1.6} u(x,y,t)}{\partial y^{1.6}} + f_3(x,y,t) \tag{7.15}$$

where $x \in [0,1]$, $y \in [0,1]$, $t \in [0,1]$, $f_1(x,y,t) = \frac{\Gamma(2.2)x^{2.8}y}{6}$, $f_2(x,y,t) = \frac{2xy^{2.6}}{\Gamma(4.6)}$ and $f_3(x,y,t) = -(1+2xy)e^{-t}x^3y^{3.6}$, under the initial condition

$$u(x,y,0) = x^3y^{3.6}, \tag{7.16}$$

Table 7.1
Maximum Absolute Error of Example 7.1 at Different Time

t	Maximum Absolute Error
0.1	3.8220e-4
0.3	1.1511e-4
0.7	1.3287e-4
1.0	1.4234e-4

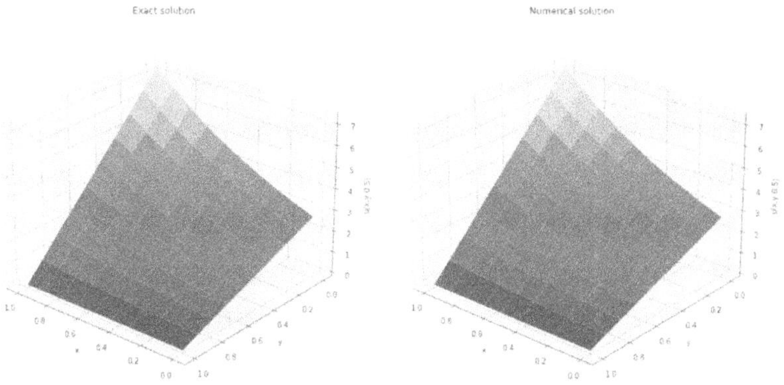

(a) (b)

Figure 7.2 Panels (a) and (b) are the exact and numerical solution of the Example 7.1 at $t = 0.5$ respectively.

and the boundary conditions

$$\begin{aligned} u(0,y,t) &= 0, \\ u(1,y,t) &= e^{-t}y^{3.6}, \\ u(x,0,t) &= 0, \\ u(x,1,t) &= e^{-t}x^{3}, \end{aligned} \tag{7.17}$$

has exact solution $u(x,y,t) = e^{-t}x^3y^{3.6}$ [23]. This example has been solved by training the FNN for $l = m = n = 5$ over 29 iterations, on 125 points as input. The points were obtained by dividing the spaces x, y, and time t by five equal points. Maximum absolute error is given at different times t in Table 7.2. Figure 7.3 has been plotted to compare the exact and numerical solution obtained with the discussed method. Figure 7.3 and Table 7.2 make it abundantly evident that the suggested method operates admirably even with a relatively small number of hidden nodes and iterations.

Table 7.2
Maximum Absolute Error of Example 7.2 at Different Time

t	Maximum Absolute Error
0.1	4.9594e-4
0.3	3.5480e-4
0.7	2.3520e-4
1.0	1.5074e-4

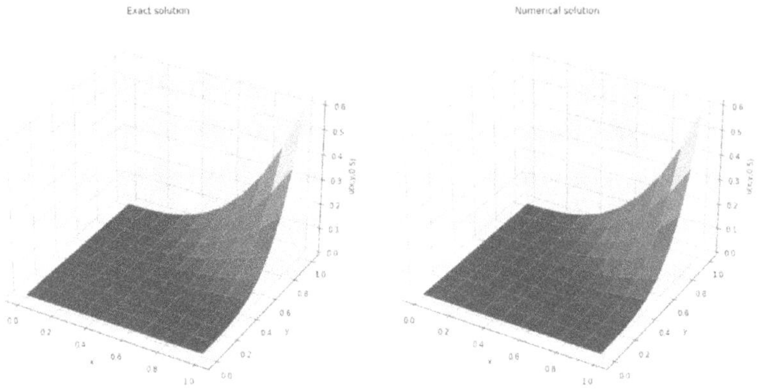

(a) (b)

Figure 7.3 Panels (a) and (b) are the exact and numerical solution of the Example 7.2 at $t = 0.5$ respectively.

Example 7.3. Consider the following two-dimensional fractional diffusion equation

$$\frac{\partial u(x,y,t)}{\partial t} = f_1(x,y,t)\frac{\partial^{1.9}u(x,y,t)}{\partial x^{1.9}} + f_2(x,y,t)\frac{\partial^{1.6}u(x,y,t)}{\partial y^{1.6}} + f_3(x,y,t), \quad (7.18)$$

where $x \in [0,1], y \in [0,1], t \in [0,1], f_1(x,y,t) = \frac{x^3 y^{1.4}}{\Gamma(3.9)}, f_2(x,y,t) = \frac{x^{1.1}y^3}{\Gamma(3.6)}, f_3(x,y,t) = -(1+2x^{1.1}y^{1.4})e^{-t}x^{2.9}y^{2.6}$, with initial condition

$$u(x,y,0) = x^{2.9}y^{2.6}, \quad (7.19)$$

and the boundary conditions

$$\begin{aligned} u(0,y,t) &= 0, \\ u(1,y,t) &= e^{-t}y^{2.6}, \\ u(x,0,t) &= 0, \\ u(x,1,t) &= e^{-t}x^{2.9}, \end{aligned} \quad (7.20)$$

has exact solution $u(x,y,t) = e^{-t}x^{2.9}y^{2.6}$ [22]. This example has been solved by training the FNN for $l = m = n = 5$ over 30 iterations on 125 points as input. The points were obtained by dividing the spaces x, y, and time t by five equal points. Maximum absolute error is given at different times t in Table 7.3. Figure 7.4 has been plotted to compare the exact and numerical solution obtained with the discussed method. It is evident from Figure 7.4 and Table 7.3 that the suggested strategy operates pretty effectively even with a very small number of hidden nodes and iterations on Example 7.3.

Example 7.4. Consider the following two-dimensional fractional diffusion equation

$$\frac{\partial u(x,y,t)}{\partial t} = f_1(x,y,t)\Big(\frac{\partial^{\alpha}u(x,y,t)}{\partial x^{\alpha}} + \frac{\partial^{\alpha}u(x,y,t)}{\partial y^{\alpha}}\Big) + f_2(x,y,t), \quad (7.21)$$

Table 7.3
Maximum Absolute Error of Example 7.3 at Different Time

t	Maximum Absolute Error
0.1	6.1622e-4
0.3	4.4530e-4
0.7	3.0786e-4
1.0	2.0095e-4

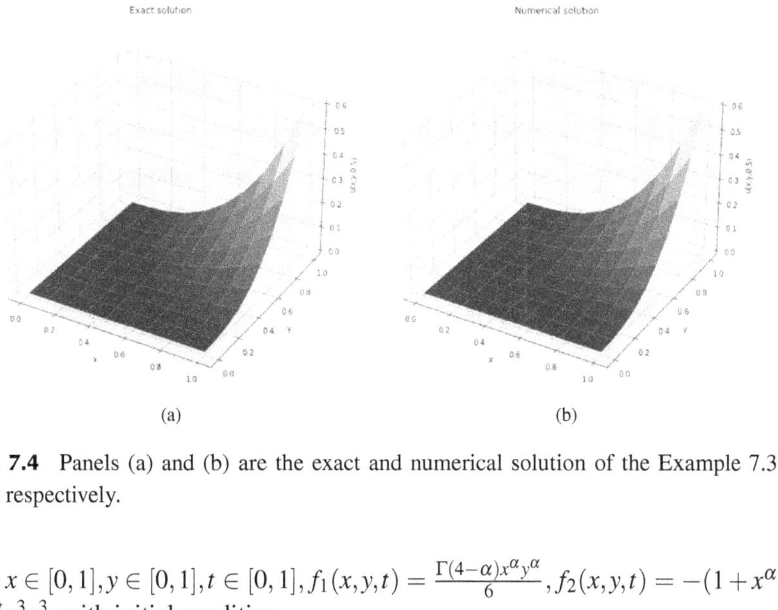

(a) (b)

Figure 7.4 Panels (a) and (b) are the exact and numerical solution of the Example 7.3 at $t = 0.5$ respectively.

where $x \in [0,1], y \in [0,1], t \in [0,1], f_1(x,y,t) = \frac{\Gamma(4-\alpha)x^\alpha y^\alpha}{6}, f_2(x,y,t) = -(1+x^\alpha + y^\alpha)e^{-t}x^3y^3$, with initial condition

$$u(x,y,0) = x^3y^3, \tag{7.22}$$

and the boundary conditions

$$\begin{aligned} u(0,y,t) &= 0, \\ u(1,y,t) &= e^{-t}y^3, \\ u(x,0,t) &= 0, \\ u(x,1,t) &= e^{-t}x^3, \end{aligned} \tag{7.23}$$

has exact solution $u(x,y,t) = e^{-t}x^3y^3$ [25]. This example has been solved by training the FNN for $l = m = n = 5$ over 31 iterations on 125 points as input. The points were obtained by dividing the spaces x, y, and time t by five equal points. Maximum absolute error is given at different times t in Table 7.4 for other orders of space derivative. Figure 7.5 has been plotted to compare the exact and numerical solution obtained with the discussed method at time 0.5 for various space order derivatives. Table 7.4

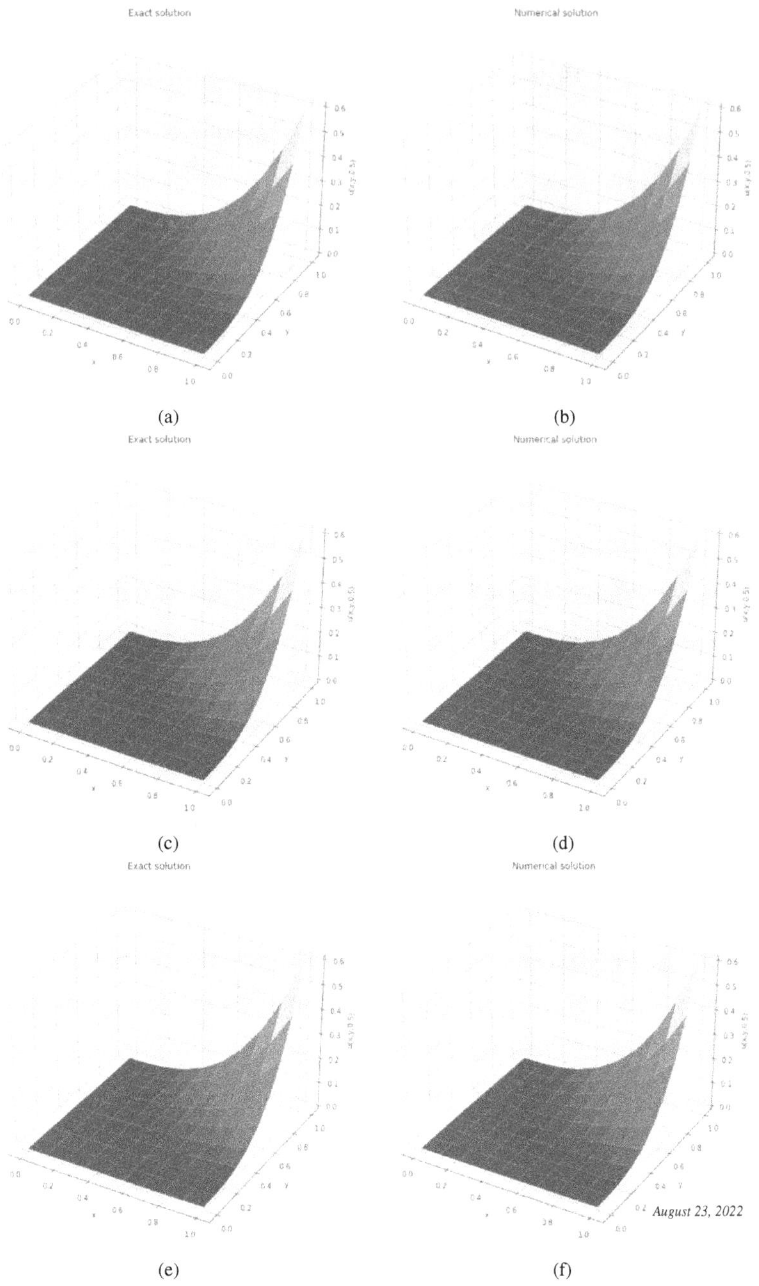

Figure 7.5 Comparison of exact and numerical solution of the Example 7.4 at at $t = 0.5$. Panels (a) and (b) for $\alpha = 1.4$, panels (c) and (d) for $\alpha = 1.7$, and panels (e) and (f) for $\alpha = 2.0$.

Table 7.4
Maximum Absolute Error of Example 7.4 for Different α and Time

	Time (t)	Maximum Absolute Error
	0.1	5.9891e-6
$\alpha = 1.4$	0.4	1.5682e-5
	0.7	1.6585e-5
	1.0	2.9095e-5
	0.1	6.4165e-6
$\alpha = 1.7$	0.4	1.2786e-5
	0.7	1.5038e-5
	1.0	2.1307e-5
	0.1	3.5027e-6
$\alpha = 2.0$	0.4	1.0044e-5
	0.7	1.2506e-5
	1.0	1.4260e-5

makes it quite evident that, after only 31 in iterations, the algorithm under discussion produces errors of the order of −5. Also, Figure 7.5 illustrates how effectively the strategies work.

Example 7.5. Consider the following time fractional order diffusion equation

$$\frac{\partial^{\gamma} u(x,y,t)}{\partial t^{\gamma}} = \frac{\partial^2 u(x,y,t)}{\partial x^2} + \frac{\partial^2 u(x,y,t)}{\partial y^2} + (t\Gamma(2+\gamma) - 2t^{1+\gamma})e^{x+y}, \tag{7.24}$$

where $x \in [0,1]$, $y \in [0,1]$ and $t \in [0,1]$ with initial condition

$$u(x,y,0) = 0 \tag{7.25}$$

and boundary conditions

$$\begin{aligned} u(0,y,t) &= e^{y}t^{1+\gamma}, \\ u(1,y,t) &= e^{1+y}t^{1+\gamma}, \\ u(x,0,t) &= e^{x}t^{1+\gamma}, \\ u(x,1,t) &= e^{1+x}t^{1+\gamma}, \end{aligned} \tag{7.26}$$

have the exact solution $u(x,y,t) = e^{x+y}t^{1+\gamma}$ [4]. By using 125 points as input and training the FNN for $l = m = n = 7$ over 50 iterations, the example has been solved. The input points were obtained by dividing the spaces x, y and time t by five equal

Table 7.5
Maximum Absolute Error of Example 7.5 for Different α and Time

	Time (t)	Maximum Absolute Error
	0.1	2.3241e-2
$\gamma = 0.4$	0.4	3.6032e-3
	0.7	2.7653e-3
	1.0	2.7240e-3
	0.1	1.0462e-2
$\gamma = 0.7$	0.4	2.1135e-3
	0.7	2.3637e-3
	1.0	2.7098e-3
	0.1	6.4943e-4
$\gamma = 1.0$	0.4	8.0814e-4
	0.7	2.0364e-3
	1.0	2.2798e-4

points. The maximum absolute error is shown in Table 7.5 for various fractional orders of time derivatives at various time t. Additionally, the exact and numerical solutions at time $t = 0.5$ for various fractional time derivatives have been compared in Figure 7.6. Figure 7.6 and Table 7.5 make it abundantly evident that the suggested method operates admirably even with a relatively small number of hidden nodes and iterations.

We can see that even with a very small number of hidden layer nodes and iterations, our mentioned method still works fairly effectively of above examples. By adding more hidden layers to the created neural network or increasing the number of nodes in the hidden layer, the method's efficiency can be improved.

7.5 SOLUTION OF TWO-DIMENSIONAL FDE

In this section, the author solves the considered two-dimensional FDE Eq. (7.1) with the above-discussed method and observes the behavior of $u(x,y,t)$ due to changes in various parameters presented in the model. Now taking the particular case of the considered model (7.1) as

$$\frac{\partial^{\gamma} u(x,y,t)}{\partial t^{\gamma}} = f_1(x,y,t)\frac{\partial^{\alpha} u(x,y,t)}{\partial x^{\alpha}} + f_2(x,y,t)\frac{\partial^{\beta} u(x,y,t)}{\partial y^{\beta}} + f_3(x,y,t) \tag{7.27}$$

where $x \in [0,1]$, $y \in [0,1]$, $t \in [0,1]$, $f_1(x,y,t) = \frac{\Gamma(2.2)x^{2.8}y}{6}$, $f_2(x,y,t) = \frac{2xy^{2.6}}{\Gamma(4.6)}$ and $f_3(x,y,t) = -(1+2xy)e^{-t}x^3y^{3.6}$, under the initial condition

$$u(x,y,0) = x^3y^{3.6}, \tag{7.28}$$

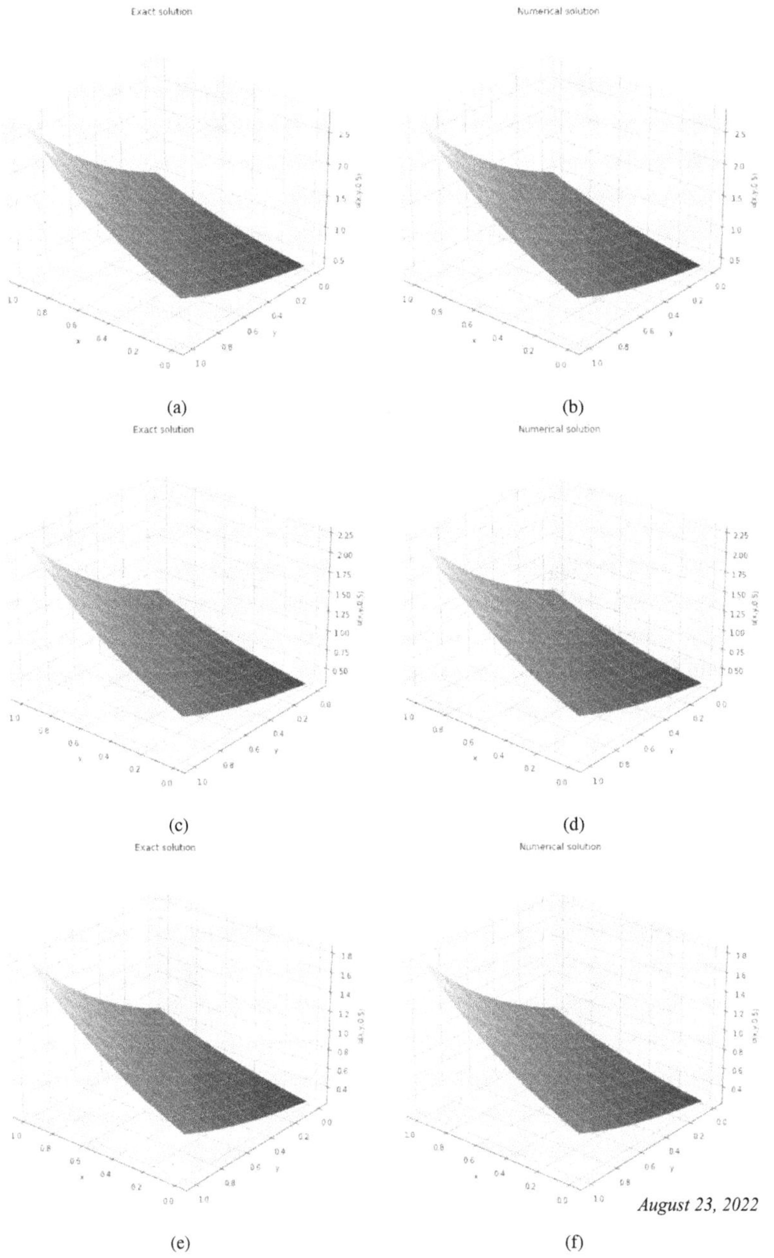

Figure 7.6 Comparison of exact and numerical solution of the Example 7.5 at at $t = 0.5$. Panels (a) and (b) for $\gamma = 0.4$, panels (c) and (d) for $\gamma = 0.7$, and panels (e) and (f) for $\gamma = 0.4$.

and the boundary conditions

$$\begin{aligned} u(0,y,t) &= 0, \\ u(1,y,t) &= e^{-t}y^{3.6}, \\ u(x,0,t) &= 0, \\ u(x,1,t) &= e^{-t}x^{3}, \end{aligned} \tag{7.29}$$

The diffusion model (7.27) with initial condition (7.28) and boundary conditions (7.29) have the exact solution for $\alpha = 1.8, \beta = 1.6, \gamma = 1$. But we don't have an exact solution for other fractional derivatives. In this section, the author has solved this problem with the discussed method for different values of fractional order derivatives and observed the behavior of the solution at different times through plot representation.

Figures 7.7–7.11 have been plotted for different values of space fractional derivatives of x, y and t to observe the behavior of the solution at different times t. From these figures, we can clearly observe that initially, the concentration was high, but it started getting low as the time passed due to diffusion. The behavior of the solution is quite similar in each figure. Also, when we observe Figures 7.7–7.11 comparatively, the different faction order derivatives of spaces are not affecting the solution much.

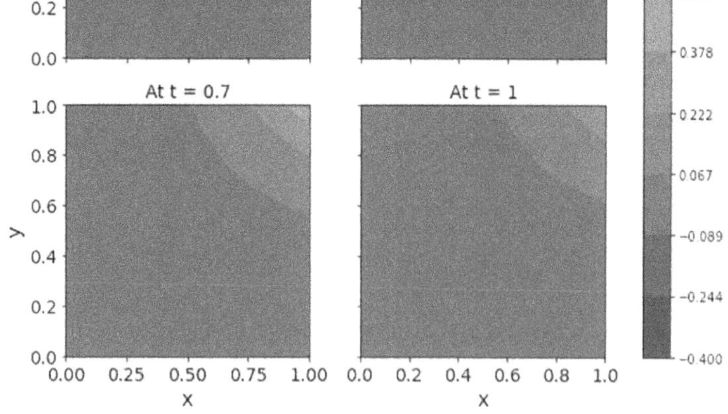

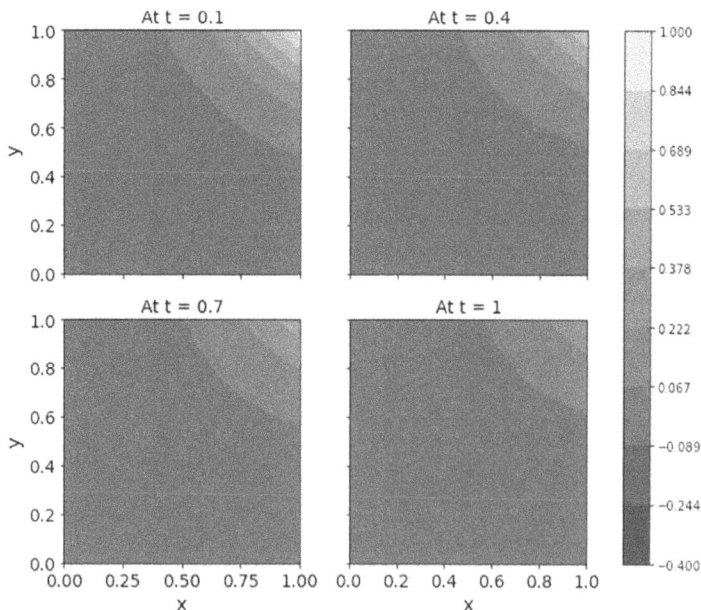

Figure 7.7 $u(x,y,t)$ at $\gamma = 1$, $\alpha = 2$, and $\beta = 2$ at different times.

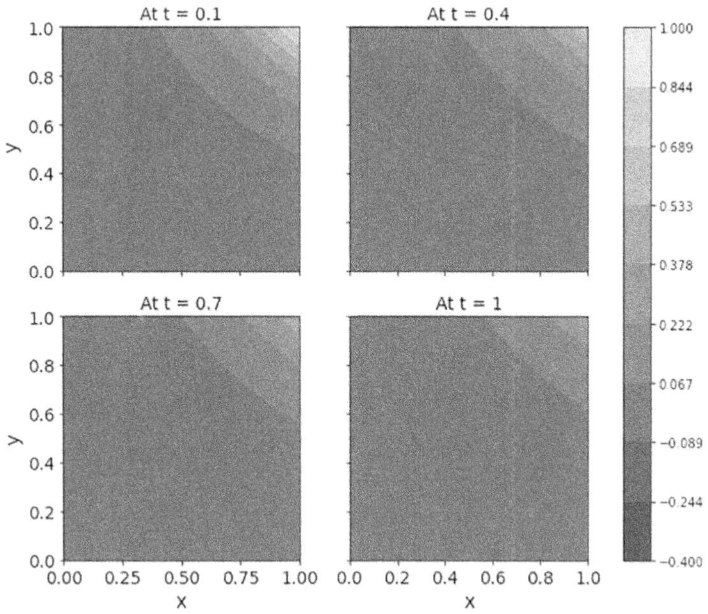

Figure 7.8 $u(x,y,t)$ at $\gamma = 1$, $\alpha = 1.6$, and $\beta = 1.6$ at different times.

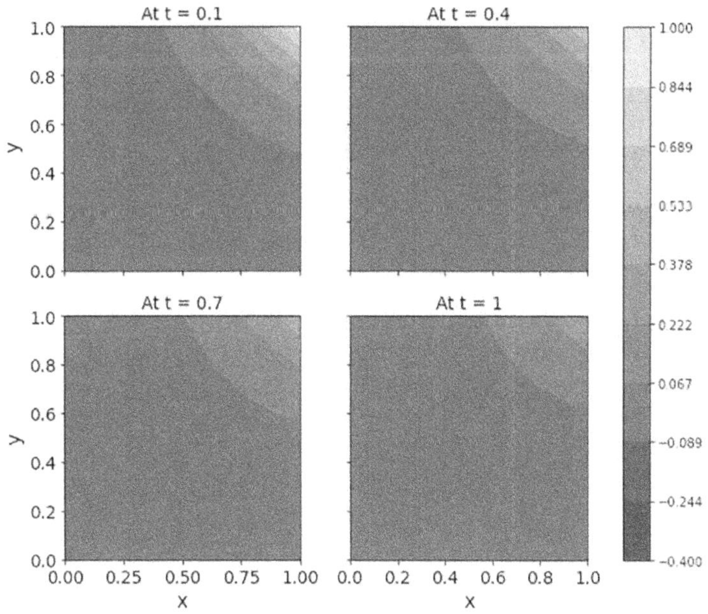

Figure 7.9 $u(x,y,t)$ at $\gamma = 1$, $\alpha = 1.6$, and $\beta = 2.0$ at different times.

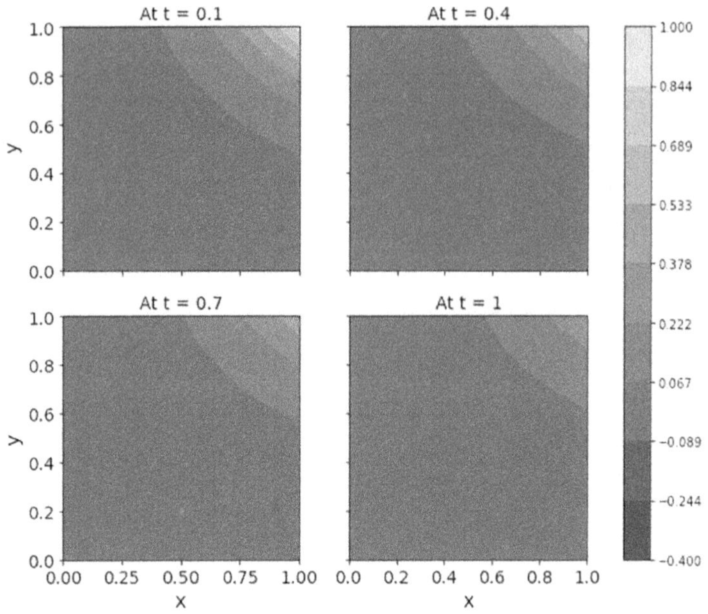

Figure 7.10 $u(x,y,t)$ at $\gamma = 1$, $\alpha = 2.0$, and $\beta = 1.6$ at different times.

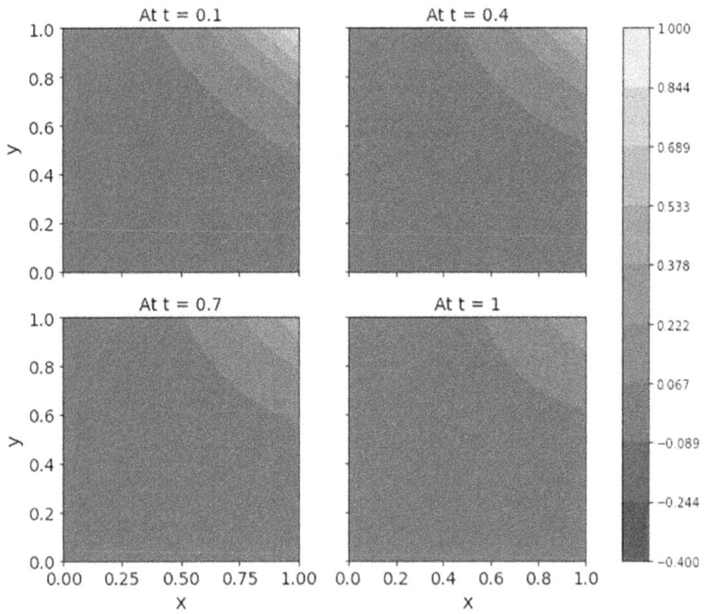

Figure 7.11 $u(x,y,t)$ at $\gamma = 0.6$, $\alpha = 2.0$, and $\beta = 2.0$ at different times.

7.6 APPLICATION OF THE METHOD IN ENGINEERING

Partial differential equations (PDEs) are used to describe physical and biological phenomena, as was mentioned in the introductory section; however, it might be difficult to determine the precise solution. Partial differential equations, in particular heat, wave, etc. are often used in engineering. This chapter is concerned with fast and effectively solving the three-dimensional PDEs. The construction of an accessible approach is emphasised in this chapter along with obtaining the precise numerical solution to PDEs.

7.7 CONCLUSION

In this chapter, the author has extended the work done by himself [5] to solve two-dimensional fractional diffusion equations. The Fibonacci neural network has been constructed by taking five layers and using different degrees of the Fibonacci polynomial as an activation function in the hidden layer. By taking the Fibonacci neural network as an initial solution to the considered diffusion problem, the authors transformed the considered problem into a non-constrained maximization problem. The method is verified by applying it to three examples. Finally, the method is used to solve unsolved problems and observe the behavior of the solution at different order space derivatives at different times. The method mentioned here can be extended to handle multidimensional differential issues in future work, or it can be used to tackle different kinds of three-dimensional differential problems. In this chapter, the author only used one hidden layer and the Fibonacci polynomial as the activation function; however, in subsequent work, the number of hidden layers may be increased for better approximation, additional hidden layers may be used, additional types of activation functions may be used, and the network may be trained with the proper activation function in order to determine whether the developed neural network performs better than this network.

Bibliography

1. Abbasbandy, S. and Shirzadi, A. (2010). A meshless method for two-dimensional diffusion equation with an integral condition. *Engineering Analysis with Boundary Elements*, 34(12):1031–1037.
2. Abbaszadeh, M. and Mohebbi, A. (2013). A fourth-order compact solution of the two-dimensional modified anomalous fractional sub-diffusion equation with a nonlinear source term. *Computers & Mathematics with Applications*, 66(8):1345–1359.
3. Berg, J. and Nyström, K. (2018). A unified deep artificial neural network approach to partial differential equations in complex geometries. *Neurocomputing*, 317:28–41.
4. Cui, M. (2013). Convergence analysis of high-order compact alternating direction implicit schemes for the two-dimensional time fractional diffusion equation. *Numerical Algorithms*, 62(3):383–409.

5. Dwivedi, K. D. et al. (2021). Numerical solution of fractional order advection reaction diffusion equation with fibonacci neural network. *Neural Processing Letters*, 53(4):2687–2699.
6. Fan, W. and Liu, F. (2018). A numerical method for solving the two-dimensional distributed order space-fractional diffusion equation on an irregular convex domain. *Applied Mathematics Letters*, 77:114–121.
7. Fang, J., Liu, C., Simos, T., and Famelis, I. T. (2020). Neural network solution of single-delay differential equations. *Mediterranean Journal of Mathematics*, 17(1):1–15.
8. Kumar, D., Singh, J., and Kumar, S. (2015). Analytical modeling for fractional multi-dimensional diffusion equations by using laplace transform. *Communications in Numerical Analysis*, 1:16–29.
9. Li, A., Chen, R., Farimani, A. B., and Zhang, Y. J. (2020). Reaction diffusion system prediction based on convolutional neural network. *Scientific Reports*, 10(1):1–9.
10. Mohebbi, A., Abbaszadeh, M., and Dehghan, M. (2014). Solution of two-dimensional modified anomalous fractional sub-diffusion equation via radial basis functions (RBF) meshless method. *Engineering Analysis with Boundary Elements*, 38:72–82.
11. Panghal, S. and Kumar, M. (2021). Optimization free neural network approach for solving ordinary and partial differential equations. *Engineering with Computers*, 37(4):2989–3002.
12. Podlubny, I. (1998). *Fractional Differential Equations: An Introduction to Fractional Derivatives, Fractional Differential Equations, to Methods of Their Solution and Some of Their Applications*. Elsevier: Amsterdam, Netherlands.
13. Ravindran, A., Reklaitis, G. V., and Ragsdell, K. M. (2006). *Engineering Optimization: Methods and Applications*. John Wiley & Sons: Boca Raton, FL.
14. Singh, H. (2019). An efficient computational method for non-linear fractional lienard equation arising in oscillating circuits. In Singh, H., Kumar, D., and Baleanu, D., editors *Methods of Mathematical Modelling*, pp. 39–50. CRC Press: : Boca Raton, FL.
15. Singh, H. (2020). Analysis for fractional dynamics of ebola virus model. *Chaos, Solitons & Fractals*, 138:109992.
16. Singh, H. (2021). Chebyshev spectral method for solving a class of local and nonlocal elliptic boundary value problems. *International Journal of Nonlinear Sciences and Numerical Simulation*, 000010151520200235.
17. Singh, H. (2022). Solving a class of local and nonlocal elliptic boundary value problems arising in heat transfer. *Heat Transfer*, 51(2):1524–1542.
18. Singh, H., Kumar, D., and Baleanu, D. (2019). *Methods of Mathematical Modelling: Fractional Differential Equations*. CRC Press: Boca Raton, FL.
19. Singh, H. and Srivastava, H. (2020). Numerical investigation of the fractional-order liénard and duffing equations arising in oscillating circuit theory. *Frontiers in Physics*, 8:120.
20. Singh, H., Srivastava, H., and Nieto, J. J. (2022). *Handbook of Fractional Calculus for Engineering and Science*. CRC Press: : Boca Raton, FL.
21. Singh, H. and Wazwaz, A.-M. (2021). Computational method for reaction diffusion-model arising in a spherical catalyst. *International Journal of Applied and Computational Mathematics*, 7(3):1–11.
22. Sweilam, N. H. and Almajbri, T. F. (2015). Large stability regions method for the two-dimensional fractional diffusion equation. *Progress in Fractional Differentiation and Applications*, 1(2):123–131.

23. Tadjeran, C. and Meerschaert, M. M. (2007). A second-order accurate numerical method for the two-dimensional fractional diffusion equation. *Journal of Computational Physics*, 220(2):813–823.
24. Tripathi, V. M., Srivastava, H. M., Singh, H., Swarup, C., and Aggarwal, S. (2021). Mathematical analysis of non-isothermal reaction–diffusion models arising in spherical catalyst and spherical biocatalyst. *Applied Sciences*, 11(21):10423.
25. Tuan, N. H., Aghdam, Y. E., Jafari, H., and Mesgarani, H. (2021). A novel numerical manner for two-dimensional space fractional diffusion equation arising in transport phenomena. *Numerical Methods for Partial Differential Equations*, 37(2):1397–1406.

8 Analysis of a Class of Reaction-Diffusion Equation Using Spectral Scheme

Prashant Pandey
KS Saket PG College

Priya Kumari
Banaras Hindu University

CONTENTS

8.1 INTRODUCTION

Recently, many scientists and researchers have analyzed different types of fractional order partial differential equations [1–5]. Due to nonlocal behavior of fractional order derivatives, the fractional extension of ordinary differential equations is also very useful for describing various physical phenomena, the fraction present in the derivative also suggests the weighting and modulation of system memory [6–11]. The fractional PDEs describe the behavior of physical phenomena such as plasma physics, electricity, propagation of waves, quantum mechanics, fluid dynamics, and other physical processes within their validity range. Even for young researchers, the

DOI: 10.1201/9781003368069-8

fractional derivatives were complicated-although it appears in many parts of sciences such as physics, engineering, bio-engineering, COVID-19 pandemic studies, and many other branches of sciences [12–14].

Fractional order derivative produces a better degree of freedom in these models. Arbitrary order derivatives are powerful tools for the discretion of the dynamical behavior of various biomaterials and systems [15]. This is why there are different fractional derivatives; Riemann and Liouville, and Caputo operators are the most conservative examples in traditional practice. Caputo presented a fractional derivative that permits the conventional initial and boundary conditions allied with the real-world problem.

One of the special types of partial differential equations (PDEs) is reaction-diffusion equations (RDEs), which has attracted the attention of many scientists and researchers. There are several numerical schemes for solving this class of PDEs such as traveling wave method, finite elements method, finite-difference methods, spectral methods, etc. On top of that, the operational matrix method based on different types of the polynomials and wavelets is one of the efficient numerical schemes, which had been widely used in solving various types of fractional order calculus problems, such as the Bernoulli operational matrix for solving fractional delay differential equations [16], operational matrix for solving fractional order partial differential equations [17], Jacobi wavelet operational matrix of fractional order integration for solving fractional differential equations [18,19], and Fibonacci wavelet operational matrix of integration for solving nonlinear integral equations, etc.

In the last few years, computation, analysis, stability analysis, and error estimations of fractional partial differential equations (FPDEs) attracted the major attention of scientists and researchers in several aspects of science. Many biological and physical phenomena have been modeled by FPDEs, and the accuracy of these models was more accurate and efficient than the integer order models. In recent times, many researchers have established various numerical algorithms to solve the FPDEs [20–24]. Recently, a few researchers also extended the application of fractional calculus in phenomena of reaction-diffusion arising in spherical catalyst [25,26], study of Ebola virus model [27], problem arising in heat transfer flow [28], problem arising in oscillating circuits [29,30], some boundary value problems [31].

The fractional nonlinear Klein-Gordon equation (FNKGE) is one of the important mathematical models in quantum mechanics. The Klein-Gordon equation is basically related to the famous Schrodinger equation. The more general form, i.e. fractional order Klein-Gordon equation, is obtained by switching integer order derivative to fractional order derivative in the Klein-Gordon equation. Here we consider generalized fractional Klein-Gordon equation as

$$\frac{\partial^{\alpha} u(x,t)}{\partial t^{\alpha}} = D\frac{\partial^2 u(x,t)}{\partial x^2} + \eta u(x,t) + \lambda u^k(x,t) + f(x,t), \quad 0 \leq x \leq 1, t \geq 0. \tag{8.1}$$

with the initial and boundary conditions

$$u(x,0) = a(x), \quad u(0,t) = b(t), \quad u(1,t) = c(t), \tag{8.2}$$

In the above general form of FNKGE D, η, λ, k are known coefficients, $f(x,t)$ is a known source function, a, b, c are known functions and the fractional order α is an arbitrary real number. Various forms of the above FNKGE are considered as test examples to assess the accuracy and applicability of our proposed collocation method.

The FNKGE is an important equation in nonlinear optics theory, solid state physics, and solitons discussion, basically in the investigations of the recurrence of initial states and the interaction of solitons for collision less plasma [32,33] and to analyze the behavior of linear wave models. The tanh method has been applied by Wazwaz to find the several exact traveling waves solutions such as solitons, compactons, and periodic solutions [33].

In the last few years, the study to find the approximate numerical solution of FNKGE has been examined considerably. Several numerical algorithms have been applied to find the numerical solution of FNKGE. The reduced differential transform method has been executed by Servi and Oturanc [34] in 2011. Hafez et al. [35] applied the novel (G'/G)-expansion method to find the exact traveling wave solutions of FNKGE.

In this chapter, we are applying an efficient computational algorithm using operational matrix method based on orthogonal Laguerre polynomials. First of all, we are using collocation method and approximation of functions to convert FNKGE into a nonlinear system of algebraic equations and on simplifying this system numerical solution of our mathematical model is obtained. To show the validity, accuracy, and efficiency of our proposed method, we have given some test examples of the FNKGE, and obtained results are compared with the analytical solutions and with the numerical solutions obtained with some existing methods. In addition, the stability analysis and convergence analysis of the proposed numerical algorithm are given to show the effectiveness of the considered numerical scheme. Numerical discussions and graphical presentations ensure that the introduced numerical method is easy to apply and reliable for nonlinear fractional order systems.

The main advantages of using this operational matrix method over other existing methods are its simplicity of implementation and programmable easily in using any computer algebra system. Besides that, if the fractional differential equations are in multi-order or having variable coefficients, operational matrix method is also efficient in finding the numerical solution.

This chapter is organized in a suitable way; definition of Caputo fractional derivative is given in Section 8.2. Section 8.3 includes some basic properties of Laguerre polynomials. Construction of operational matrix is discussed in Section 8.4. Section 8.5 includes the brief discussion of approximation of unknown functions. Convergence analysis of the scheme and error analysis is given in Sections 8.6 and 8.7, respectively. In Section 8.8, obtained results have been discussed, which is followed by "Conclusion" section.

8.2 PRELIMINARIES

Many definitions of fractional derivatives such as Riemann Liouville fractional derivative, Atangana-Baleanu derivative, Caputo derivative, etc., are given in Refs. [36,37]. Here, we have provided some basic definitions and properties of the fractional calculus which are important for this chapter.

Caputo Definition of Fractional Derivative

The fractional order derivative in the Caputo sense of any order α is given by Refs. [36,38]

$$(D^{\alpha}\omega)(\theta) = \begin{cases} \frac{d^k\omega(\theta)}{d\theta^k}, & \text{when } \alpha = k \in Z^+, \\ \frac{1}{\Gamma(k-\alpha)}\int_0^{\theta}(\theta-\rho)^{k-\alpha-1}\omega^{(k)}(\rho)\,d\rho, & \text{when } k-1<\alpha<k. \end{cases} \quad (8.3)$$

One can find more details of fractional calculus in Refs. [36,38]

8.3 BASIC PROPERTIES OF LAGUERRE POLYNOMIALS

Due to orthogonality properties of Laguerre polynomials, it can be applicable in various aspects of science such as homotopy analysis, quantum physics complex analytic number theory, etc.

Now, we define the Laguerre polynomials of degree l in the interval I as

$$P_l(x) = \frac{1}{l!}e^x\partial_x^l(x^l e^{-x}), \quad l = 0,1,\ldots; \quad (8.4)$$

Some recurrence relation for the Laguerre polynomials is given by

$$\partial_x(xe^{-x}\partial_x P_l(x)) + le^{-x}P_l(x) = 0, \quad x \in I, \quad (8.5)$$

and

$$P_l(x) = \partial_x P_l(x) - \partial_x P_{l+1}(x), \quad l \geq 0. \quad (8.6)$$

The set Laguerre polynomials form the $L_r^2(I)$-orthogonal system with the weight function $r(x) = e^{-x}$, i.e.,

$$\int_I P_h(x)P_k(x)r(x)dx = \delta_{hk}, \quad \forall h,k \geq 0, \quad (8.7)$$

In the above expression, δ_{hk} is the well-known kronecher delta function.

Thus, on the interval I the analytical expression of the Laguerre polynomials of degree l is defined by

$$P_l(x) = \sum_{\varepsilon=0}^{l}\frac{l!(-1)^{\varepsilon}}{(-\varepsilon+l)!(\varepsilon!)^2}x^{\varepsilon}, \quad l = 0,1,\ldots; \quad (8.8)$$

8.4 FORMULATION OF OPERTIONAL MATRIX

Any function $A(x) \in L_r^2(I)$ can be expressed in the combination of Laguerre polynomials as

$$A(x) = \sum_{\varepsilon=0}^{\infty} \sigma_\varepsilon P_\varepsilon(x), \quad \sigma_\varepsilon = \int_0^\infty A(x) P_\varepsilon(x) r(x) dx, \quad \varepsilon = 0,1,2,\ldots; \tag{8.9}$$

Usually, one can use only the initial (N+1)-terms of Laguerre polynomials. Thus, the above equation can be reduced as

$$A_N(x) = \sum_{\varepsilon=0}^{N} \sigma_\varepsilon P_\varepsilon(x) = K^T B(x). \tag{8.10}$$

where $B(x)$ is the Laguerre vector and K is the coefficients of Laguerre vector and expressed as

$$K^T = [k_0, k_1, \ldots, k_N], \quad B(x) = [P_0(x), P_1(x), \ldots, P_N(x)]^T. \tag{8.11}$$

The derivative of the Laguerre vector $B(x)$ is

$$\frac{dB(x)}{dx} = H^{(1)} B(x). \tag{8.12}$$

where $H^{(1)}$ is an $(N+1) \times (N+1)$ order operational matrix for the differentiation and written as

$$H^{(1)} = -\begin{bmatrix} 0 & 0 & 0 & 0 & 0 & \ldots & 0 & 0 \\ 1 & 0 & 0 & 0 & 0 & \ldots & 0 & 0 \\ 1 & 1 & 0 & 0 & 0 & \ldots & 0 & 0 \\ 1 & 1 & 1 & 0 & 0 & \ldots & 0 & 0 \\ 1 & 1 & 1 & 1 & 0 & \ldots & 0 & 0 \\ \vdots & \vdots & \vdots & \vdots & \vdots & \ldots & \vdots & \vdots \\ 1 & 1 & 1 & 1 & 1 & \ldots & 1 & 0 \end{bmatrix}. \tag{8.13}$$

Thus, in view of Eq. (8.12), we can also write

$$\frac{d^n B(x)}{dx^n} = (H^{(1)})^n B(x), \tag{8.14}$$

The superscript in $H^{(1)}$ has the property

$$H^{(m)} = (H^{(1)})^m, \quad m = 1,2,\ldots \;\; ; \tag{8.15}$$

Lemma 8.1. Let $P_m(x)$ be the Laguerre polynomial of degree m; then α order fractional derivative of $P_m(x)$ has the property

$$H^\alpha P_m(x) = 0, \quad m = 0,1,\ldots,\lceil \alpha \rceil - 1, \quad \alpha > 0, \tag{8.16}$$

Proof. Lemma 8.1 can be easily proved by using properties of fractional calculus in Eq. (8.8).

Now we generalize the operational matrix given in Eq. (8.13) for arbitrary fractional derivatives by the following theorem.

Theorem 8.1. *Let $\alpha > 0$ be an arbitrary real number and $B(x)$ be the Laguerre vector as in Eq. (8.10) then*

$$H^{\alpha}B(x) = H^{(\alpha)}B(x), \tag{8.17}$$

In the above equation $H^{(\alpha)}$ is $(N+1)\times(N+1)$ order operational matrix for α order fractional differentiation in Caputo sense and expressed as

$$H^{(\alpha)} = \begin{bmatrix} 0 & 0 & 0 & \dots & 0 \\ \vdots & \vdots & \vdots & \dots & \vdots \\ 0 & 0 & 0 & \dots & 0 \\ E_\alpha(\lceil\alpha\rceil,0) & E_\alpha(\lceil\alpha\rceil,1) & E_\alpha(\lceil\alpha\rceil,2) & \dots & E_\alpha(\lceil\alpha\rceil,N) \\ \vdots & \vdots & \vdots & \dots & \vdots \\ E_\alpha(i,0) & E_\alpha(i,1) & E_\alpha(i,2) & \dots & E_\alpha(i,N) \\ \vdots & \vdots & \vdots & \dots & \vdots \\ E_\alpha(N,0) & E_\alpha(N,1) & E_\alpha(N,2) & \dots & E_\alpha(N,N) \end{bmatrix}, \tag{8.18}$$

where

$$E_\alpha(\zeta,\eta) = \sum_{h=\lceil\alpha\rceil}^{\zeta}\sum_{k=0}^{\eta}\frac{(-1)^{h+k}\zeta!\eta!\Gamma(h-\alpha+k+1)}{(\zeta-h)!\Gamma(h-\alpha+1)h!(\eta-k)!(k!)^2}. \tag{8.19}$$

One can clearly see that in $H^{(\alpha)}$, the initial $\lceil\alpha\rceil$ rows are zero.

Proof. To prove this, Eqs. (8.8–8.10) solved together, we have

$$H^{\alpha}P_\phi(x) = \sum_{h=0}^{\phi}\frac{(-1)^h\phi!H^{\alpha}x^h}{(\phi-h)!(h!)^2} = \sum_{h=\lceil\alpha\rceil}^{\phi}\frac{(-1)^h\phi!x^{h-\alpha}}{(\phi-h)!\Gamma(h-\alpha+1)h!};\quad \phi=\lceil\alpha\rceil,\dots,N. \tag{8.20}$$

The term $x^{h-\alpha}$ in the above equation can be written in terms of Laguerre polynomials as

$$x^{h-\alpha} = \sum_{\psi=0}^{N}\sigma_\psi P_\psi(x), \tag{8.21}$$

and σ_ψ can be found from Eq. (8.10) using $A(x) = x^{h-\alpha}$ as

$$\sigma_\psi = \sum_{k=0}^{\psi}\frac{(-1)^k\psi!\Gamma(h-\alpha+k+1)}{(\psi-k)!(k!)^2} \tag{8.22}$$

By Eqs. (8.20) – (8.22), we get

$$H^{\alpha}P_\zeta(x) = \sum_{\eta=0}^{N}E_\alpha(\zeta,\eta)P_\eta(x),\quad \zeta=\lceil\alpha\rceil,\dots,N, \tag{8.23}$$

where $E_\alpha(\zeta,\eta) = \sum_{h=\lceil\alpha\rceil}^{\zeta} \sum_{k=0}^{\eta} \frac{(-1)^{h+k}\zeta!\eta!\Gamma(h-\alpha+k+1)}{(\zeta-h)!\Gamma(h-\alpha+1)h!(\eta-k)!(k!)^2}$.
The vector form of Eq. (8.23) is

$$H^\alpha P_\zeta(x) = [E_\alpha(\zeta,0), E_\alpha(\zeta,1), \ldots, E_\alpha(\zeta,N)]B(x), \quad \zeta = \lceil\alpha\rceil, \ldots, N. \tag{8.24}$$

In view of Lemma 8.1, we get

$$H^\alpha P_\zeta(x) = [0,0,\ldots,0]B(x), \quad \zeta = 0,1,\ldots,\lceil\alpha\rceil - 1. \tag{8.25}$$

Thus the required proof can be done by using Eqs. (8.24) – (8.25).

8.5 APPROXIMATION OF FUNCTION

This part of the chapter includes the implementation of our proposed numerical collocation scheme to FNKGE in an efficient way. The approximation of the function $u(x,t)$ in terms of Laguerre polynomials is given as

$$u(x,t) = \sum_{h=0}^{N} \sum_{k=0}^{N} \sigma_{hk} P_h(x) P_k(t), \tag{8.26}$$

where σ_{hk} is the unknown coefficients to be determine latter for $h,k = 1,2,3,\ldots$; In the vector form the above Eq. (8.26) is written as

$$u(x,t) = B^T(x).K.B(t), \tag{8.27}$$

where $K = [\sigma_{hk}]$ be the matrix of unknown coefficients of order $(N+1)\times(N+1)$ and $B(t) = (P_0(t), P_1(t), \ldots, P_N(t))^T$ is a column vector.

In view of Theorem 8.1, the fractional derivatives of order β with respect to x and of order α with respect to t of Eq. (8.27) is given by

$$\frac{\partial^\alpha u(x,t)}{\partial t^\alpha} = H^\alpha u(x,t) = B^T(x).K.H^\alpha B(t), \tag{8.28}$$

$$\frac{\partial^\beta u(x,t)}{\partial x^\beta} = H^\beta u(x,t) = H^\beta B^T(x).K.B(t), \tag{8.29}$$

With the help of Eq. (8.27) the boundary conditions (8.2) can be written as

$$B^T(x).K.B(0) = x(x-1), \quad B^T(0).K.B(t) = 0, \quad B^T(1).K.B(t) = 0. \tag{8.30}$$

Using collocation method to our proposed FNKGE (Eq. 8.1) with the aid of Eq. (8.30) at collocation points $x_\varepsilon = \frac{\varepsilon}{N}$ for $\varepsilon = 0,1,2,\ldots,N$ and $t_\varepsilon = \frac{\varepsilon}{N}$ for $\varepsilon = 0,1,2,\ldots,N$, we get a nonlinear system of algebraic equations. On solving this nonlinear system of algebraic equations, we can find the matrix K of unknown coefficients, and thus we obtain approximate numerical solution of our proposed FNKGE by substituting K.

8.6 STABILITY ANALYSIS

This section of the chapter includes the estimation of upper bound of the error arises in the approximation which is necessary to ensure the convergence and efficiency of the method.

Theorem 8.2. *For the generalized Laguerre polynomials $P_\gamma^{(\tau)}(t)$ the global uniform bounds is estimated as*

$$| P_\gamma^{(\tau)}(t) | \leq \begin{cases} \frac{(\tau+1)_\gamma}{\gamma!} e^{\frac{t}{2}}, & \text{if } \tau \geq 0, t \geq 0, \gamma = 0,1,2,\ldots; \\ (2 - \frac{(\tau+1)_\gamma}{\gamma!}) e^{\frac{t}{2}}, & \text{if } -1 < \tau \leq 0, t \geq 0, \gamma = 0,1,2,\ldots \end{cases} \tag{8.31}$$

where $(\sigma)_\gamma := \sigma(\sigma+1)(\sigma+2)\ldots(\sigma+\gamma-1), \gamma = 1,2,3,\ldots$.

Proof. The authors of [39–41] have been proved and discussed these global uniform bounds.

Remark. For $\tau = 0$ the generalized Laguerre polynomials $P_\gamma^{(\tau)}(t)$ is reduced the usual Laguerre polynomials $P_\gamma(t)$.

The global uniform bounds (Eq. 8.31) for usual Laguerre polynomials is reduced to

$$| P_\gamma(t) | \leq \frac{1}{\gamma!} e^{\frac{t}{2}}, \quad t \geq 0, \quad \gamma = 0,1,2,\ldots \tag{8.32}$$

Theorem 8.3. *Consider a sufficiently smooth functions $u(x,t)$ on the region D. The error $M_r(N)$ arising in the approximation of $(\frac{\partial^\alpha u(x,t)}{\partial t^\alpha})$ by $(\frac{\partial^\alpha u(x,t)}{\partial t^\alpha})_N$ has the bound*

$$| M_r(N) | \leq \sum_{\delta=N+1}^{\infty} \sum_{\omega=N+1}^{\infty} \frac{c_{\delta\omega}}{\delta!.\gamma!} \chi_{\gamma\omega N} e^{\frac{x+t}{2}}, \quad x,t \geq 0, \quad \delta, \omega = 0,1,2,\ldots \tag{8.33}$$

where $\chi_{\gamma\omega N} = \sum_{\gamma=0}^{N} E_\alpha(\omega, \gamma)$.

Proof. Using Eq. (8.26), we can write

$$u(x,t) = \sum_{\delta=0}^{\infty} \sum_{\omega=0}^{\infty} c_{\delta\omega} P_\delta(x) P_\omega(t), \quad u_N(x,t) = \sum_{\delta=0}^{N} \sum_{\omega=0}^{N} c_{\delta\omega} P_\delta(x) P_\omega(t), \tag{8.34}$$

Operating α order fractional partial derivative with respect to t in the above equation, we have

$$\frac{\partial^\alpha u(x,t)}{\partial t^\alpha} = \sum_{\delta=0}^{\infty} \sum_{\omega=0}^{\infty} c_{\delta\omega} P_\delta(x) \frac{\partial^\alpha P_\omega(t)}{\partial t^\alpha}, \quad (\frac{\partial^\alpha u}{\partial t^\alpha})_N = \sum_{\delta=0}^{N} \sum_{\omega=0}^{N} c_{\delta\omega} P_\delta(x) \frac{\partial^\alpha P_\omega(t)}{\partial t^\alpha}. \tag{8.35}$$

or,

$$M_r(N) = \frac{\partial^\alpha u}{\partial t^\alpha} - (\frac{\partial^\alpha u}{\partial t^\alpha})_N = \sum_{\delta=N+1}^{\infty} \sum_{\omega=N+1}^{\infty} c_{\delta\omega} P_\delta(x) \frac{\partial^\alpha P_\omega(t)}{\partial t^\alpha} \tag{8.36}$$

Applying Eq. (8.23) in the above equation, we have

$$| M_r(N) |=| \sum_{\delta=N+1}^{\infty} \sum_{\omega=N+1}^{\infty} c_{\delta\omega} P_\delta(x) (\sum_{\gamma=0}^{N} E_\alpha(\omega,\gamma) P_\gamma(t)) | \tag{8.37}$$

or,

$$| M_r(N) |= \sum_{\delta=N+1}^{\infty} \sum_{\omega=N+1}^{\infty} c_{\delta\omega} \cdot \chi_{\gamma\omega N} | P_\gamma(t) | . | P_\delta(x) | \tag{8.38}$$

Using Eq. (8.32), we have

$$| M_r(N) |\leq \sum_{\delta=N+1}^{\infty} \sum_{\omega=N+1}^{\infty} c_{\delta\omega} \cdot \chi_{\gamma\omega N} \cdot \frac{1}{\gamma!} e^{\frac{t}{2}} . \frac{1}{\delta!} e^{\frac{x}{2}} \tag{8.39}$$

or,

$$| M_r(N) |\leq \sum_{\delta=N+1}^{\infty} \sum_{\omega=N+1}^{\infty} \frac{c_{\delta\omega}}{\delta!.\gamma!} \chi_{\gamma\omega N} e^{\frac{x+t}{2}}, \quad x,t \geq 0, \quad \delta,\omega = 0,1,2,\ldots \tag{8.40}$$

which is the required proof.

8.7 NUMERICAL EXAMINATION OF FNKGE

This section of this chapter consists of the experimental test of our proposed numerical scheme by testing some different forms of nonlinear Klein-Gordon equation by giving some particular value to unknown coefficients and functions in generalized FNKGE (8.1) and boundary conditions (8.2). The results obtained by the proposed numerical algorithm are compared by the exact solution and with numerical solution obtained by some existing methods.

Example 8.1 If we put $D = 1, \eta = 0, \lambda = -1, k = 2, \alpha = 2$ and $f(x,t) = 6xt(x^2 - t^2) + x^6t^6$ then the generalized FNKGE is reduced to the following form

$$\frac{\partial^2 u(x,t)}{\partial t^2} = \frac{\partial^2 u(x,t)}{\partial x^2} - u^2(x,t) + 6xt(x^2 - t^2) + x^6t^6, \quad 0 \leq x \leq 1, t \geq 0. \tag{8.41}$$

with suitable boundary conditions the exact solution of the above nonlinear Klein-Gordon equation is $u(x,t) = x^3t^3$. The root mean square error and the L_∞ error for the above example are calculated and compared with various existing method in Table 8.1.

The error analysis for various values of t of Example 8.1 in both root mean square and L_∞ errors are compared with existing methods through Table 8.1. This table clearly ensure that our proposed numerical scheme is more accurate and computationally authenticate than the method given in Refs. [42,43].

Table 8.1
Comparison of L_∞ and L_2 Errors for $u(x,t)$

t	Proposed Method		Zeytinoglu et al. [42]		Sarboland et al. [43]	
	L_2	L_∞	L_2	L_∞	L_2	L_∞
1	3.53×10^{-8}	4.01×10^{-8}	6.36×10^{-6}	1.50×10^{-5}	4.86×10^{-6}	7.80×10^{-6}
2	4.31×10^{-8}	4.89×10^{-8}	2.54×10^{-5}	6.00×10^{-5}	7.77×10^{-5}	1.23×10^{-4}
3	7.34×10^{-8}	3.21×10^{-7}	5.73×10^{-5}	1.35×10^{-4}	2.45×10^{-5}	5.30×10^{-4}
4	1.70×10^{-7}	1.86×10^{-7}	1.02×10^{-4}	2.40×10^{-4}	9.70×10^{-4}	1.86×10^{-3}
5	5.54×10^{-7}	2.43×10^{-6}	1.47×10^{-4}	3.50×10^{-4}	1.90×10^{-3}	3.52×10^{-3}

Example 8.2 Similarly, if we put $D=\frac{5}{2}, \eta=-1, \lambda=-\frac{3}{2}, k=3, \alpha=2$ and $f(x,t)=0$ then the generalized FNKGE (8.1) is reduced to the following form

$$\frac{\partial^2 u(x,t)}{\partial t^2}=\frac{5}{2}\frac{\partial^2 u(x,t)}{\partial x^2}-u(x,t)-\frac{3}{2}u^3(x,t),\ \ 0\leq x\leq 1, t\geq 0. \tag{8.42}$$

with suitable boundary conditions the exact solution of above nonlinear Klein-Gordon equation is $u(x,t)=\sqrt{\frac{2}{3}}tan(\sqrt{\frac{2}{9}}(x+\frac{t}{2}))$. The root mean square error and the L_∞ error for above example are calculated and compared with the various existing method in Table 8.2.

The error analysis for various value of t of Example 8.2 in both root mean square and L_∞ errors are compared with existing methods through the Table 8.2. This table clearly ensures that our proposed numerical scheme is more accurate and computationally authenticate than the method given in Refs. [44,45].

Table 8.2
Comparison of L_∞ and L_2 Errors for $u(x,t)$

t	Proposed Method		Yin et al. [44]		Dehghan et al. [45]	
	L_2	L_∞	L_2	L_∞	L_2	L_∞
1	6.21×10^{-7}	6.56×10^{-7}	3.56×10^{-6}	6.12×10^{-6}	4.06×10^{-6}	4.08×10^{-5}
2	7.23×10^{-7}	8.01×10^{-7}	1.35×10^{-5}	2.22×10^{-5}	1.57×10^{-5}	1.58×10^{-4}
3	9.87×10^{-7}	2.04×10^{-6}	5.42×10^{-5}	9.13×10^{-5}	6.45×10^{-5}	6.48×10^{-4}
4	3.75×10^{-6}	4.86×10^{-6}	4.48×10^{-4}	7.79×10^{-4}	5.33×10^{-3}	5.36×10^{-3}

8.8 DISCUSSION OF OUTCOMES

An efficient accuracy and validation of proposed scheme by some various form of Klein-Gordon equation, the author has been inspired to solve generalized fractional order Klein-Gordon equation (8.1) with boundary conditions (8.2) for different fractional order i.e. for different value of α and known constants. The mobility of field variable $u(x,t)$ at time $t = 0.5$ and $t = 1$ with the space variable x for the various value of α is numerically computed and demonstrated through the graphical presentation in Figures 8.1 and 8.2.

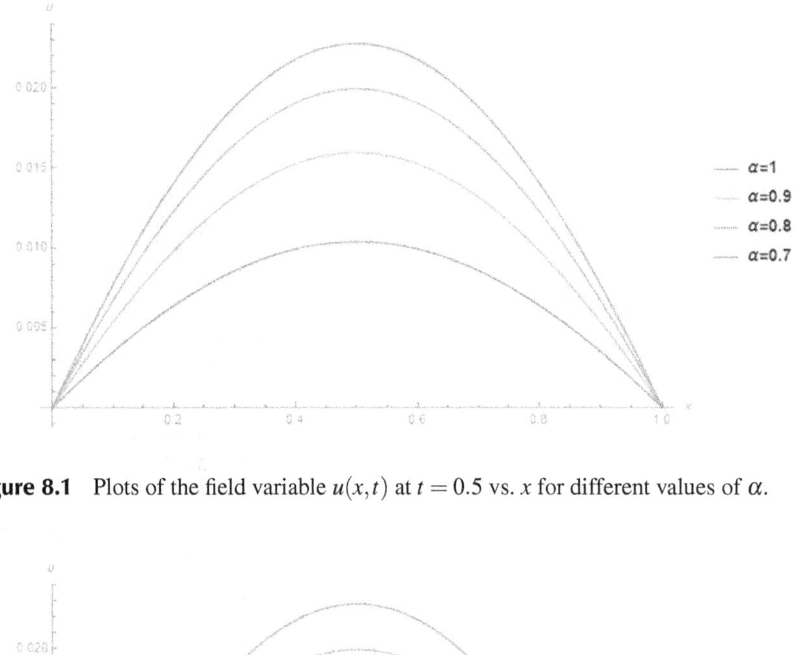

Figure 8.1 Plots of the field variable $u(x,t)$ at $t = 0.5$ vs. x for different values of α.

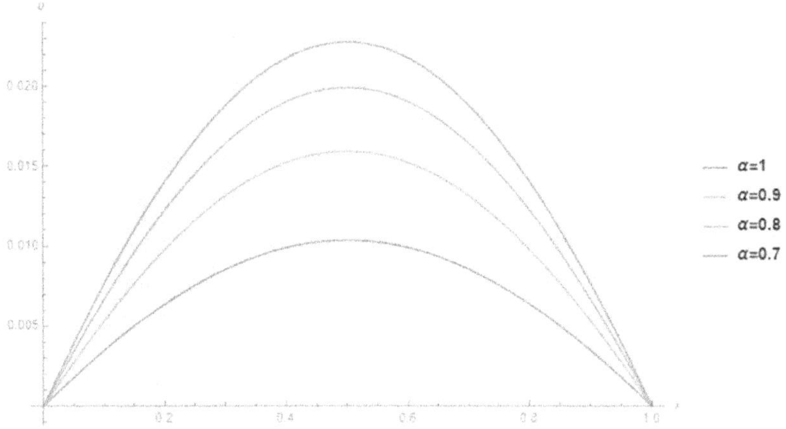

Figure 8.2 Plots of the field variable $u(x,t)$ at $t = 1$ vs. x for different values of α.

The variation of field variable for different values of α at $t = 0.5$ and $t = 1$ is shown in Figures 8.1 and 8.2, respectively. In both cases, it is clearly seen that as the proposed system moves toward integer order from fractional order the graph of solute concentration for different fractional systems decreases and advances toward the graph of integer systems.

8.9 APPLICATION OF MODEL

The Klein-Gordon equations are relativistic type of wave equations and are mainly related to the famous Schrödinger equations. The main objective of these types of relativistic wave equations is to analyze the dynamic behavior of particles with the existence of negative energy states with negative mass, with energies approaches to $-\infty$ implying the instability of the corresponding considered systems. Especially, the Klein-Gordon equations describe and analyze the spin-zero particles and models various phenomena such as the propagation of dislocations in crystals and the behavior of elementary particles of the system.

8.10 CONCLUSION

In this chapter, important aspects of nonlinear fractional Klein-Gordon equation have been considered. First one is to find the approximate analytical solution of FNKGE efficiently for different cases of fractional order at different time variable and stability of the unknown variable approximation. Second one is to introduce the efficient computational technique of operational matrix method based on orthogonal Laguerre polynomials. Last one is the pictorial demonstration of field variable $u(x,t)$ at different time and different fractional order. To more study of the Klein-Gordon equation our future task is to deal the multi dimensional variable order Klein-Gordon equation. Also in future authors are oriented to analyze and study of fractional Klein-Gordon equations with non-local and non-singular kernels. The discussion of stability analysis of solitary wave solutions for coupled and (2+ 1)-dimensional cubic Klein-Gordon equations and its various applications is also the main task of authors for future research.

References

1. Prashant Pandey, Sachin Kumar, José Francisco Gómez-Aguilar, and Dumitru Baleanu. An efficient technique for solving the space-time fractional reaction-diffusion equation in porous media. *Chinese Journal of Physics*, 68:483–492, 2020.
2. Prashant Pandey, Sachin Kumar, and Subir Das. Approximate analytical solution of coupled fractional order reaction-advection-diffusion equations. *The European Physical Journal Plus*, 134(7):364, 2019.
3. Xia Tian, Sergiy Yu Reutskiy, and Zhuo-Jia Fu. A novel meshless collocation solver for solving multi-term variable-order time fractional pdes. *Engineering with Computers*, 38(2):1527–1538, 2022.

4. Arezou Rezazadeh and Zakieh Avazzadeh. Numerical approach for solving two dimensional fractal-fractional pdes using peridynamic method. *International Journal of Computer Mathematics*, 99(3):486–505, 2022.
5. Hussam Aljarrah, Mohammad Alaroud, Anuar Ishak, and Maslina Darus. Approximate solution of nonlinear time-fractional pdes by laplace residual power series method. *Mathematics*, 10(12):1980, 2022.
6. Kolade M Owolabi and Abdon Atangana. Mathematical analysis and computational experiments for an epidemic system with nonlocal and nonsingular derivative. *Chaos, Solitons & Fractals*, 126:41–49, 2019.
7. Lanre Akinyemi, Kottakkaran Sooppy Nisar, Ahamed Saleel, Hadi Rezazadeh, Pundikala Veeresha, Mostafa MA Khater, and Mustafa Inc. Novel approach to the analysis of fifth-order weakly nonlocal fractional schrödinger equation with caputo derivative. *Results in Physics*, 31:104958, 2021.
8. Siwei Duo, Hong Wang, and Yanzhi Zhang. A comparative study on nonlocal diffusion operators related to the fractional laplacian. arXiv preprint arXiv:1711.06916, 2017.
9. Harendra Singh, Hari Mohan Srivastava, and Juan J Nieto. *Handbook of Fractional Calculus for Engineering and Science*. CRC Press: Boca Raton, FL, 2022.
10. Harendra Singh, Devendra Kumar, and Dumitru Baleanu. *Methods of Mathematical Modelling: Fractional Differential Equations*. CRC Press: Boca Raton, FL, 2019.
11. Harendra Singh, Hari Mohan Srivastava, and Dumitru Baleanu. *Methods of Mathematical Modeling Infectious Diseases*. Elsevier Science: Amsterdam, Netherlands, 2022.
12. Santanu Saha Ray and B Sagar. Numerical soliton solutions of fractional modified (2+1)-dimensional konopelchenko–dubrovsky equations in plasma physics. *Journal of Computational and Nonlinear Dynamics*, 17(1):011007, 2022.
13. Mahmoud AE Abdelrahman, Samia Zaki Hassan, Reem Alomair, and DM Alsaleh. The new wave structures to the fractional ion sound and langmuir waves equation in plasma physics. *Fractal and Fractional*, 6(5):227, 2022.
14. Prashant Pandey, José Francisco Gómez-Aguilar, Mohammed KA Kaabar, Zailan Siri, and Abd Allah Mousa. Mathematical modeling of covid-19 pandemic in india using caputo-fabrizio fractional derivative. *Computers in Biology and Medicine*, 145:105518, 2022.
15. Mati ur Rahman, Muhammad Arfan, Zahir Shah, and Ebraheem Alzahrani. Evolution of fractional mathematical model for drinking under atangana-baleanu caputo derivatives. *Physica Scripta*, 96(11):115203, 2021.
16. Parisa Rahimkhani, Yadollah Ordokhani, and Esmail Babolian. A new operational matrix based on bernoulli wavelets for solving fractional delay differential equations. *Numerical Algorithms*, 74(1):223–245, 2017.
17. Prashant Pandey, Subir Das, Eduard-Marius Craciun, and Tomasz Sadowski. Two-dimensional nonlinear time fractional reaction–diffusion equation in application to sub-diffusion process of the multicomponent fluid in porous media. *Meccanica*, 56(1):99–115, 2021.
18. A Mahmoud, Ibrahem G Ameen, and Abdelkawy Mohamed. A new operational matrix based on jacobi wavelets for a class of variable-order fractional differential equations. *Proceedings of the Romanian Academy Series A*, 18(4):315–322, 2017.
19. Shyam Lal and Priya Kumari. Approximation of functions with bounded derivative and solution of riccati differential equations by jacobi wavelet operational matrix. *Applied Mathematics and Computation*, 394:125834, 2021.

20. José Francisco Gómez-Aguilar and Abdon Atangana. Time-fractional variable-order telegraph equation involving operators with mittag-leffler kernel. *Journal of Electromagnetic Waves and Applications*, 33(2):165–177, 2019.
21. Khaled M Saad and José Francisco Gómez-Aguilar. Analysis of reaction–diffusion system via a new fractional derivative with non-singular kernel. *Physica A: Statistical Mechanics and Its Applications*, 509:703–716, 2018.
22. Prashant Pandey and Jagdev Singh. An efficient computational approach for nonlinear variable order fuzzy fractional partial differential equations. *Computational and Applied Mathematics*, 41(1):1–21, 2022.
23. Prashant Pandey and José Francisco Gómez-Aguilar. On solution of a class of nonlinear variable order fractional reaction–diffusion equation with mittag–leffler kernel. *Numerical Methods for Partial Differential Equations*, 37(2):998–1011, 2021.
24. Ndolane Sene and Abdon Atangana. Integral-balance methods for the fractional diffusion equation described by the caputo-generalized fractional derivative. In: Harendra Singh, Devendra Kumar, and Dumitru Baleanu (Eds.), *Methods of Mathematical Modelling: Fractional Differential Equations*, pp. 83. CRC Press: Boca Raton, FL, 2019.
25. Harendra Singh and Abdul-Majid Wazwaz. Computational method for reaction diffusion-model arising in a spherical catalyst. *International Journal of Applied and Computational Mathematics*, 7(3):1–11, 2021.
26. Vivek Mani Tripathi, Hari Mohan Srivastava, Harendra Singh, Chetan Swarup, and Sudhanshu Aggarwal. Mathematical analysis of non-isothermal reaction–diffusion models arising in spherical catalyst and spherical biocatalyst. *Applied Sciences*, 11(21):10423, 2021.
27. Harendra Singh. Analysis for fractional dynamics of ebola virus model. *Chaos, Solitons & Fractals*, 138:109992, 2020.
28. Harendra Singh. Solving a class of local and nonlocal elliptic boundary value problems arising in heat transfer. *Heat Transfer*, 51(2):1524–1542, 2022.
29. Harendra Singh. An efficient computational method for non-linear fractional lienard equation arising in oscillating circuits. In: Harendra Singh, Devendra Kumar, and Dumitru Baleanu (Eds.), *Methods of Mathematical Modelling*, pp. 39–50. CRC Press: Boca Raton, FL, 2019.
30. Harendra Singh and Hari Mohan Srivastava. Numerical investigation of the fractional-order liénard and duffing equations arising in oscillating circuit theory. *Frontiers in Physics*, 8:120, 2020.
31. Harendra Singh. Chebyshev spectral method for solving a class of local and nonlocal elliptic boundary value problems. *International Journal of Nonlinear Sciences and Numerical Simulation*, 000010151520200235, 2021.
32. Salah M El-Sayed. The decomposition method for studying the klein–gordon equation. *Chaos, Solitons & Fractals*, 18(5):1025–1030, 2003.
33. Abdul-Majid Wazwaz. Compactons, solitons and periodic solutions for some forms of nonlinear klein–gordon equations. *Chaos, Solitons & Fractals*, 28(4):1005–1013, 2006.
34. Yıldıray Keskin, Sema Servi, and Galip Oturanc. Reduced differential transform method for solving klein gordon equations. In *Proceedings of the World Congress on Engineering (WCE)*, London, 2011.
35. Mohammad Golam Hafez, Md Nur Alam, and Mohammad Ali Akbar. Exact traveling wave solutions to the klein–gordon equation using the novel (G'/G)-expansion method. *Results in Physics*, 4:177–184, 2014.

36. Igor Podlubny. *Fractional Differential Equations: An Introduction to Fractional Derivatives, Fractional Differential Equations, to Methods of Their Solution and Some of Their Applications*, volume 198. Elsevier: Amsterdam, Netherlands, 1998.
37. Huitzilin Yépez-Martínez and José Francisco Gómez-Aguilar. A new modified definition of caputo–fabrizio fractional-order derivative and their applications to the multi step homotopy analysis method (mham). *Journal of Computational and Applied Mathematics*, 346:247–260, 2019.
38. Anatoly A Kilbas, Hari Mohan Srivastava, and Juan J Trujillo. *Theory and applications of fractional differential equations*. Elsevier: Amsterdam, 2006.
39. Milton Abramowitz and Irene Ann Stegun. *Handbook of Mathematical Functions*. Dover: New York, 1964.
40. Mohamed Meabed Khader. The use of generalized laguerre polynomials in spectral methods for solving fractional delay differential equations. *Journal of Computational and Nonlinear Dynamics*, 8(4):041018, 2013.
41. MM Khader, Talaat S El Danaf, and Ahmed S Hendy. Efficient spectral collocation method for solving multi-term fractional differential equations based on the generalized laguerre polynomials. *Fractional Calculus and Applied Analysis*, 3(13):1–14, 2012.
42. Asuman Zeytinoglu and Murat Sari. Some difference algorithms for nonlinear klein-gordon equations. In *International Conference on Numerical Methods and Applications*, Borovets, Bulgaria, pp. 491–498. Springer, 2018.
43. M Sarboland and Azim Aminataei. Numerical solution of the nonlinear klein–gordon equation using multiquadric quasi-interpolation scheme. *Universal Journal of Applied Mathematics*, 3(3):40–49, 2015.
44. Fukang Yin, Tian Tian, Junqiang Song, and Min Zhu. Spectral methods using legendre wavelets for nonlinear klein\sine-gordon equations. *Journal of Computational and Applied Mathematics*, 275:321–334, 2015.
45. Mehdi Dehghan and Ali Shokri. Numerical solution of the nonlinear klein–gordon equation using radial basis functions. *Journal of Computational and Applied Mathematics*, 230(2):400–410, 2009.

9 New Fractional Calculus Results for the Families of Extended Hurwitz-Lerch Zeta Function

Rakesh K. Parmar
Pondicherry University

Arjun K. Rathie and S. D. Purohit
Rajasthan Technical University

CONTENTS

9.1 INTRODUCTION AND PRELIMINARIES

Fractional Calculus goes back to the beginning of the theory of differential calculus. During the second half of the twentieth century, considerable amount of research in fractional calculus was published in science and engineering literature. It has been recognized the advantageous use of this mathematical tool in the modeling and control of many dynamical systems. There is no doubt that fractional calculus has become an exciting recent mathematical tool of the solution of diverse problems in mathematics, science, and engineering [29–37,48]. This chapter is devoted to the study of novel and current applications of fractional calculus in science and engineering in order to generate more interest in the topic and demonstrate its applicability.

DOI: 10.1201/9781003368069-9

In 1978, Saigo [22] defined a pair of fractional integral and differential operators involving the Gauss hypergeometric function ${}_2F_1$ as kernel and are defined for $\varsigma,\ \varpi,\ \eta \in \mathbb{C}$ $(\Re(\varsigma) > 0)$ and $x > 0$ (see, *e.g.*, [1,17,19,45]):

$$\left(\mathscr{I}_{0+}^{\varsigma,\varpi,\eta} f\right)(x) = \frac{x^{-\varsigma-\varpi}}{\Gamma(\varsigma)} \int_0^x (x-t)^{\varsigma-1}\, {}_2F_1\left[\begin{matrix}\varsigma+\varpi, -\eta \\ \varsigma\end{matrix} \middle| 1-\frac{t}{x}\right] f(t)\, \mathrm{d}t, \tag{9.1}$$

$$\left(\mathscr{I}_{-}^{\varsigma,\varpi,\eta} f\right)(x) = \frac{1}{\Gamma(\varsigma)} \int_x^\infty (t-x)^{\varsigma-1} t^{-\varsigma-\varpi}\, {}_2F_1\left[\begin{matrix}\varsigma+\varpi, -\eta \\ \varsigma\end{matrix} \middle| 1-\frac{x}{t}\right] f(t)\, \mathrm{d}t \tag{9.2}$$

and

$$\begin{aligned}\left(\mathscr{D}_{0+}^{\varsigma,\varpi,\eta} f\right)(x) &= \left(\mathscr{I}_{0+}^{-\varsigma,-\varpi,\varsigma+\eta} f\right)(x) \\ &= \left(\frac{d}{dx}\right)^n \left(\mathscr{I}_{0+}^{-\varsigma+n,-\varpi-n,\varsigma+\eta-n} f\right)(x) \quad (n = [\Re(\varsigma)]+1),\end{aligned} \tag{9.3}$$

$$\begin{aligned}\left(\mathscr{D}_{-}^{\varsigma,\varpi,\eta} f\right)(x) &= \left(\mathscr{I}_{-}^{-\varsigma,-\varpi,\varsigma+\eta} f\right)(x) \\ &= \left(-\frac{d}{dx}\right)^n \left(\mathscr{I}_{-}^{-\varsigma+n,-\varpi-n,\varsigma+\eta} f\right)(x) \quad (n = [\Re(\varsigma)]+1),\end{aligned} \tag{9.4}$$

respectively. On putting $\varpi = -\varsigma$, Eqs. (9.1)–(9.4) reduces to the well-known Riemann-Liouville(R-L) fractional integration and derivatives of order $\varsigma \in \mathbb{C}$ $(\Re(\varsigma) > 0)$ and $x > 0$ (see, for example, [6,7,17,19]):

$$\left(\mathscr{I}_{0+}^{\varsigma,-\varsigma,\eta} f\right)(x) = \left(\mathscr{I}_{0+}^{\varsigma} f\right)(x) = \frac{1}{\Gamma(\varsigma)} \int_0^x \frac{f(t)}{(x-t)^{1-\varsigma}}\, \mathrm{d}t, \tag{9.5}$$

$$\left(\mathscr{I}_{-}^{\varsigma,-\varsigma,\eta} f\right)(x) = \left(\mathscr{I}_{-}^{\varsigma} f\right)(x) = \frac{1}{\Gamma(\varsigma)} \int_x^\infty \frac{f(t)}{(t-x)^{1-\varsigma}}\, \mathrm{d}t \tag{9.6}$$

and

$$\begin{aligned}\left(\mathscr{D}_{0+}^{\varsigma,-\varsigma,\eta} f\right)(x) &= \left(\mathscr{D}_{0+}^{\varsigma} f\right)(x) \\ &= \left(\frac{d}{dx}\right)^n \left(\mathscr{I}_{0+}^{n-\varsigma} f\right)(x) \qquad (n = [\Re(\varsigma)]+1),\end{aligned} \tag{9.7}$$

$$\begin{aligned}\left(\mathscr{D}_{-}^{\varsigma,-\varsigma,\eta} f\right)(x) &= \left(\mathscr{D}_{-}^{\varsigma} f\right)(x) \\ &= \left(-\frac{d}{dx}\right)^n \left(\mathscr{I}_{-}^{n-\varsigma} f\right)(x) \qquad (n = [\Re(\varsigma)]+1),\end{aligned} \tag{9.8}$$

respectively, where $[\Re(\varsigma)]$ is the integral part of $\Re(\varsigma)$.

If $\varpi = 0$, Eqs. (9.1)–(9.4) reduce to the so-called Erdélyi-Kober fractional integrals and derivatives defined for $\varsigma \in \mathbb{C}$ ($\Re(\varsigma) > 0$) and $x > 0$ (see, *e.g.*, [6,7,17,19]):

$$\left(\mathscr{I}_{0+}^{\varsigma,0,\eta} f\right)(x) = \left(\mathscr{I}_{\eta,\varsigma}^{+} f\right)(x) = \frac{x^{-\varsigma-\eta}}{\Gamma(\varsigma)} \int_0^x \frac{t^{\eta} f(t)}{(x-t)^{1-\varsigma}}\, \mathrm{d}t, \tag{9.9}$$

$$\left(\mathscr{I}_{-}^{\varsigma,0,\eta} f\right)(x) = \left(K_{\eta,\varsigma}^{-} f\right)(x) = \frac{x^{\eta}}{\Gamma(\varsigma)} \int_x^{\infty} \frac{t^{-\varsigma-\eta} f(t)}{(t-x)^{1-\varsigma}}\, \mathrm{d}t \tag{9.10}$$

and

$$\begin{aligned}\left(\mathscr{D}_{0+}^{\varsigma,0,\eta} f\right)(x) &= \left(\mathscr{D}_{\eta,\varsigma}^{+} f\right)(x)\\ &= \left(\frac{d}{dx}\right)^{n} \left(\mathscr{I}_{0+}^{-\varsigma+n,-\varsigma,\varsigma+\eta-n} f\right)(x) \quad (n = [\Re(\varsigma)] + 1),\end{aligned} \tag{9.11}$$

$$\begin{aligned}\left(\mathscr{D}_{-}^{\varsigma,0,\eta} f\right)(x) &= \left(\mathscr{D}_{\eta,\varsigma}^{-} f\right)(x)\\ &= \left(-\frac{d}{dx}\right)^{n} \left(\mathscr{I}_{-}^{-\varsigma+n,-\varsigma,\varsigma+\eta} f\right)(x) \quad (n = [\Re(\varsigma)] + 1),\end{aligned} \tag{9.12}$$

respectively.

In addition to this, the *Hurwitz-Lerch zeta function* $\Phi(z,s,a)$ is defined by (see, *e.g.*, [10, p. 27, Eq. 1.11(1)]; see also [39, p. 121] and [40, p. 194]):

$$\Phi(z,s,a) := \sum_{n=0}^{\infty} \frac{z^n}{(n+a)^s} \tag{9.13}$$

$$\left(a \in \mathbb{C}\setminus\mathbb{Z}_0^-;\ s \in \mathbb{C} \text{ when } |z| < 1;\ \Re(s) > 1 \text{ when } |z| = 1\right).$$

Various properties and the special cases of the Hurwitz-Lerch zeta function $\Phi(z,s,a)$, one may refer to the standard texts ([39], [40]). Various generalizations of the Hurwitz-Lerch Zeta function $\Phi(z,s,a)$ have been investigated by many authors (see, *e.g.*, [2,4,5,9–12,15,16,20,24–26,38,41,43,46]). In particular, Srivastava et al. [46, p. 491, Eq. (1.20)] introduced and investigated the extended Hurwitz-Lerch zeta function as follows:

$$\Phi_{\lambda,\mu;\nu}^{(\rho,\tau;\kappa)}(z,s,a) := \sum_{n=0}^{\infty} \frac{(\lambda)_{\rho n}(\mu)_{\tau n}}{(\nu)_{\kappa n}\, n!} \frac{z^n}{(n+a)^s} \tag{9.14}$$

$$\begin{gathered}\big(\lambda,\mu \in \mathbb{C};\ a,\ \nu \in \mathbb{C}\setminus\mathbb{Z}_0^-;\ \rho,\ \tau,\ \kappa \in \mathbb{R}^+;\ \kappa-\rho-\tau > -1 \text{ when } s,\ z \in \mathbb{C};\\ \kappa-\rho-\tau = -1 \text{ and } s \in \mathbb{C} \text{ when } |z| < \delta^* := \rho^{-\rho}\tau^{-\tau}\kappa^{\kappa};\\ \kappa-\rho-\tau = -1 \text{ and } \Re(s+\nu-\lambda-\mu) > 1 \text{ when } |z| = \delta^*\big).\end{gathered}$$

In the same paper, they systematically studied various integrals and computational representations of several families of generalized Hurwitz-Lerch zeta functions, Mellin-Barnes integral representations of almost all generalized and specialized

Hurwitz-Lerch zeta functions. They also derived another analytic continuation formula, which is an elegant extension of the well-known analytic continuation formula for Gaussian hypergeometric functions. Finally, they have examined various special or boundary cases of the extended Hurwitz-Lerch zeta function (9.14)$\Phi_{\lambda,\mu;\nu}^{(\rho,\sigma,\kappa)}(z,s,a)$ presented in [46, p. 491–492]. However, in our present investigations, we draw attention to some known results for particular choices of parameters given below

Case 1. For $\rho=\tau=\kappa=1$ in Eq. (9.14), yield another known generalized Hurwitz-Lerch Zeta function introduced by Garg et al. [11, p. 313, Eq. (1.7)]:

$$\Phi_{\lambda,\mu;\nu}^{(1,1,1)}(z,s,a):=\Phi_{\lambda,\mu;\nu}(z,s,a):=\sum_{n=0}^{\infty}\frac{(\lambda)_n(\mu)_n}{(\nu)_n n!}\frac{z^n}{(n+a)^s} \tag{9.15}$$

$$\left(\lambda,\mu\in\mathbb{C};\ \nu,a\in\mathbb{C}\setminus\mathbb{Z}_0^-;\ s\in\mathbb{C}\text{ when }|z|<1;\ \Re(s+\nu-\lambda-\mu)>1\text{ when }|z|=1\right).$$

Case 2. For $\lambda=1=\rho$ in Eq. (9.14), yields generalized Hurwitz-Lerch Zeta function studied earlier by Lin and Srivastava [16, p. 727, Eq. (8)]:

$$\Phi_{\mu;\nu}^{(\tau,\kappa)}(z,s,a):=\Phi_{1,\mu;\nu}^{(1,\tau,\kappa)}(z,s,a):=\sum_{n=0}^{\infty}\frac{(\mu)_{\tau n}}{(\nu)_{\kappa n}}\frac{z^n}{(n+a)^s} \tag{9.16}$$

$$\left(\mu\in\mathbb{C};\ \nu,a\in\mathbb{C}\setminus\mathbb{Z}_0^-;\ s\in\mathbb{C}\text{ when }|z|<1;\ \Re(s+\nu-\mu)>1\text{ when }|z|=1\right).$$

Case 3. For $\rho=\tau=\kappa=1$ and $\lambda=\nu$ in Eq. (9.14), reduces to the known definition of generalized Hurwitz-Lerch Zeta function defined by Goyal and Laddha [12, p. 100, Eq. (1.5)]:

$$\Phi_{\nu,\mu;\nu}^{(1,1,1)}(z,s,a):=\Phi_{\mu}^{*}(z,s,a):=\sum_{n=0}^{\infty}\frac{(\mu)_n}{n!}\frac{z^n}{(n+a)^s} \tag{9.17}$$

$$\left(\mu\in\mathbb{C};\ a\in\mathbb{C}\setminus\mathbb{Z}_0^-;\ s\in\mathbb{C}\text{ when }|z|<1;\Re(s-\mu)>1\text{ when }|z|=1\right)$$

Case 4. The following two limiting cases have also been discussed in Eq. (9.14) [46, p. 492, Eq. (1.25) and (1.26)]:

$$\Phi_{\mu;\nu}^{*(\tau,\kappa)}(z,s,a):=\lim_{|\lambda|\to\infty}\Phi_{\lambda,\mu;\nu}^{(\rho,\sigma,\kappa)}\left(\frac{z}{\lambda^{\rho}},s,a\right):=\sum_{n=0}^{\infty}\frac{(\mu)_{\tau n}}{(\nu)_{\kappa n}n!}\frac{z^n}{(n+a)^s} \tag{9.18}$$

$$\left(\mu\in\mathbb{C};\ \nu,a\in\mathbb{C}\setminus\mathbb{Z}_0^-;\tau,\ \kappa\in\mathbb{R}^+;\ s\in\mathbb{C}\text{ when }|z|<\tau^{-\tau}\kappa^{-\kappa};\ \Re(s+\nu-\mu)>1\right.$$
$$\left.\text{when }|z|=\tau^{-\tau}\kappa^{-\kappa}\right)$$

and

$$\Phi_{\mu}^{*(\tau)}(z,s,a):=\lim_{\min|\lambda|,|\nu|\to\infty}\left\{\Phi_{\lambda,\mu;\nu}^{(\rho,\tau,\kappa)}\left(\frac{z\nu^{\kappa}}{\lambda^{\rho}},s,a\right)\right\}:=\sum_{n=0}^{\infty}\frac{(\mu)_{\tau n}}{n!}\frac{z^n}{(n+a)^s} \tag{9.19}$$

$\left(\mu \in \mathbb{C}; a \in \mathbb{C} \setminus \mathbb{Z}_0^-; 0 < \tau < 1 \text{ and } s, z \in \mathbb{C} \text{ when } |z| < \tau^{-\tau}; \tau = 1 \text{ and } \Re(s-\mu) > 1\right.$
$\left.\text{when } |z| = \tau^{-\tau}\right)$.

Finally the Mellin-Barnes-type contour integral representation for extended Hurwitz-Lerch Zeta function $\Phi_{\lambda,\mu;\nu}^{(\rho,\tau;\kappa)}(t,s,a)$ has been obtained by Srivastava et al. [46, p. 498, Eq. (3.4)] and is defined as follows:

$$\begin{aligned}\Phi_{\lambda,\mu;\nu}^{(\rho,\tau,\kappa)}&(z,s,a)\\&=\frac{\Gamma(\nu)}{\Gamma(\lambda)\Gamma(\mu)}\frac{1}{2\pi i}\int_{-i\infty}^{+i\infty}\frac{\Gamma(-\xi)\Gamma(\lambda+\rho\xi)\Gamma(\mu+\tau\xi)\{\Gamma(\xi+a)\}^s}{\Gamma(\nu+\kappa\xi)\{\Gamma(\xi+a+1)\}^s}(-z)^\xi\,\mathrm{d}\xi\,,\\&(|\arg(-z)|<\pi,\ \min\{\Re(a),\Re(s),\Re(\nu)\}>0).\end{aligned}\tag{9.20}$$

We conclude this section by mentioning that, our aim is to study the compositions formulas of the generalized fractional integration and differentiation operators (9.1)–(9.4) with the extended Hurwitz-Lerch zeta function (9.14) in terms of the I–function introduced by Rathie [28]. Since the extended Hurwitz-Lerch zeta function and generalized Saigo's hypergeometric fractional calculus operators are of very general nature, therefore on specializing the parameters, a large number of special cases involving Riemann-Liouville and Erdélyi-Kober fractional integral and differential operators and Hurwitz-Lerch zeta function have been expressed in terms of I–function and $\overline{H}$–function, as our main findings.

9.2 FRACTIONAL INTEGRAL OPERATORS OF THE $\Phi_{\lambda,\mu;\nu}^{(\rho,\tau;\kappa)}(t,s,a)$

We begin by recalling the definition of the generalized hypergeometric function, that is I–function involving fractional powers of gamma functions introduced by Rathie [28] in the following Mellin-Barnes-type contour integral:

$$\begin{aligned}I_{p,q}^{m,n}(z)&=I_{p,q}^{m,n}\left[z\left|\begin{matrix}(a_j,A_j,\varsigma_j)_{\overline{1,p}}\\(b_j,B_j,\varpi_j)_{\overline{1,q}}\end{matrix}\right.\right]=I_{p,q}^{m,n}\left[z\left|\begin{matrix}(a_1,A_1,\varsigma_1),\ldots,(a_p,A_p,\varsigma_p)\\(b_1,B_1,\varpi_1),\ldots,(b_q,B_q,\varpi_q)\end{matrix}\right.\right]\\&=\frac{1}{2\pi i}\int_{\pounds}\phi_{p,q}^{m,n}(s)\,z^s\,ds,\end{aligned}\tag{9.21}$$

for all $z \neq 0$, where

$$\phi_{p,q}^{m,n}(s)=\frac{\prod\limits_{j=1}^{m}\{\Gamma(b_j-B_js)\}^{\varpi_j}\prod\limits_{j=1}^{n}\{\Gamma(1-a_j+A_js)\}^{\varsigma_j}}{\prod\limits_{j=m+1}^{q}\{\Gamma(1-b_j+B_js)\}^{\varpi_j}\prod\limits_{j=n+1}^{p}\{\Gamma(a_j-A_js)\}^{\varsigma_j}},\tag{9.22}$$

with $a_j \in \mathbb{C}\ (j=1,\ldots,p), b_j \in \mathbb{C}\ (j=1,\ldots,q),\ A_j \in \mathbb{R}^+\ (j=1,\ldots,p)$ and $B_j \in \mathbb{R}^+\ (j=1,\ldots,q)$. and the exponents $\varsigma_j\ (j=1,\ldots,n)$ and $\varpi_j\ (j=m+1,\ldots,q)$ can take non-integer values. $\pounds$ is a suitable contour of the Mellin-Barnes type separating the poles of $\{\Gamma(b_j-B_js)\}^{\varpi_j}\ (j=1,\ldots,m)$ from those of $\{\Gamma(1-a_j+A_js)\}^{\varsigma_j}$

$(j = 1, \ldots, n)$ with usual indentations. An empty product is interpreted as 1, the integers m, n, p, q satisfy the inequalities $0 \leq m \leq q$ and $0 \leq n \leq p$. Also, the Mellin-Barnes contour integral representing in the I-function converges absolutely and defines an analytic function for $|\arg(z)| < \frac{\pi}{2}\triangle$, where

$$\triangle = \sum_{j=1}^{m} |\varpi_j| B_j + \sum_{j=1}^{n} |\varsigma_j| A_j - \sum_{j=m+1}^{q} |\varpi_j| B_j - \sum_{j=n+1}^{p} |\varsigma_j| A_j > 0.$$

However, we use the following power function formulae in our investigation:

Lemma 9.1. *[21] Let $\varsigma, \varpi, \eta \in \mathbb{C}$. Then there exist the power function formulas*

(a) *If $\Re(\varsigma) > 0$ and $\Re(\sigma) > \max[0, \Re(\varpi - \eta)]$, then for $x > 0$, we have*

$$(\mathscr{I}_{0+}^{\varsigma,\varpi,\eta} t^{\sigma-1})(x) = \frac{\Gamma(\sigma)\Gamma(\sigma+\eta-\varpi)}{\Gamma(\sigma-\varpi)\Gamma(\sigma+\varsigma+\eta)} x^{\sigma-\varpi-1}. \tag{9.23}$$

In particular for $\varpi = -\varsigma$ and $\varpi = 0$, we get

$$(\mathscr{I}_{0+}^{\varsigma} t^{\sigma-1})(x) = \frac{\Gamma(\sigma)}{\Gamma(\sigma+\varsigma)} x^{\sigma+\varsigma-1} \qquad (\Re(\varsigma) > 0,\ \Re(\sigma) > 0) \tag{9.24}$$

and

$$(\mathscr{I}_{\eta,\varsigma}^{+} t^{\sigma-1})(x) = \frac{\Gamma(\sigma+\eta)}{\Gamma(\sigma+\varsigma+\eta)} x^{\sigma-1} \qquad (\Re(\varsigma) > 0,\ \Re(\sigma) > -\Re(\eta)). \tag{9.25}$$

(b) *If $\Re(\varsigma) > 0$ and $\Re(\sigma) < 1 + \min[\Re(\varpi), \Re(\eta)]$, then for $x > 0$, we have*

$$(\mathscr{I}_{-}^{\varsigma,\varpi,\eta} t^{\sigma-1})(x) = \frac{\Gamma(1-\sigma+\varpi)\Gamma(1-\sigma+\eta)}{\Gamma(1-\sigma)\Gamma(1-\sigma+\varsigma+\varpi+\eta)} x^{\sigma-\varpi-1}. \tag{9.26}$$

In particular for $\varpi = -\varsigma$ and $\varpi = 0$, we get

$$(\mathscr{I}_{-}^{\varsigma} t^{\sigma-1})(x) = \frac{\Gamma(1-\varsigma-\sigma)}{\Gamma(1-\sigma)} x^{\sigma+\varsigma-1} \qquad (0 < \Re(\varsigma) < 1 - \Re(\sigma)) \tag{9.27}$$

and

$$(\mathscr{K}_{\eta,\varsigma}^{-} t^{\sigma-1})(x) = \frac{\Gamma(1-\sigma+\eta)}{\Gamma(1-\sigma+\varsigma+\eta)} x^{\sigma-1} \qquad (\Re(\sigma) < 1 + \Re(\sigma)). \tag{9.28}$$

We begin the main results exposition by presenting the composition formulas of generalized fractional integrals (9.1) and (9.2) involving the extended Hurwitz-Lerch Zeta function in terms of the I–function (9.21).

Theorem 9.1. *Let $\varsigma, \varpi, \eta, \sigma, \lambda, \mu, s \in \mathbb{C}$ and $a, \nu \in \mathbb{C} \setminus \mathbb{Z}_0^-$ with $\rho, \rho, \tau, \kappa \in \mathbb{R}^+$, such that $\Re(\varsigma) > 0$ and $\Re(\sigma) > \max[0, \Re(\varpi - \eta)]$. Then for $\Re(a) > 0$ and $\Re(s) > 0$, following Saigo's fractional integral $\mathscr{I}_{0+}^{\varsigma,\varpi,\eta}$ of $\Phi_{\lambda,\mu;\nu}^{(\rho,\tau;\kappa)}(t,s,a)$ holds true:*

$$\left(\mathscr{I}_{0+}^{\varsigma,\varpi,\eta}\left\{t^{\sigma-1}\Phi_{\lambda,\mu;\nu}^{(\rho,\tau;\kappa)}(t^{\rho},s,a)\right\}\right)(x) = x^{\sigma-\varpi-1}\frac{\Gamma(\nu)}{\Gamma(\lambda)\Gamma(\mu)}$$
$$\times I_{5,5}^{1,5}\left[-x^{\rho}\left|\begin{matrix}(1-\lambda,\rho;1),(1-\mu,\tau;1),(1-a,1;s),(1-\sigma,\rho;1),(1-\sigma+\varpi-\eta,\rho;1)\\(0,1;1),(1-\nu,\kappa;1),(-a,1;s),(1-\sigma+\varpi,\rho;1),(1-\sigma-\varsigma-\eta,\rho;1)\end{matrix}\right.\right]. \tag{9.29}$$

Proof. Using the definitions (9.20) and (9.1), by changing the order of integration and applying the relation (9.23)

$$\begin{aligned}
&\left(\mathscr{I}_{0+}^{\varsigma,\varpi,\eta}\left\{t^{\sigma-1}\Phi_{\lambda,\mu;\nu}^{(\rho,\tau;\kappa)}(t^{\rho},s,a)\right\}\right)(x)\\
&=\frac{\Gamma(\nu)}{\Gamma(\lambda)\Gamma(\mu)}\frac{1}{2\pi i}\int_{\mathscr{L}}\frac{\Gamma(-\xi)\Gamma(\lambda+\rho\xi)\Gamma(\mu+\tau\xi)\{\Gamma(\xi+a)\}^{s}(-1)^{\xi}}{\Gamma(\nu+\kappa\xi)\{\Gamma(\xi+a+1)\}^{s}}\\
&\quad\left(\mathscr{I}_{0+}^{\varsigma,\varpi,\eta}t^{\sigma+\rho\xi-1}\right)(x)\,\mathrm{d}\xi\\
&=x^{\sigma-\varpi-1}\frac{\Gamma(\nu)}{\Gamma(\lambda)\Gamma(\mu)}\frac{1}{2\pi i}\int_{\mathscr{L}}\frac{\Gamma(-\xi)\Gamma(\lambda+\rho\xi)\Gamma(\mu+\tau\xi)\{\Gamma(a+\xi)\}^{s}}{\Gamma(\nu+\kappa\xi)\{\Gamma(a+1+\xi)\}^{s}}\\
&\quad\times\frac{\Gamma(\sigma+\rho\xi)\Gamma(\sigma+\eta-\varpi+\rho\xi)}{\Gamma(\sigma-\varpi+\rho\xi)\Gamma(\sigma+\varsigma+\eta+\rho\xi)}(-x^{\rho})^{\xi}\,\mathrm{d}\xi,
\end{aligned}$$

which upon using the definition (9.21), yields the desired formula (9.29). □

Theorem 9.2. *Let* $\varsigma,\varpi,\eta,\sigma,\lambda,\mu,s\in\mathbb{C}$ *and* $a,\nu\in\mathbb{C}\setminus\mathbb{Z}_0^-$ *with* $\rho,\rho,\tau,\kappa\in\mathbb{R}^+$, *such that* $\Re(\varsigma)>0$ *and* $\Re(\sigma)<1+\min[\Re(\varpi),\Re(\eta)]$. *Then for* $\Re(a)>0$ *and* $\Re(s)>0$, *following Saigo's fractional integral* $\mathscr{I}_{-}^{\varsigma,\varpi,\eta}$ *of* $\Phi_{\lambda,\mu;\nu}^{(\rho,\tau;\kappa)}(t,s,a)$ *holds true:*

$$\begin{aligned}
&\left(\mathscr{I}_{-}^{\varsigma,\varpi,\eta}\left\{t^{\sigma-1}\Phi_{\lambda,\mu;\nu}^{(\rho,\tau;\kappa)}(t^{\rho},s,a)\right\}\right)(x)=x^{\sigma-\varpi-1}\frac{\Gamma(\nu)}{\Gamma(\lambda)\Gamma(\mu)}\\
&\times I_{5,5}^{3,3}\left[-x^{\rho}\,\middle|\,\begin{matrix}(1-\lambda,\rho;1),(1-\mu,\tau;1),(1-a,1;s),(1-\sigma,\rho;1),(1-\sigma+\varsigma+\varpi+\eta,\rho;1)\\(0,1;1),(1-\sigma+\varpi,\rho;1),(1-\sigma+\eta,\rho;1),(1-\nu,\kappa;1),(-a,1;s)\end{matrix}\right].
\end{aligned}\tag{9.30}$$

Proof. Using the definitions (9.20) and (9.2), by changing the order of integration and applying the relation (9.26)

$$\begin{aligned}
&\left(\mathscr{I}_{-}^{\varsigma,\varpi,\eta}\left\{t^{\sigma-1}\Phi_{\lambda,\mu;\nu}^{(\rho,\tau;\kappa)}(t^{\rho},s,a)\right\}\right)(x)\\
&=\frac{\Gamma(\nu)}{\Gamma(\lambda)\Gamma(\mu)}\frac{1}{2\pi i}\int_{\mathscr{L}}\frac{\Gamma(-\xi)\Gamma(\lambda+\rho\xi)\Gamma(\mu+\tau\xi)\{\Gamma(\xi+a)\}^{s}(-1)^{\xi}}{\Gamma(\nu+\kappa\xi)\{\Gamma(\xi+a+1)\}^{s}}\\
&\quad\left(\mathscr{I}_{-}^{\varsigma,\varpi,\eta}t^{\sigma+\rho\xi-1}\right)(x)\,\mathrm{d}\xi\\
&=x^{\sigma-\varpi-1}\frac{\Gamma(\nu)}{\Gamma(\lambda)\Gamma(\mu)}\frac{1}{2\pi i}\int_{\mathscr{L}}\frac{\Gamma(-\xi)\Gamma(\lambda+\rho\xi)\Gamma(\mu+\tau\xi)\{\Gamma(a+\xi)\}^{s}}{\Gamma(\nu+\kappa\xi)\{\Gamma(a+1+\xi)\}^{s}}\\
&\quad\times\frac{\Gamma(1-\sigma+\varpi-\rho\xi)\Gamma(1-\sigma+\eta-\rho\xi)}{\Gamma(1-\sigma-\rho\xi)\Gamma(1-\sigma+\varsigma+\varpi+\eta-\rho\xi)}(-x^{\rho})^{\xi}\,\mathrm{d}\xi,
\end{aligned}$$

which upon using the definition (9.21), yields the desired formula (9.30). □

Now we deduce the fractional integral formulas for the classical Riemann-Liouville and Erdélyi-Kober by letting $\varpi=-\varsigma$ and $\varpi=0$ respectively, which are asserted by Corollaries 8–11 below.

Corollary 8. *Let* $\varsigma, \varpi, \eta, \sigma, \lambda, \mu, s \in \mathbb{C}$ *and* $a, \nu \in \mathbb{C}\setminus\mathbb{Z}_0^-$ *with* $\rho, \rho, \tau, \kappa \in \mathbb{R}^+$, *such that* $\Re(\varsigma) > 0$ *and* $\Re(\sigma) > 0$. *Then for* $\Re(a) > 0$ *and* $\Re(s) > 0$, *following Riemann-Liouville fractional integral* $\mathscr{I}_{0+}^{\varsigma}$ *of* $\Phi_{\lambda,\mu;\nu}^{(\rho,\tau;\kappa)}(t,s,a)$ *holds true:*

$$\left(\mathscr{I}_{0+}^{\varsigma}\left\{t^{\sigma-1}\Phi_{\lambda,\mu;\nu}^{(\rho,\tau;\kappa)}(t^{\rho},s,a)\right\}\right)(x) = x^{\sigma+\varsigma-1}\frac{\Gamma(\nu)}{\Gamma(\lambda)\Gamma(\mu)}$$
$$\times I_{4,4}^{1,4}\left[-x^{\sigma}\left|\begin{array}{c}(1-\lambda,\rho;1),(1-\mu,\tau;1),(1-a,1;s),(1-\sigma,\rho;1)\\(0,1;1),(1-\nu,\kappa;1),(-a,1;s),(1-\sigma-\varsigma,\rho;1)\end{array}\right.\right]. \quad (9.31)$$

Corollary 9. *Let* $\varsigma, \varpi, \eta, \sigma, \lambda, \mu, s \in \mathbb{C}$ *and* $a, \nu \in \mathbb{C}\setminus\mathbb{Z}_0^-$ *with* $\rho, \rho, \tau, \kappa \in \mathbb{R}^+$, *such that* $\Re(\varsigma) > 0$ *and* $\Re(\sigma) > -\Re(\eta)$. *Then for* $\Re(a) > 0$ *and* $\Re(s) > 0$, *following Erdélyi-Kober fractional integral* $\mathscr{I}_{\eta,\varsigma}^{+}$ *of* $\Phi_{\lambda,\mu;\nu}^{(\rho,\tau;\kappa)}(t,s,a)$ *holds true:*

$$\left(\mathscr{I}_{\eta,\varsigma}^{+}\left\{t^{\sigma-1}\Phi_{\lambda,\mu;\nu}^{(\rho,\tau;\kappa)}(t^{\rho},s,a)\right\}\right)(x) = x^{\sigma-1}\frac{\Gamma(\nu)}{\Gamma(\lambda)\Gamma(\mu)}$$
$$\times I_{4,4}^{1,4}\left[-x^{\rho}\left|\begin{array}{c}(1-\lambda,\rho;1),(1-\mu,\tau;1),(1-a,1;s),(1-\sigma-\eta,\rho;1)\\(0,1;1),(1-\nu,\kappa;1),(-a,1;s),(1-\sigma-\varsigma-\eta,\rho;1)\end{array}\right.\right]. \quad (9.32)$$

Corollary 10. *Let* $\varsigma, \varpi, \eta, \sigma, \lambda, \mu, s \in \mathbb{C}$ *and* $a, \nu \in \mathbb{C}\setminus\mathbb{Z}_0^-$ *with* $\rho, \rho, \tau, \kappa \in \mathbb{R}^+$, *such that* $0 < \Re(\varsigma) < 1 - \Re(\sigma)$. *Then for* $\Re(a) > 0$ *and* $\Re(s) > 0$, *following Riemann-Liouville fractional integral* $\mathscr{I}_{-}^{\varsigma}$ *of* $\Phi_{\lambda,\mu;\nu}^{(\rho,\tau;\kappa)}(t,s,a)$ *holds true:*

$$\left(\mathscr{I}_{-}^{\varsigma}\left\{t^{\sigma-1}\Phi_{\lambda,\mu;\nu}^{(\rho,\tau;\kappa)}(t^{\rho},s,a)\right\}\right)(x) = x^{\sigma+\varsigma-1}\frac{\Gamma(\nu)}{\Gamma(\lambda)\Gamma(\mu)}$$
$$\times I_{4,4}^{2,3}\left[-x^{\rho}\left|\begin{array}{c}(1-\lambda,\rho;1),(1-\mu,\tau;1),(1-a,1;s),(1-\sigma,\rho;1)\\(0,1;1),(1-\sigma-\varsigma,\rho;1),(1-\nu,\kappa;1),(-a,1;s)\end{array}\right.\right]. \quad (9.33)$$

Corollary 11. *Let* $\varsigma, \varpi, \eta, \sigma, \lambda, \mu, s \in \mathbb{C}$ *and* $a, \nu \in \mathbb{C}\setminus\mathbb{Z}_0^-$ *with* $\rho, \rho, \tau, \kappa \in \mathbb{R}^+$, *such that* $\Re(\varsigma) > 0$ *and* $\Re(\sigma) < 1 + \Re(\eta)$. *Then for* $\Re(a) > 0$ *and* $\Re(s) > 0$, *following Erdélyi-Kober fractional integral* $\mathscr{K}_{\eta,\varsigma}^{-}$ *of* $\Phi_{\lambda,\mu;\nu}^{(\rho,\tau;\kappa)}(t,s,a)$ *holds true:*

$$\left(\mathscr{K}_{\eta,\varsigma}^{-}\left\{t^{\sigma-1}\Phi_{\lambda,\mu;\nu}^{(\rho,\tau;\kappa)}(t^{\rho},s,a)\right\}\right)(x) = x^{\sigma-1}\frac{\Gamma(\nu)}{\Gamma(\lambda)\Gamma(\mu)}I_{4,4}^{2,3}$$
$$\left[-x^{\rho}\left|\begin{array}{c}(1-\lambda,\rho;1),(1-\mu,\tau;1),(1-a,1;s),(1-\sigma+\varsigma+\eta,\rho;1)\\(0,1;1),(1-\sigma+\eta,\rho;1),(1-\nu,\kappa;1),(-a,1;s)\end{array}\right.\right]. \quad (9.34)$$

9.3 FRACTIONAL DIFFERENTIAL OPERATORS OF THE $\Phi_{\lambda,\mu;\nu}^{(\rho,\tau;\kappa)}(t,s,a)$

In this section, we begin the main results exposition by presenting the composition formulas of generalized fractional integrals (9.3) and (9.4) involving the extended Hurwitz-Lerch Zeta function in terms of the I–function (9.21). The following power function formulas are useful in our investigation:

Lemma 9.2. *[21] Let $\varsigma, \varpi, \eta \in \mathbb{C}$. Then there exists the power function formulas*

(a) *If $\Re(\varsigma) > 0$ and $\Re(\sigma) > -\min[0, \Re(\varsigma+\varpi+\eta)]$, then for $x > 0$, we have*

$$(\mathscr{D}_{0+}^{\varsigma,\varpi,\eta} t^{\sigma-1})(x) = \frac{\Gamma(\sigma)\Gamma(\sigma+\varsigma+\varpi+\eta)}{\Gamma(\sigma+\varpi)\Gamma(\sigma+\eta)} x^{\sigma+\varpi-1} \tag{9.35}$$

In particular for $\varpi = -\varsigma$ and $\varpi = 0$, we get

$$(\mathscr{D}_{0+}^{\varsigma} t^{\sigma-1})(x) = \frac{\Gamma(\sigma)}{\Gamma(\sigma-\varsigma)} x^{\sigma-\varsigma-1} \qquad (\Re(\varsigma) > 0,\ \Re(\sigma) > 0) \tag{9.36}$$

and

$$(\mathscr{D}_{\eta,\varsigma}^{+} t^{\sigma-1})(x) = \frac{\Gamma(\sigma+\varsigma+\eta)}{\Gamma(\sigma+\eta)} x^{\sigma-1} \qquad (\Re(\varsigma) > 0,\ \Re(\sigma) > -\Re(\varsigma+\eta)). \tag{9.37}$$

(b) *If $\Re(\varsigma) > 0$, $\Re(\sigma) < 1 + \min[\Re(-\varpi-n), \Re(\varsigma+\eta)]$ and $n = [\Re(\varsigma)] + 1$, then for $x > 0$, we have*

$$(\mathscr{D}_{-}^{\varsigma,\varpi,\eta} t^{\sigma-1})(x) = \frac{\Gamma(1-\sigma-\varpi)\Gamma(1-\sigma+\varsigma+\eta)}{\Gamma(1-\sigma)\Gamma(1-\sigma+\eta-\varpi)} x^{\sigma+\varpi-1}. \tag{9.38}$$

In particular for $\varpi = -\varsigma$ and $\varpi = 0$, we get

$$(\mathscr{D}_{-}^{\varsigma} t^{\sigma-1})(x) = \frac{\Gamma(1-\sigma+\varsigma)}{\Gamma(1-\sigma)} x^{\sigma-\varsigma-1} \qquad (\Re(\varsigma) > 0,\ \Re(\sigma) < 1 + \Re(\varsigma) - n) \tag{9.39}$$

and

$$(\mathscr{D}_{\eta,\varsigma}^{-} t^{\sigma-1})(x) = \frac{\Gamma(1-\sigma+\varsigma+\eta)}{\Gamma(1-\sigma-\eta)} x^{\sigma-1} \qquad (\Re(\varsigma) > 0,\ \Re(\sigma) < 1 + \Re(\varsigma+\eta) - n). \tag{9.40}$$

Theorem 9.3. *Let $\varsigma, \varpi, \eta, \sigma, \lambda, \mu, s \in \mathbb{C}$ and $a, \nu \in \mathbb{C} \setminus \mathbb{Z}_0^-$ with $\rho, \rho, \tau, \kappa \in \mathbb{R}^+$, such that $\Re(\varsigma) \geq 0$ and $\Re(\sigma) > -\min[0, \Re(\varsigma+\varpi+\eta)]$. Then for $\Re(a) > 0$ and $\Re(s) > 0$, following Saigo's fractional derivative $\mathscr{D}_{0+}^{\varsigma,\varpi,\eta}$ of $\Phi_{\lambda,\mu;\nu}^{(\rho,\tau;\kappa)}(t,s,a)$ holds true:*

$$\left(\mathscr{D}_{0+}^{\varsigma,\varpi,\eta}\left\{t^{\sigma-1}\Phi_{\lambda,\mu;\nu}^{(\rho,\tau;\kappa)}(t^{\rho},s,a)\right\}\right)(x) = x^{\sigma+\varpi-1}\frac{\Gamma(\nu)}{\Gamma(\lambda)\Gamma(\mu)}$$
$$\times I_{5,5}^{1,5}\left[-x^{\rho}\,\middle|\, \begin{matrix}(1-\lambda,\rho;1),(1-\mu,\tau;1),(1-a,1;s),(1-\sigma,\rho;1),(1-\sigma-\varsigma-\varpi-\eta,\rho;1)\\(0,1;1),(1-\nu,\kappa;1),(-a,1;s),(1-\sigma-\varpi,\rho;1),(1-\sigma-\eta,\rho;1)\end{matrix}\right]. \tag{9.41}$$

Proof. Using the definitions (9.20) and (9.3), by changing the order of integration and applying the relation (9.35)

$$\left(\mathscr{D}_{0+}^{\varsigma,\varpi,\eta}\left\{t^{\sigma-1}\Phi_{\lambda,\mu;\nu}^{(\rho,\tau;\kappa)}(t^{\rho},s,a)\right\}\right)(x)$$

$$=\frac{\Gamma(\nu)}{\Gamma(\lambda)\Gamma(\mu)}\frac{1}{2\pi i}\int_{\mathscr{L}}\frac{\Gamma(-\xi)\Gamma(\lambda+\rho\xi)\Gamma(\mu+\tau\xi)\{\Gamma(\xi+a)\}^{s}(-1)^{\xi}}{\Gamma(\nu+\kappa\xi)\{\Gamma(\xi+a+1)\}^{s}}\left(\mathscr{D}_{0+}^{\varsigma,\varpi,\eta}t^{\sigma+\rho\xi-1}\right)(x)\,\mathrm{d}\xi$$

$$=x^{\sigma+\varpi-1}\frac{\Gamma(\nu)}{\Gamma(\lambda)\Gamma(\mu)}\frac{1}{2\pi i}\int_{\mathscr{L}}\frac{\Gamma(-\xi)\Gamma(\lambda+\rho\xi)\Gamma(\mu+\tau\xi)\{\Gamma(a+\xi)\}^{s}}{\Gamma(\nu+\kappa\xi)\{\Gamma(a+1+\xi)\}^{s}}$$

$$\times\frac{\Gamma(\sigma+\rho\xi)\Gamma(\sigma+\varsigma+\varpi+\eta+\rho\xi)}{\Gamma(\sigma+\varpi+\rho\xi)\Gamma(\sigma+\eta+\rho\xi)}(-x^{\rho})^{\xi}\,\mathrm{d}\xi,$$

which upon using the definition (9.21), yields the desired formula (9.41). □

Theorem 9.4. *Let* $\varsigma,\varpi,\eta,\sigma,\lambda,\mu,s\in\mathbb{C}$ *and* $a,\nu\in\mathbb{C}\setminus\mathbb{Z}_0^-$ *with* $\rho,\rho,\tau,\kappa\in\mathbb{R}^+$, *such that* $\Re(\varsigma)\geq 0$ *and* $\Re(\sigma)<1+\min[\Re(-\varpi-n),\Re(\varsigma+\eta)]$, $n=[\Re(\varsigma)]+1$. *Then for* $\Re(a)>0$ *and* $\Re(s)>0$, *following Saigo's fractional derivative* $\mathscr{D}_{-}^{\varsigma,\varpi,\eta}$ *of* $\Phi_{\lambda,\mu;\nu}^{(\rho,\tau;\kappa)}(t,s,a)$ *holds true:*

$$\left(\mathscr{D}_{-}^{\varsigma,\varpi,\eta}\left\{t^{\sigma-1}\Phi_{\lambda,\mu;\nu}^{(\rho,\tau;\kappa)}(t^{\rho},s,a)\right\}\right)(x)=x^{\sigma+\varpi-1}\frac{\Gamma(\nu)}{\Gamma(\lambda)\Gamma(\mu)}$$

$$\times I_{5,5}^{3,3}\left[-x^{\rho}\;\middle|\;\begin{array}{l}(1-\lambda,\rho;1),(1-\mu,\tau;1),(1-a,1;s),(1-\sigma,\rho;1),(1-\sigma+\eta-\varpi,\rho;1)\\(0,1;1),(1-\sigma-\varpi,\rho;1),(1-\sigma+\varsigma+\eta,\rho;1),(1-\nu,\kappa;1),(-a,1;s)\end{array}\right].\tag{9.42}$$

Proof. Using the definitions (9.20) and (9.4), by changing the order of integration and applying the relation (9.38)

$$\left(\mathscr{D}_{-}^{\varsigma,\varpi,\eta}\left\{t^{\sigma-1}\Phi_{\lambda,\mu;\nu}^{(\rho,\tau;\kappa)}(t^{\rho},s,a)\right\}\right)(x)$$

$$=\frac{\Gamma(\nu)}{\Gamma(\lambda)\Gamma(\mu)}\frac{1}{2\pi i}\int_{\mathscr{L}}\frac{\Gamma(-\xi)\Gamma(\lambda+\rho\xi)\Gamma(\mu+\tau\xi)\{\Gamma(\xi+a)\}^{s}(-1)^{\xi}}{\Gamma(\nu+\kappa\xi)\{\Gamma(\xi+a+1)\}^{s}}\left(\mathscr{D}_{-}^{\varsigma,\varpi,\eta}t^{\sigma+\xi-1}\right)(x)\,\mathrm{d}\xi$$

$$=x^{\sigma+\varpi-1}\frac{\Gamma(\nu)}{\Gamma(\lambda)\Gamma(\mu)}\frac{1}{2\pi i}\int_{\mathscr{L}}\frac{\Gamma(-\xi)\Gamma(\lambda+\rho\xi)\Gamma(\mu+\tau\xi)\{\Gamma(a+\xi)\}^{s}}{\Gamma(\nu+\kappa\xi)\{\Gamma(a+1+\xi)\}^{s}}$$

$$\times\frac{\Gamma(1-\sigma-\varpi-\rho\xi)\Gamma(1-\sigma+\varsigma+\eta-\rho\xi)}{\Gamma(1-\sigma-\rho\xi)\Gamma(1-\sigma+\eta-\varpi-\rho\xi)}(-x^{\rho})^{\xi}\,\mathrm{d}\xi,$$

which upon using the definition (9.21), yields the desired formula (9.42). □

Now we deduce the fractional derivative formulas for the classical Riemann-Liouville and Erdélyi-Kober by letting $\varpi=-\varsigma$ and $\varpi=0$ respectively, which are asserted by Corollaries 12–15 below.

Corollary 12. *Let* $\varsigma, \varpi, \eta, \sigma, \lambda, \mu, s \in \mathbb{C}$ *and* $a, \nu \in \mathbb{C}\setminus\mathbb{Z}_0^-$ *with* $\rho, \rho, \tau, \kappa \in \mathbb{R}^+$, *such that* $\Re(\varsigma) \geq 0$ *and* $\Re(\sigma) > 0$. *Then for* $\Re(a) > 0$ *and* $\Re(s) > 0$, *following Riemann-Liouville fractional differentiation* $\mathscr{D}_{0+}^{\varsigma}$ *of* $\Phi_{\lambda,\mu;\nu}^{(\rho,\tau;\kappa)}(t,s,a)$ *holds true:*

$$\left(\mathscr{D}_{0+}^{\varsigma}\left\{t^{\sigma-1}\Phi_{\lambda,\mu;\nu}^{(\rho,\tau;\kappa)}(t^{\rho},s,a)\right\}\right)(x)$$
$$= x^{\sigma+\varsigma-1}\frac{\Gamma(\nu)}{\Gamma(\lambda)\Gamma(\mu)} I_{4,4}^{1,4}\left[-x^{\rho}\left|\begin{matrix}(1-\lambda,\rho;1),(1-\mu,\tau;1),(1-a,1;s),(1-\sigma,\rho;1)\\(0,1;1),(1-\nu,\kappa;1),(-a,1;s),(1-\sigma+\varsigma,\rho;1)\end{matrix}\right.\right]. \tag{9.43}$$

Corollary 13. *Let* $\varsigma, \varpi, \eta, \sigma, \lambda, \mu, s \in \mathbb{C}$ *and* $a, \nu \in \mathbb{C}\setminus\mathbb{Z}_0^-$ *with* $\rho, \rho, \tau, \kappa \in \mathbb{R}^+$, *such that* $\Re(\varsigma) \geq 0$ *and* $\Re(\sigma) > -\Re(\varsigma+\eta)$. *Then for* $\Re(a) > 0$ *and* $\Re(s) > 0$, *following Erdélyi-Kober fractional derivative* $\mathscr{D}_{\eta,\varsigma}^{+}$ *of* $\Phi_{\lambda,\mu;\nu}^{(\rho,\tau;\kappa)}(t,s,a)$ *holds true:*

$$\left(\mathscr{D}_{\eta,\varsigma}^{+}\left\{t^{\sigma-1}\Phi_{\lambda,\mu;\nu}^{(\rho,\tau;\kappa)}(t^{\rho},s,a)\right\}\right)(x)$$
$$= x^{\sigma-1}\frac{\Gamma(\nu)}{\Gamma(\lambda)\Gamma(\mu)} I_{4,4}^{1,4}\left[-x^{\rho}\left|\begin{matrix}(1-\lambda,\rho;1),(1-\mu,\tau;1),(1-a,1;s),(1-\sigma-\varsigma-\eta,\rho;1)\\(0,1;1),(1-\nu,\kappa;1),(-a,1;s),(1-\sigma-\eta,\rho;1)\end{matrix}\right.\right]. \tag{9.44}$$

Corollary 14. *Let* $\varsigma, \varpi, \eta, \sigma, \lambda, \mu, s \in \mathbb{C}$ *and* $a, \nu \in \mathbb{C}\setminus\mathbb{Z}_0^-$ *with* $\rho, \rho, \tau, \kappa \in \mathbb{R}^+$, *such that* $\Re(\varsigma) \geq 0$ *and* $\Re(\sigma) < \Re(\varsigma) - [\Re(\varsigma)]$. *Then for* $\Re(a) > 0$ *and* $\Re(s) > 0$, *following Riemann-Liouville fractional differentiation* $\mathscr{D}_{-}^{\varsigma}$ *of* $\Phi_{\lambda,\mu;\nu}^{(\rho,\tau;\kappa)}(t,s,a)$ *holds true:*

$$\left(\mathscr{D}_{-}^{\varsigma}\left\{t^{\sigma-1}\Phi_{\lambda,\mu;\nu}^{(\rho,\tau;\kappa)}(t^{\rho},s,a)\right\}\right)(x)$$
$$= x^{\sigma-\varsigma-1}\frac{\Gamma(\nu)}{\Gamma(\lambda)\Gamma(\mu)} I_{4,4}^{2,3}\left[-x^{\rho}\left|\begin{matrix}(1-\lambda,\rho;1),(1-\mu,\tau;1),(1-a,1;s),(1-\sigma,\rho;1)\\(0,1;1),(1-\sigma+\varsigma,\rho;1),(1-\nu,\kappa;1),(-a,1;s)\end{matrix}\right.\right]. \tag{9.45}$$

Corollary 15. *Let* $\varsigma, \varpi, \eta, \sigma, \lambda, \mu, s \in \mathbb{C}$ *and* $a, \nu \in \mathbb{C}\setminus\mathbb{Z}_0^-$ *with* $\rho, \rho, \tau, \kappa \in \mathbb{R}^+$, *such that* $\Re(\varsigma) \geq 0$ *and* $\Re(\sigma) < \Re(\varsigma+\eta) - [\Re(\varsigma)]$. *Then for* $\Re(a) > 0$ *and* $\Re(s) > 0$, *following Erdélyi-Kober fractional differentiation* $\mathscr{D}_{\eta,\varsigma}^{-}$ *of* $\Phi_{\lambda,\mu;\nu}^{(\rho,\tau;\kappa)}(t,s,a)$ *holds true:*

$$\left(\mathscr{D}_{\eta,\varsigma}^{-}\left\{t^{\sigma-1}\Phi_{\lambda,\mu;\nu}^{(\rho,\tau;\kappa)}(t^{\rho},s,a)\right\}\right)(x)$$
$$= x^{\sigma-1}\frac{\Gamma(\nu)}{\Gamma(\lambda)\Gamma(\mu)} I_{4,4}^{2,3}\left[-x^{\rho}\left|\begin{matrix}(1-\lambda,\rho;1),(1-\mu,\tau;1),(1-a,1;s),(1-\sigma+\eta,\rho;1)\\(0,1;1),(1-\sigma+\varsigma+\eta,\rho;1),(1-\nu,\kappa;1),(-a,1;s)\end{matrix}\right.\right]. \tag{9.46}$$

Next, we obtained various fractional calculus results for the families of extended Hurwitz-Lerch Zeta function defined by Garg et al. [11, p. 313, Eq. (1.7)], Lin and Srivastava [16, p. 727, Eq. (8)] and Goyal and Laddha [12, p. 100, Eq. (1.5)]. These results are given in the following subsections.

9.3.1 FRACTIONAL CALCULUS OPERATORS OF THE $\Phi_{\lambda,\mu;\nu}(t,s,a)$

In this subsection, we shall derive generalized Saigo's fractional calculus results involving the generalized Hurwitz-Lerch zeta function defined Garg *et al.* [11, p. 313, Eq. (1.7)] for particular choices of parameters in extended Hurwitz-Lerch Zeta function (9.14) given in Case 1. Further, special cases for the classical Riemann-Liouville and Erdélyi-Kober fractional integral and differential operators are deduced. The results are given in the following subsections as Corollaries.

9.3.1.1 Fractional Integration of the $\Phi_{\lambda,\mu;\nu}(t,s,a)$

Corollary 16. *Let $\varsigma, \varpi, \eta, \sigma, \lambda, \mu, s \in \mathbb{C}$ and $a, \nu \in \mathbb{C}\setminus\mathbb{Z}_0^-$ with $\rho \in \mathbb{R}^+$, such that $\Re(\varsigma) > 0$ and $\Re(\sigma) > \max[0, \Re(\varpi - \eta)]$. Then for $\Re(a) > 0$ and $\Re(s) > 0$, following Saigo's fractional integral $\mathscr{I}_{0+}^{\varsigma,\varpi,\eta}$ of $\Phi_{\lambda,\mu;\nu}(t,s,a)$ holds true:*

$$\left(\mathscr{I}_{0+}^{\varsigma,\varpi,\eta}\left\{t^{\sigma-1}\Phi_{\lambda,\mu;\nu}(t^{\rho},s,a)\right\}\right)(x) = x^{\sigma-\varpi-1}\frac{\Gamma(\nu)}{\Gamma(\lambda)\Gamma(\mu)}$$
$$\times I_{5,5}^{1,5}\left[-x^{\rho}\left|\begin{matrix}(1-\lambda,1;1),(1-\mu,1;1),(1-a,1;s),(1-\sigma,\rho;1),(1-\sigma+\varpi-\eta,\rho;1)\\(0,1;1),(1-\nu,1;1),(-a,1;s),(1-\sigma+\varpi,\rho;1),(1-\sigma-\varsigma-\eta,\rho;1)\end{matrix}\right.\right]. \tag{9.47}$$

Corollary 17. *Let $\varsigma, \varpi, \eta, \sigma, \lambda, \mu, s \in \mathbb{C}$ and $a, \nu \in \mathbb{C}\setminus\mathbb{Z}_0^-$ with $\rho \in \mathbb{R}^+$, such that $\Re(\varsigma) > 0$ and $\Re(\sigma) < 1+\min[\Re(\varpi), \Re(\eta)]$. Then for $\Re(a) > 0$ and $\Re(s) > 0$, following Saigo's fractional integral $\mathscr{I}_{-}^{\varsigma,\varpi,\eta}$ of $\Phi_{\lambda,\mu;\nu}(t,s,a)$ holds true:*

$$\left(\mathscr{I}_{-}^{\varsigma,\varpi,\eta}\left\{t^{\sigma-1}\Phi_{\lambda,\mu;\nu}(t^{\rho},s,a)\right\}\right)(x) = x^{\sigma-\varpi-1}\frac{\Gamma(\nu)}{\Gamma(\lambda)\Gamma(\mu)}$$
$$\times I_{5,5}^{3,3}\left[-x^{\rho}\left|\begin{matrix}(1-\lambda,1;1),(1-\mu,1;1),(1-a,1;s),(1-\sigma,\rho;1),(1-\sigma+\varsigma+\varpi+\eta,\rho;1)\\(0,1;1),(1-\sigma+\varpi,\rho;1),(1-\sigma+\eta,\rho;1,(1-\nu,1;1),(-a,1;s))\end{matrix}\right.\right]. \tag{9.48}$$

Corollary 18. *Let $\varsigma, \varpi, \eta, \sigma, \lambda, \mu, s \in \mathbb{C}$ and $a, \nu \in \mathbb{C}\setminus\mathbb{Z}_0^-$ with $\rho \in \mathbb{R}^+$, such that $\Re(\varsigma) > 0$ and $\Re(\sigma) > 0$. Then for $\Re(a) > 0$ and $\Re(s) > 0$, following Riemann-Liouville fractional integral $\mathscr{I}_{0+}^{\varsigma}$ of $\Phi_{\lambda,\mu;\nu}(t,s,a)$ holds true:*

$$\left(\mathscr{I}_{0+}^{\varsigma}\left\{t^{\sigma-1}\Phi_{\lambda,\mu;\nu}^{(1,1;1)}(t^{\rho},s,a)\right\}\right)(x)$$
$$= x^{\sigma+\varsigma-1}\frac{\Gamma(\nu)}{\Gamma(\lambda)\Gamma(\mu)}I_{4,4}^{1,4}\left[-x^{\sigma}\left|\begin{matrix}(1-\lambda,1;1),(1-\mu,1;1),(1-a,1;s),(1-\sigma,\rho;1)\\(0,1;1),(1-\nu,1;1),(-a,1;s),(1-\sigma-\varsigma,\rho;1)\end{matrix}\right.\right]. \tag{9.49}$$

Corollary 19. *Let $\varsigma, \varpi, \eta, \sigma, \lambda, \mu, s \in \mathbb{C}$ and $a, \nu \in \mathbb{C}\setminus\mathbb{Z}_0^-$ with $\rho \in \mathbb{R}^+$, such that $\Re(\varsigma) > 0$ and $\Re(\sigma) > -\Re(\eta)$. Then for $\Re(a) > 0$ and $\Re(s) > 0$, following*

Erdélyi-Kober fractional integral $\mathscr{I}^{+}_{\eta,\varsigma}$ *of* $\Phi_{\lambda,\mu;\nu}(t,s,a)$ *holds true:*

$$\left(\mathscr{I}^{+}_{\eta,\varsigma}\left\{t^{\sigma-1}\Phi_{\lambda,\mu;\nu}(t^{\rho},s,a)\right\}\right)(x)$$
$$= x^{\sigma-1}\frac{\Gamma(\nu)}{\Gamma(\lambda)\Gamma(\mu)} I^{1,4}_{4,4}\left[-x^{\rho}\,\middle|\, \begin{matrix}(1-\lambda,1;1),(1-\mu,1;1),(1-a,1;s),(1-\sigma-\eta,\rho;1)\\(0,1;1),(1-\nu,1;1),(-a,1;s),(1-\sigma-\varsigma-\eta,\rho;1)\end{matrix}\right]. \tag{9.50}$$

Corollary 20. *Let* $\varsigma,\varpi,\eta,\sigma,\lambda,\mu,s\in\mathbb{C}$ *and* $a,\nu\in\mathbb{C}\setminus\mathbb{Z}_0^-$ *with* $\rho\in\mathbb{R}^+$, *such that* $0<\Re(\varsigma)<1-\Re(\sigma)$. *Then for* $\Re(a)>0$ *and* $\Re(s)>0$, *following Riemann-Liouville fractional integral* $\mathscr{I}^{\varsigma}_{-}$ *of* $\Phi_{\lambda,\mu;\nu}(t,s,a)$ *holds true:*

$$\left(\mathscr{I}^{\varsigma}_{-}\left\{t^{\sigma-1}\Phi_{\lambda,\mu;\nu}(t^{\rho},s,a)\right\}\right)(x)$$
$$= x^{\sigma+\varsigma-1}\frac{\Gamma(\nu)}{\Gamma(\lambda)\Gamma(\mu)} I^{2,3}_{4,4}\left[-x^{\rho}\,\middle|\, \begin{matrix}(1-\lambda,1;1),(1-\mu,1;1),(1-a,1;s),(1-\sigma,\rho;1)\\(0,1;1),(1-\sigma-\varsigma,\rho),(1-\nu,1;1),(-a,1;s)\end{matrix}\right]. \tag{9.51}$$

Corollary 21. *Let* $\varsigma,\varpi,\eta,\sigma,\lambda,\mu,s\in\mathbb{C}$ *and* $a,\nu\in\mathbb{C}\setminus\mathbb{Z}_0^-$ *with* $\rho\in\mathbb{R}^+$, *such that* $\Re(\varsigma)>0$ *and* $\Re(\sigma)<1+\Re(\eta)$. *Then for* $\Re(a)>0$ *and* $\Re(s)>0$, *following Erdélyi-Kober fractional integral* $K^{-}_{\eta,\varsigma}$ *of* $\Phi_{\lambda,\mu;\nu}(t,s,a)$ *holds true:*

$$\left(K^{-}_{\eta,\varsigma}\left\{t^{\sigma-1}\Phi_{\lambda,\mu;\nu}(t^{\rho},s,a)\right\}\right)(x)$$
$$= x^{\sigma-1}\frac{\Gamma(\nu)}{\Gamma(\lambda)\Gamma(\mu)} I^{2,3}_{4,4}\left[-x^{\rho}\,\middle|\, \begin{matrix}(1-\lambda,1;1),(1-\mu,1;1),(1-a,1;s),(1-\sigma+\varsigma+\eta,\rho;1)\\(0,1;1),(1-\sigma+\eta,\rho;1),(1-\nu,1;1),(-a,1;s)\end{matrix}\right]. \tag{9.52}$$

9.3.1.2 Fractional Differentiation of the $\Phi_{\lambda,\mu;\nu}(t,s,a)$

Corollary 22. *Let* $\varsigma,\varpi,\eta,\sigma,\lambda,\mu,s\in\mathbb{C}$ *and* $a,\nu\in\mathbb{C}\setminus\mathbb{Z}_0^-$ *with* $\rho\in\mathbb{R}^+$, *such that* $\Re(\varsigma)\geq 0$ *and* $\Re(\sigma)>-\min[0,\Re(\varsigma+\varpi+\eta)]$. *Then for* $\Re(a)>0$ *and* $\Re(s)>0$, *following Saigo's fractional derivative* $\mathscr{D}^{\varsigma,\varpi,\eta}_{0+}$ *of* $\Phi_{\lambda,\mu;\nu}(t,s,a)$ *holds true:*

$$\left(\mathscr{D}^{\varsigma,\varpi,\eta}_{0+}\left\{t^{\sigma-1}\Phi_{\lambda,\mu;\nu}(t^{\rho},s,a)\right\}\right)(x) = x^{\sigma+\varpi-1}\frac{\Gamma(\nu)}{\Gamma(\lambda)\Gamma(\mu)}$$
$$\times I^{1,5}_{5,5}\left[-x^{\rho}\,\middle|\, \begin{matrix}(1-\sigma,\rho;1),(1-\sigma-\varsigma-\varpi-\eta,\rho;1),(1-\lambda,1;1),(1-\mu,1;1),(1-a,1;s)\\(0,1;1),(1-\nu,1;1),(-a,1;s),(1-\sigma-\varpi,\rho;1),(1-\sigma-\eta,\rho;1)\end{matrix}\right] \tag{9.53}$$

Corollary 23. *Let* $\varsigma,\varpi,\eta,\sigma,\lambda,\mu,s\in\mathbb{C}$ *and* $a,\nu\in\mathbb{C}\setminus\mathbb{Z}_0^-$ *with* $\rho\in\mathbb{R}^+$, *such that* $\Re(\varsigma)\geq 0$ *and* $\Re(\sigma)<1+\min[\Re(-\varpi-n),\Re(\varsigma+\eta)]$, $n=[\Re(\varsigma)]+1$. *Then for* $\Re(a)>0$ *and* $\Re(s)>0$, *following Saigo's fractional derivative* $\mathscr{D}^{\varsigma,\varpi,\eta}_{-}$

of $\Phi_{\lambda,\mu;\nu}(t,s,a)$ *holds true:*

$$\left(\mathscr{D}_{-}^{\varsigma,\varpi,\eta}\left\{t^{\sigma-1}\Phi_{\lambda,\mu;\nu}(t^{\rho},s,a)\right\}\right)(x)=x^{\sigma+\varpi-1}\frac{\Gamma(\nu)}{\Gamma(\lambda)\Gamma(\mu)}$$
$$\times I_{5,5}^{3,3}\left[-x^{\rho}\left|\begin{matrix}(1-\lambda,1;1),(1-\mu,1;1),(1-a,1;s),(1-\sigma,\rho;1),(1-\sigma+\eta-\varpi,\rho;1)\\(0,1;1),(1-\nu,1;1),(-a,1;s),(1-\sigma-\varpi,\rho;1),(1-\sigma+\varsigma+\eta,\rho;1)\end{matrix}\right.\right] \tag{9.54}$$

Corollary 24. *Let* $\varsigma,\varpi,\eta,\sigma,\lambda,\mu,s\in\mathbb{C}$ *and* $a,\nu\in\mathbb{C}\setminus\mathbb{Z}_0^-$ *with* $\rho\in\mathbb{R}^+$, *such that* $\Re(\varsigma)\geq 0$ *and* $\Re(\sigma)>0$. *Then for* $\Re(a)>0$ *and* $\Re(s)>0$, *following Riemann-Liouville fractional differentiation* $\mathscr{D}_{0+}^{\varsigma}$ *of* $\Phi_{\lambda,\mu;\nu}(t,s,a)$ *holds true:*

$$\left(\mathscr{D}_{0+}^{\varsigma}\left\{t^{\sigma-1}\Phi_{\lambda,\mu;\nu}(t^{\rho},s,a)\right\}\right)(x)$$
$$=x^{\sigma+\varsigma-1}\frac{\Gamma(\nu)}{\Gamma(\lambda)\Gamma(\mu)}I_{4,4}^{1,4}\left[-x^{\rho}\left|\begin{matrix}(1-\lambda,1;1),(1-\mu,1;1),(1-a,1;s),(1-\sigma,\rho;1)\\(0,1;1),(1-\nu,1;1),(-a,1;s),(1-\sigma+\varsigma,\rho;1)\end{matrix}\right.\right]. \tag{9.55}$$

Corollary 25. *Let* $\varsigma,\varpi,\eta,\sigma,\lambda,\mu,s\in\mathbb{C}$ *and* $a,\nu\in\mathbb{C}\setminus\mathbb{Z}_0^-$ *with* $\rho\in\mathbb{R}^+$, *such that* $\Re(\varsigma)\geq 0$ *and* $\Re(\sigma)>-\Re(\varsigma+\eta)$. *Then for* $\Re(a)>0$ *and* $\Re(s)>0$, *following Erdélyi-Kober fractional derivative* $\mathscr{D}_{\eta,\varsigma}^{+}$ *of* $\Phi_{\lambda,\mu;\nu}(t,s,a)$ *holds true:*

$$\left(\mathscr{D}_{\eta,\varsigma}^{+}\left\{t^{\sigma-1}\Phi_{\lambda,\mu;\nu}(t^{\rho},s,a)\right\}\right)(x)$$
$$=x^{\sigma-1}\frac{\Gamma(\nu)}{\Gamma(\lambda)\Gamma(\mu)}I_{4,4}^{1,4}\left[-x^{\rho}\left|\begin{matrix}(1-\lambda,1;1),(1-\mu,1;1),(1-a,1;s),(1-\sigma-\varsigma-\eta,\rho;1)\\(0,1;1),(1-\nu,1;1),(-a,1;s),(1-\sigma-\eta,\rho;1)\end{matrix}\right.\right]. \tag{9.56}$$

Corollary 26. *Let* $\varsigma,\varpi,\eta,\sigma,\lambda,\mu,s\in\mathbb{C}$ *and* $a,\nu\in\mathbb{C}\setminus\mathbb{Z}_0^-$ *with* $\rho\in\mathbb{R}^+$, *such that* $\Re(\varsigma)\geq 0$ *and* $\Re(\sigma)<\Re(\varsigma)-[\Re(\varsigma)]$. *Then for* $\Re(a)>0$ *and* $\Re(s)>0$, *following Riemann-Liouville fractional differentiation* $\mathscr{D}_{-}^{\varsigma}$ *of* $\Phi_{\lambda,\mu;\nu}(t,s,a)$ *holds true:*

$$\left(\mathscr{D}_{-}^{\varsigma}\left\{t^{\sigma-1}\Phi_{\lambda,\mu;\nu}^{(1,1;1)}(t^{\rho},s,a)\right\}\right)(x)$$
$$=x^{\sigma-\varsigma-1}\frac{\Gamma(\nu)}{\Gamma(\lambda)\Gamma(\mu)}I_{4,4}^{2,3}\left[-x^{\rho}\left|\begin{matrix}(1-\lambda,1;1),(1-\mu,1;1),(1-a,1;s),(1-\sigma,\rho;1)\\(0,1;1),(1-\sigma+\varsigma,\rho;1),(1-\nu,1;1),(-a,1;s)\end{matrix}\right.\right]. \tag{9.57}$$

Corollary 27. *Let* $\varsigma,\varpi,\eta,\sigma,\lambda,\mu,s\in\mathbb{C}$ *and* $a,\nu\in\mathbb{C}\setminus\mathbb{Z}_0^-$ *with* $\rho\in\mathbb{R}^+$, *such that* $\Re(\varsigma)\geq 0$ *and* $\Re(\sigma)<\Re(\varsigma+\eta)-[\Re(\varsigma)]$. *Then for* $\Re(a)>0$ *and* $\Re(s)>0$, *following Erdélyi-Kober fractional differentiation* $\mathscr{D}_{\eta,\varsigma}^{-}$ *of* $\Phi_{\lambda,\mu;\nu}(t,s,a)$ *holds true:*

$$\left(\mathscr{D}_{\eta,\varsigma}^{-}\left\{t^{\sigma-1}\Phi_{\lambda,\mu;\nu}(t^{\rho},s,a)\right\}\right)(x)$$
$$=x^{\sigma-1}\frac{\Gamma(\nu)}{\Gamma(\lambda)\Gamma(\mu)}I_{4,4}^{2,3}\left[-x^{\rho}\left|\begin{matrix}(1-\lambda,1;1),(1-\mu,1;1),(1-a,1;s),(1-\sigma+\eta,\rho;1)\\(0,1;1),(1-\sigma+\varsigma+\eta,\rho;1),(1-\nu,1;1),(-a,1;s)\end{matrix}\right.\right]. \tag{9.58}$$

9.3.2 FRACTIONAL CALCULUS OPERATORS OF THE $\Phi_{\mu;\nu}^{(\tau;\kappa)}(t,s,a)$

In this section, we derive generalized Saigo's fractional calculus results involving the generalized Hurwitz-Lerch zeta function defined Lin and Srivastava [16, p. 727, Eq. (8)] for particular choices of parameters in extended Hurwitz-Lerch Zeta function (9.14) given in Case 2. Further, special cases for the classical Riemann-Liouville and Erdélyi-Kober fractional integral and differential operators are deduced. The results are given in the following subsections as Corollaries.

9.3.2.1 Fractional Integration of the $\Phi_{\mu;\nu}^{(\tau;\kappa)}(t,s,a)$

Corollary 28. *Let $\varsigma, \varpi, \eta, \sigma, \mu, s \in \mathbb{C}$ and $a, \nu \in \mathbb{C}\setminus\mathbb{Z}_0^-$ with $\rho,\ \tau,\ \kappa \in \mathbb{R}^+$, such that $\Re(\varsigma) > 0$ and $\Re(\sigma) > \max[0, \Re(\varpi-\eta)]$. Then for $\Re(a) > 0$ and $\Re(s) > 0$, following Saigo's fractional integral $\mathscr{I}_{0+}^{\varsigma,\varpi,\eta}$ of $\Phi_{\mu;\nu}^{(\tau;\kappa)}(t,s,a)$ holds true:*

$$\left(\mathscr{I}_{0+}^{\varsigma,\varpi,\eta}\left\{t^{\sigma-1}\Phi_{\mu;\nu}^{(\tau;\kappa)}(t^{\rho},s,a)\right\}\right)(x) = x^{\sigma-\varpi-1}\frac{\Gamma(\nu)}{\Gamma(\mu)}$$
$$\times I_{5,5}^{1,5}\left[-x^{\rho}\left|\begin{matrix}(0,1;1),(1-\mu,\tau;1),(1-a,1;s),(1-\sigma,\rho;1),(1-\sigma+\varpi-\eta,\rho;1)\\(0,1;1),(1-\nu,\kappa;1),(-a,1;s),(1-\sigma+\varpi,\rho;1),(1-\sigma-\varsigma-\eta,\rho;1)\end{matrix}\right.\right]. \tag{9.59}$$

Corollary 29. *Let $\varsigma, \varpi, \eta, \sigma, \mu, s \in \mathbb{C}$ and $a, \nu \in \mathbb{C}\setminus\mathbb{Z}_0^-$ with $\rho,\ \tau,\ \kappa \in \mathbb{R}^+$, such that $\Re(\varsigma) > 0$ and $\Re(\sigma) < 1+\min[\Re(\varpi), \Re(\eta)]$. Then for $\Re(a) > 0$ and $\Re(s) > 0$, following Saigo's fractional integral $\mathscr{I}_{-}^{\varsigma,\varpi,\eta}$ of $\Phi_{\mu;\nu}^{(\tau;\kappa)}(t,s,a)$ holds true:*

$$\left(\mathscr{I}_{-}^{\varsigma,\varpi,\eta}\left\{t^{\sigma-1}\Phi_{\mu;\nu}^{(\tau;\kappa)}(t^{\rho},s,a)\right\}\right)(x) = x^{\sigma-\varpi-1}\frac{\Gamma(\nu)}{\Gamma(\mu)}$$
$$\times I_{5,5}^{3,3}\left[-x^{\rho}\left|\begin{matrix}(0,1;1),(1-\mu,\tau;1),(1-a,1;s),(1-\sigma,\rho;1),(1-\sigma+\varsigma+\varpi+\eta,\rho;1)\\(0,1;1),(1-\sigma+\varpi,\rho;1),(1-\sigma+\eta,\rho;1),(1-\nu,\kappa;1),(-a,1;s)\end{matrix}\right.\right]. \tag{9.60}$$

Corollary 30. *Let $\varsigma, \varpi, \eta, \sigma, \mu, s \in \mathbb{C}$ and $a, \nu \in \mathbb{C}\setminus\mathbb{Z}_0^-$ with $\rho,\ \tau,\ \kappa \in \mathbb{R}^+$, such that $\Re(\varsigma) > 0$ and $\Re(\sigma) > 0$. Then for $\Re(a) > 0$ and $\Re(s) > 0$, following Riemann-Liouville fractional integral $\mathscr{I}_{0+}^{\varsigma}$ of $\Phi_{\mu;\nu}^{(\tau;\kappa)}(t,s,a)$ holds true:*

$$\left(\mathscr{I}_{0+}^{\varsigma}\left\{t^{\sigma-1}\Phi_{\mu;\nu}^{(\tau;\kappa)}(t^{\rho},s,a)\right\}\right)(x)$$
$$= x^{\sigma+\varsigma-1}\frac{\Gamma(\nu)}{\Gamma(\mu)}I_{4,4}^{1,4}\left[-x^{\sigma}\left|\begin{matrix}(0,1;1),(1-\mu,\tau;1),(1-a,1;s),(1-\sigma,\rho;1)\\(0,1;1),(1-\nu,\kappa;1),(-a,1;s),(1-\sigma-\varsigma,\rho;1)\end{matrix}\right.\right]. \tag{9.61}$$

Corollary 31. *Let $\varsigma, \varpi, \eta, \sigma, \mu, s \in \mathbb{C}$ and $a, \nu \in \mathbb{C}\setminus\mathbb{Z}_0^-$ with $\rho, \tau, \kappa \in \mathbb{R}^+$, such that $\Re(\varsigma) > 0$ and $\Re(\sigma) > -\Re(\eta)$. Then for $\Re(a) > 0$ and $\Re(s) > 0$, following Erdélyi-Kober fractional integral $\mathscr{I}^+_{\eta,\varsigma}$ of $\Phi^{(\tau;\kappa)}_{\mu;\nu}(t,s,a)$ holds true:*

$$\left(\mathscr{I}^+_{\eta,\varsigma}\left\{t^{\sigma-1}\Phi^{(\tau;\kappa)}_{\mu;\nu}(t^\rho,s,a)\right\}\right)(x) = x^{\sigma-1}\frac{\Gamma(\nu)}{\Gamma(\mu)} I^{1,4}_{4,4}\left[-x^\rho \left| \begin{matrix} (0,1;1),(1-\mu,\tau;1),(1-a,1;s),(1-\sigma-\eta,\rho;1) \\ (0,1;1),(1-\nu,\kappa;1),(-a,1;s),(1-\sigma-\varsigma-\eta,\rho;1) \end{matrix}\right.\right]. \tag{9.62}$$

Corollary 32. *Let $\varsigma, \varpi, \eta, \sigma, \mu, s \in \mathbb{C}$ and $a, \nu \in \mathbb{C}\setminus\mathbb{Z}_0^-$ with $\rho, \tau, \kappa \in \mathbb{R}^+$, such that $0 < \Re(\varsigma) < 1 - \Re(\sigma)$. Then for $\Re(a) > 0$ and $\Re(s) > 0$, following Riemann-Liouville fractional integral $\mathscr{I}^\varsigma_-$ of $\Phi^{(\tau;\kappa)}_{\mu;\nu}(t,s,a)$ holds true:*

$$\left(\mathscr{I}^\varsigma_-\left\{t^{\sigma-1}\Phi^{(\tau;\kappa)}_{\mu;\nu}(t^\rho,s,a)\right\}\right)(x) = x^{\sigma+\varsigma-1}\frac{\Gamma(\nu)}{\Gamma(\mu)} I^{2,3}_{4,4}\left[-x^\rho \left| \begin{matrix} (0,1;1),(1-\mu,\tau;1),(1-a,1;s),(1-\sigma,\rho;1) \\ (0,1;1),(1-\sigma-\varsigma,\rho;1),(1-\nu,\kappa;1),(-a,1;s) \end{matrix}\right.\right]. \tag{9.63}$$

Corollary 33. *Let $\varsigma, \varpi, \eta, \sigma, \mu, s \in \mathbb{C}$ and $a, \nu \in \mathbb{C}\setminus\mathbb{Z}_0^-$ with $\rho, \tau, \kappa \in \mathbb{R}^+$, such that $\Re(\varsigma) > 0$ and $\Re(\sigma) < 1 + \Re(\eta)$. Then for $\Re(a) > 0$ and $\Re(s) > 0$, following Erdélyi-Kober fractional integral $K^-_{\eta,\varsigma}$ of $\Phi^{(\tau;\kappa)}_{\mu;\nu}(t,s,a)$ holds true:*

$$\left(K^-_{\eta,\varsigma}\left\{t^{\sigma-1}\Phi^{(\tau;\kappa)}_{\mu;\nu}(t^\rho,s,a)\right\}\right)(x) = x^{\sigma-1}\frac{\Gamma(\nu)}{\Gamma(\mu)} I^{2,3}_{4,4}\left[-x^\rho \left| \begin{matrix} (0,1;1),(1-\mu,\tau;1),(1-a,1;s),(1-\sigma+\varsigma+\eta,\rho;1) \\ (0,1;1),(1-\sigma+\eta,\rho;1),(1-\nu,\kappa;1),(-a,1;s) \end{matrix}\right.\right]. \tag{9.64}$$

9.3.2.2 Fractional Differentiation of the $\Phi^{(\tau;\kappa)}_{\mu;\nu}(t,s,a)$

Corollary 34. *Let $\varsigma, \varpi, \eta, \sigma, \mu, s \in \mathbb{C}$ and $a, \nu \in \mathbb{C}\setminus\mathbb{Z}_0^-$ with $\rho, \tau, \kappa \in \mathbb{R}^+$, such that $\Re(\varsigma) \geq 0$ and $\Re(\sigma) > -\min[0, \Re(\varsigma+\varpi+\eta)]$. Then for $\Re(a) > 0$ and $\Re(s) > 0$, following Saigo's fractional derivative $\mathscr{D}^{\varsigma,\varpi,\eta}_{0+}$ of $\Phi^{(\tau;\kappa)}_{\mu;\nu}(t,s,a)$ holds true:*

$$\left(\mathscr{D}^{\varsigma,\varpi,\eta}_{0+}\left\{t^{\sigma-1}\Phi^{(\tau;\kappa)}_{\mu;\nu}(t^\rho,s,a)\right\}\right)(x) = x^{\sigma+\varpi-1}\frac{\Gamma(\nu)}{\Gamma(\mu)} \times I^{1,5}_{5,5}\left[-x^\rho \left| \begin{matrix} (0,1;1),(1-\mu,\tau;1),(1-a,1;s),(1-\sigma,\rho;1),(1-\sigma-\varsigma-\varpi-\eta,\rho;1) \\ (0,1;1),(1-\nu,\kappa;1),(-a,1;s),(1-\sigma-\varpi,\rho;1),(1-\sigma-\eta,\rho;1) \end{matrix}\right.\right] \tag{9.65}$$

Corollary 35. *Let $\varsigma, \varpi, \eta, \sigma, \mu, s \in \mathbb{C}$ and $a, \nu \in \mathbb{C}\setminus\mathbb{Z}_0^-$ with $\rho, \tau, \kappa \in \mathbb{R}^+$, such that $\Re(\varsigma) \geq 0$ and $\Re(\sigma) < 1 + \min[\Re(-\varpi - n), \Re(\varsigma+\eta)]$, $n = [\Re(\varsigma)] + 1$.*

Then for $\Re(a) > 0$ *and* $\Re(s) > 0$, *following Saigo's fractional derivative* $\mathscr{D}_{-}^{\varsigma,\varpi,\eta}$ *of* $\Phi_{\mu;\nu}^{(\tau;\kappa)}(t,s,a)$ *holds true:*

$$\left(\mathscr{D}_{-}^{\varsigma,\varpi,\eta}\left\{t^{\sigma-1}\Phi_{\mu;\nu}^{(\tau;\kappa)}(t^{\rho},s,a)\right\}\right)(x) = x^{\sigma+\varpi-1}\frac{\Gamma(\nu)}{\Gamma(\mu)} \times I_{5,5}^{3,3}\left[-x^{\rho}\,\middle|\, \begin{matrix}(0,1;1),(1-\mu,\tau;1),(1-a,1;s),(1-\sigma,\rho;1),(1-\sigma+\eta-\varpi,\rho;1)\\(0,1;1),(1-\sigma-\varpi,\rho;1),(1-\sigma+\varsigma+\eta,\rho;1),(1-\nu,\kappa;1),(-a,1;s)\end{matrix}\right] \tag{9.66}$$

Corollary 36. *Let* $\varsigma, \varpi, \eta, \sigma, \mu, s \in \mathbb{C}$ *and* $a, \nu \in \mathbb{C}\setminus\mathbb{Z}_0^-$ *with* $\rho, \tau, \kappa \in \mathbb{R}^+$, *such that* $\Re(\varsigma) \geq 0$ *and* $\Re(\sigma) > 0$. *Then for* $\Re(a) > 0$ *and* $\Re(s) > 0$, *following Riemann-Liouville fractional differentiation* $\mathscr{D}_{0+}^{\varsigma}$ *of* $\Phi_{\mu;\nu}^{(\tau;\kappa)}(t,s,a)$ *holds true:*

$$\left(\mathscr{D}_{0+}^{\varsigma}\left\{t^{\sigma-1}\Phi_{\mu;\nu}^{(\tau;\kappa)}(t^{\rho},s,a)\right\}\right)(x) = x^{\sigma+\varsigma-1}\frac{\Gamma(\nu)}{\Gamma(\mu)} I_{4,4}^{1,4}\left[-x^{\rho}\,\middle|\, \begin{matrix}(0,1;1),(1-\mu,\tau;1),(1-a,1;s),(1-\sigma,\rho;1)\\(0,1;1),(1-\nu,\kappa;1),(-a,1;s),(1-\sigma+\varsigma,\rho;1)\end{matrix}\right]. \tag{9.67}$$

Corollary 37. *Let* $\varsigma, \varpi, \eta, \sigma, \mu, s \in \mathbb{C}$ *and* $a, \nu \in \mathbb{C}\setminus\mathbb{Z}_0^-$ *with* $\rho, \tau, \kappa \in \mathbb{R}^+$, *such that* $\Re(\varsigma) \geq 0$ *and* $\Re(\sigma) > -\Re(\varsigma+\eta)$. *Then for* $\Re(a) > 0$ *and* $\Re(s) > 0$, *following Erdélyi-Kober fractional derivative* $\mathscr{D}_{\eta,\varsigma}^{+}$ *of* $\Phi_{\mu;\nu}^{(\tau;\kappa)}(t,s,a)$ *holds true:*

$$\left(\mathscr{D}_{\eta,\varsigma}^{+}\left\{t^{\sigma-1}\Phi_{\mu;\nu}^{(\tau;\kappa)}(t^{\rho},s,a)\right\}\right)(x) = x^{\sigma-1}\frac{\Gamma(\nu)}{\Gamma(\mu)} I_{4,4}^{1,4}\left[-x^{\rho}\,\middle|\, \begin{matrix}(0,1;1),(1-\mu,\tau;1),(1-a,1;s),(1-\sigma-\varsigma-\eta,\rho;1)\\(0,1;1),(1-\nu,\kappa;1),(-a,1;s),(1-\sigma-\eta,\rho;1)\end{matrix}\right]. \tag{9.68}$$

Corollary 38. *Let* $\varsigma, \varpi, \eta, \sigma, \mu, s \in \mathbb{C}$ *and* $a, \nu \in \mathbb{C}\setminus\mathbb{Z}_0^-$ *with* $\rho, \tau, \kappa \in \mathbb{R}^+$, *such that* $\Re(\varsigma) \geq 0$ *and* $\Re(\sigma) < \Re(\varsigma) - [\Re(\varsigma)]$. *Then for* $\Re(a) > 0$ *and* $\Re(s) > 0$, *following Riemann-Liouville fractional differentiation* $\mathscr{D}_{-}^{\varsigma}$ *of* $\Phi_{\mu;\nu}^{(\tau;\kappa)}(t,s,a)$ *holds true:*

$$\left(\mathscr{D}_{-}^{\varsigma}\left\{t^{\sigma-1}\Phi_{\mu;\nu}^{(\tau;\kappa)}(t^{\rho},s,a)\right\}\right)(x) = x^{\sigma-\varsigma-1}\frac{\Gamma(\nu)}{\Gamma(\mu)} I_{4,4}^{2,3}\left[-x^{\rho}\,\middle|\, \begin{matrix}(0,1;1),(1-\mu,\tau;1),(1-a,1;s),(1-\sigma,\rho;1)\\(0,1;1),(1-\sigma+\varsigma,\rho;1),(1-\nu,\kappa;1),(-a,1;s)\end{matrix}\right]. \tag{9.69}$$

Corollary 39. *Let* $\varsigma, \varpi, \eta, \sigma, \mu, s \in \mathbb{C}$ *and* $a, \nu \in \mathbb{C}\setminus\mathbb{Z}_0^-$ *with* $\rho, \tau, \kappa \in \mathbb{R}^+$, *such that* $\Re(\varsigma) \geq 0$ *and* $\Re(\sigma) < \Re(\varsigma+\eta) - [\Re(\varsigma)]$. *Then for* $\Re(a) > 0$ *and* $\Re(s) > 0$, *following Erdélyi-Kober fractional differentiation* $\mathscr{D}_{\eta,\varsigma}^{-}$ *of* $\Phi_{\mu;\nu}^{(\tau;\kappa)}(t,s,a)$ *holds true:*

$$\left(\mathscr{D}_{\eta,\varsigma}^{-}\left\{t^{\sigma-1}\Phi_{\mu;\nu}^{(\tau;\kappa)}(t^{\rho},s,a)\right\}\right)(x) = x^{\sigma-1}\frac{\Gamma(\nu)}{\Gamma(\mu)} I_{4,4}^{2,3}\left[-x^{\rho}\,\middle|\, \begin{matrix}(0,1;1),(1-\mu,\tau;1),(1-a,1;s),(1-\sigma+\eta,\rho;1)\\(0,1;1),(1-\sigma+\varsigma+\eta,\rho;1),(1-\nu,\kappa;1),(-a,1;s)\end{matrix}\right]. \tag{9.70}$$

9.3.3 FRACTIONAL CALCULUS OPERATORS OF THE $\Phi_{\mu}^{*}(t,s,a)$

In this subsection, we shall derive generalized Saigo's fractional calculus results involving the generalized Hurwitz-Lerch Zeta function defined Goyal and Laddha [12, p. 100, Eq. (1.5)] for particular choices of parameters in extended Hurwitz-Lerch zeta function (9.14) given in Case 3. Further, special cases for the classical Riemann-Liouville and Erdélyi-Kober fractional integral and differential operators are deduced. The results are given in the following subsections as Corollaries.

9.3.3.1 Fractional Integration of the $\Phi_{\mu}^{*}(t,s,a)$

Corollary 40. *Let $\varsigma, \varpi, \eta, \sigma, \mu, s \in \mathbb{C}$ and $a, \in \mathbb{C}\setminus\mathbb{Z}_0^-$ with $\rho \in \mathbb{R}^+$, such that $\Re(\varsigma) > 0$ and $\Re(\sigma) > \max[0, \Re(\varpi - \eta)]$. Then for $\Re(a) > 0$ and $\Re(s) > 0$, following Saigo's fractional integral $\mathscr{I}_{0+}^{\varsigma,\varpi,\eta}$ of $\Phi_{\mu}^{*}(t,s,a)$ holds true:*

$$\left(\mathscr{I}_{0+}^{\varsigma,\varpi,\eta}\left\{t^{\sigma-1}\Phi_{\mu}^{*}(t^{\rho},s,a)\right\}\right)(x)$$

$$= x^{\sigma-\varpi-1}\frac{1}{\Gamma(\mu)}I_{4,4}^{1,4}\left[-x^{\rho}\,\middle|\, \begin{matrix}(1-\sigma,\rho;1),(1-\sigma+\varpi-\eta,\rho;1),(1-\mu,1;1),(1-a,1;s)\\(0,1),(-a,1;s),(1-\sigma+\varpi,\rho;1),(1-\sigma-\varsigma-\eta,\rho;1)\end{matrix}\right]. \tag{9.71}$$

Corollary 41. *Let $\varsigma, \varpi, \eta, \sigma, \mu, s \in \mathbb{C}$ and $a \in \mathbb{C}\setminus\mathbb{Z}_0^-$ with $\rho \in \mathbb{R}^+$, such that $\Re(\varsigma) > 0$ and $\Re(\sigma) < 1 + \min[\Re(\varpi), \Re(\eta)]$. Then for $\Re(a) > 0$ and $\Re(s) > 0$, following Saigo's fractional integral $\mathscr{I}_{-}^{\varsigma,\varpi,\eta}$ of $\Phi_{\mu}^{*}(t,s,a)$ holds true:*

$$\left(\mathscr{I}_{-}^{\varsigma,\varpi,\eta}\left\{t^{\sigma-1}\Phi_{\mu}^{*}(t^{\rho},s,a)\right\}\right)(x)$$

$$= x^{\sigma-\varpi-1}\frac{1}{\Gamma(\mu)}I_{4,4}^{3,2}\left[-x^{\rho}\,\middle|\, \begin{matrix}(1-\mu,1;1),(1-a,1;s),(1-\sigma,\rho;1),(1-\sigma+\varsigma+\varpi+\eta,\rho;1)\\(0,1;1),(1-\sigma+\varpi,\rho;1),(1-\sigma+\eta,\rho;1),(-a,1;s)\end{matrix}\right]. \tag{9.72}$$

Corollary 42. *Let $\varsigma, \varpi, \eta, \sigma, \mu, s \in \mathbb{C}$ and $a \in \mathbb{C}\setminus\mathbb{Z}_0^-$ with $\rho \in \mathbb{R}^+$, such that $\Re(\varsigma) > 0$ and $\Re(\sigma) > 0$. Then for $\Re(a) > 0$ and $\Re(s) > 0$, following Riemann-Liouville fractional integral $\mathscr{I}_{0+}^{\varsigma}$ of $\Phi_{\mu}^{*}(t,s,a)$ holds true:*

$$\left(\mathscr{I}_{0+}^{\varsigma}\left\{t^{\sigma-1}\Phi_{\mu}^{*}(t^{\rho},s,a)\right\}\right)(x)$$

$$= x^{\sigma+\varsigma-1}\frac{1}{\Gamma(\mu)}I_{3,3}^{1,3}\left[-x^{\sigma}\,\middle|\, \begin{matrix}(1-\mu,1;1),(1-a,1;s),(1-\sigma,\rho;1)\\(0,1;1),(-a,1;s),(1-\sigma-\varsigma,\rho;1)\end{matrix}\right]. \tag{9.73}$$

Corollary 43. *Let $\varsigma, \varpi, \eta, \sigma, \mu, s \in \mathbb{C}$ and $a \in \mathbb{C}\setminus\mathbb{Z}_0^-$ with $\rho \in \mathbb{R}^+$, such that $\Re(\varsigma) > 0$ and $\Re(\sigma) > -\Re(\eta)$. Then for $\Re(a) > 0$ and $\Re(s) > 0$, following Erdélyi-Kober fractional integral $\mathscr{I}_{\eta,\varsigma}^{+}$ of $\Phi_{\mu}^{*}(t,s,a)$ holds true:*

$$\left(\mathscr{I}_{\eta;\varsigma}^{+}\left\{t^{\sigma-1}\Phi_{\mu}^{*}(t^{\rho},s,a)\right\}\right)(x)$$

$$= x^{\sigma-1}\frac{1}{\Gamma(\mu)}I_{3,3}^{1,3}\left[-x^{\rho}\,\middle|\, \begin{matrix}(1-\mu,1;1),(1-a,1;s),(1-\sigma-\eta,\rho;1)\\(0,1;1),(-a,1;s),(1-\sigma-\varsigma-\eta,\rho;1)\end{matrix}\right]. \tag{9.74}$$

Corollary 44. *Let $\varsigma, \varpi, \eta, \sigma, \mu, s \in \mathbb{C}$ and $a \in \mathbb{C}\setminus\mathbb{Z}_0^-$ with $\rho \in \mathbb{R}^+$, such that $0 < \Re(\varsigma) < 1 - \Re(\sigma)$. Then for $\Re(a) > 0$ and $\Re(s) > 0$, following Riemann-Liouville fractional integral $\mathscr{I}_-^{\varsigma}$ of $\Phi_\mu^*(t,s,a)$ holds true:*

$$\left(\mathscr{I}_-^{\varsigma}\left\{t^{\sigma-1}\Phi_\mu^*(t^\rho,s,a)\right\}\right)(x)$$
$$= x^{\sigma+\varsigma-1}\frac{1}{\Gamma(\mu)} I_{3,3}^{2,2}\left[-x^\rho \,\middle|\, \begin{matrix}(1-\mu,1;1),(1-a,1;s),(1-\sigma,\rho;1)\\(0,1;1),(1-\sigma-\varsigma,\rho),(-a,1;s)\end{matrix}\right]. \tag{9.75}$$

Corollary 45. *Let $\varsigma, \varpi, \eta, \sigma, \mu, s \in \mathbb{C}$ and $a \in \mathbb{C}\setminus\mathbb{Z}_0^-$ with $\rho \in \mathbb{R}^+$, such that $\Re(\varsigma) > 0$ and $\Re(\sigma) < 1 + \Re(\eta)$. Then for $\Re(a) > 0$ and $\Re(s) > 0$, following Erdélyi-Kober fractional integral $K_{\eta,\varsigma}^-$ of $\Phi_\mu^*(t,s,a)$ holds true:*

$$\left(K_{\eta,\varsigma}^-\left\{t^{\sigma-1}\Phi_\mu^*(t^\rho,s,a)\right\}\right)(x)$$
$$= x^{\sigma-1}\frac{1}{\Gamma(\mu)} I_{3,3}^{2,2}\left[-x^\rho \,\middle|\, \begin{matrix}(1-\mu,1;1),(1-a,1;s),(1-\sigma+\varsigma+\eta,\rho;1)\\(0,1;1),(1-\sigma+\eta,\rho;1),(-a,1;s)\end{matrix}\right]. \tag{9.76}$$

9.3.3.2 Fractional Differentiation of the $\Phi_\mu^*(t,s,a)$

Corollary 46. *Let $\varsigma, \varpi, \eta, \sigma, \mu, s \in \mathbb{C}$ and $a \in \mathbb{C}\setminus\mathbb{Z}_0^-$ with $\rho \in \mathbb{R}^+$, such that $\Re(\varsigma) \geq 0$ and $\Re(\sigma) > -\min[0, \Re(\varsigma+\varpi+\eta)]$. Then for $\Re(a) > 0$ and $\Re(s) > 0$, following Saigo's fractional derivative $\mathscr{D}_{0+}^{\varsigma,\varpi,\eta}$ of $\Phi_\mu^*(t,s,a)$ holds true:*

$$\left(\mathscr{D}_{0+}^{\varsigma,\varpi,\eta}\left\{t^{\sigma-1}\Phi_\mu^*(t^\rho,s,a)\right\}\right)(x) = x^{\sigma+\varpi-1}\frac{1}{\Gamma(\mu)}$$
$$\times I_{4,4}^{1,4}\left[-x^\rho \,\middle|\, \begin{matrix}(1-\mu,1;1),(1-a,1;s),(1-\sigma,\rho;1),(1-\sigma-\varsigma-\varpi-\eta,\rho;1)\\(0,1;1),(-a,1;s),(1-\sigma-\varpi,\rho;1),(1-\sigma-\eta,\rho;1)\end{matrix}\right] \tag{9.77}$$

Corollary 47. *Let $\varsigma, \varpi, \eta, \sigma, \mu, s \in \mathbb{C}$ and $a \in \mathbb{C}\setminus\mathbb{Z}_0^-$ with $\rho \in \mathbb{R}^+$, such that $\Re(\varsigma) \geq 0$ and $\Re(\sigma) < 1 + \min[\Re(-\varpi-n), \Re(\varsigma+\eta)]$, $n = [\Re(\varsigma)] + 1$. Then for $\Re(a) > 0$ and $\Re(s) > 0$, following Saigo's fractional derivative $\mathscr{D}_-^{\varsigma,\varpi,\eta}$ of $\Phi_\mu^*(t,s,a)$ holds true:*

$$\left(\mathscr{D}_-^{\varsigma,\varpi,\eta}\left\{t^{\sigma-1}\Phi_\mu^*(t^\rho,s,a)\right\}\right)(x) = x^{\sigma+\varpi-1}\frac{1}{\Gamma(\mu)}$$
$$\times I_{4,4}^{3,2}\left[-x^\rho \,\middle|\, \begin{matrix}(1-\mu,1;1),(1-a,1;s),(1-\sigma,\rho;1),(1-\sigma+\eta-\varpi,\rho;1)\\(0,1;1),(-a,1;s),(1-\sigma-\varpi,\rho;1),(1-\sigma+\varsigma+\eta,\rho;1)\end{matrix}\right] \tag{9.78}$$

Corollary 48. *Let $\varsigma, \varpi, \eta, \sigma, \mu, s \in \mathbb{C}$ and $a \in \mathbb{C} \setminus \mathbb{Z}_0^-$ with $\rho \in \mathbb{R}^+$, such that $\Re(\varsigma) \geq 0$ and $\Re(\sigma) > 0$. Then for $\Re(a) > 0$ and $\Re(s) > 0$, following Riemann-Liouville fractional differentiation $\mathscr{D}_{0+}^{\varsigma}$ of $\Phi_\mu^*(t,s,a)$ holds true:*

$$\left(\mathscr{D}_{0+}^{\varsigma}\left\{t^{\sigma-1}\Phi_\mu^*(t^\rho,s,a)\right\}\right)(x)$$
$$= x^{\sigma+\varsigma-1}\frac{1}{\Gamma(\mu)}I_{3,3}^{1,3}\left[-x^\rho \,\middle|\, \begin{matrix}(1-\mu,1;1),(1-a,1;s),(1-\sigma,\rho;1)\\(0,1;1),(-a,1;s),(1-\sigma+\varsigma,\rho;1)\end{matrix}\right]. \tag{9.79}$$

Corollary 49. *Let $\varsigma, \varpi, \eta, \sigma, \mu, s \in \mathbb{C}$ and $a \in \mathbb{C} \setminus \mathbb{Z}_0^-$ with $\rho \in \mathbb{R}^+$, such that $\Re(\varsigma) \geq 0$ and $\Re(\sigma) > -\Re(\varsigma+\eta)$. Then for $\Re(a) > 0$ and $\Re(s) > 0$, following Erdélyi-Kober fractional derivative $\mathscr{D}_{\eta,\varsigma}^{+}$ of $\Phi_\mu^*(t,s,a)$ holds true:*

$$\left(\mathscr{D}_{\eta,\varsigma}^{+}\left\{t^{\sigma-1}\Phi_\mu^*(t^\rho,s,a)\right\}\right)(x)$$
$$= x^{\sigma-1}\frac{1}{\Gamma(\mu)}I_{3,3}^{1,3}\left[-x^\rho \,\middle|\, \begin{matrix}(1-\mu,1;1),(1-a,1;s),(1-\sigma-\varsigma-\eta,\rho;1)\\(0,1;1),(-a,1;s),(1-\sigma-\eta,\rho;1)\end{matrix}\right]. \tag{9.80}$$

Corollary 50. *Let $\varsigma, \varpi, \eta, \sigma, \mu, s \in \mathbb{C}$ and $a \in \mathbb{C} \setminus \mathbb{Z}_0^-$ with $\rho \in \mathbb{R}^+$, such that $\Re(\varsigma) \geq 0$ and $\Re(\sigma) < \Re(\varsigma) - [\Re(\varsigma)]$. Then for $\Re(a) > 0$ and $\Re(s) > 0$, following Riemann-Liouville fractional differentiation $\mathscr{D}_{-}^{\varsigma}$ of $\Phi_\mu^*(t,s,a)$ holds true:*

$$\left(\mathscr{D}_{-}^{\varsigma}\left\{t^{\sigma-1}\Phi_\mu^*(t^\rho,s,a)\right\}\right)(x)$$
$$= x^{\sigma-\varsigma-1}\frac{1}{\Gamma(\mu)}I_{3,3}^{2,2}\left[-x^\rho \,\middle|\, \begin{matrix}(1-\mu,1;1),(1-a,1;s),(1-\sigma,\rho;1)\\(0,1;1),(1-\sigma+\varsigma,\rho;1),(-a,1;s)\end{matrix}\right]. \tag{9.81}$$

Corollary 51. *Let $\varsigma, \varpi, \eta, \sigma, \mu, s \in \mathbb{C}$ and $a \in \mathbb{C} \setminus \mathbb{Z}_0^-$ with $\rho \in \mathbb{R}^+$, such that $\Re(\varsigma) \geq 0$ and $\Re(\sigma) < \Re(\varsigma+\eta) - [\Re(\varsigma)]$. Then for $\Re(a) > 0$ and $\Re(s) > 0$, following Erdélyi-Kober fractional differentiation $\mathscr{D}_{\eta,\varsigma}^{-}$ of $\Phi_\mu^*(t,s,a)$ holds true:*

$$\left(\mathscr{D}_{\eta,\varsigma}^{-}\left\{t^{\sigma-1}\Phi_\mu^*(t^\rho,s,a)\right\}\right)(x)$$
$$= x^{\sigma-1}\frac{1}{\Gamma(\mu)}I_{3,3}^{2,2}\left[-x^\rho \,\middle|\, \begin{matrix}(1-\mu,1;1),(1-a,1;s),(1-\sigma+\eta,\rho;1)\\(0,1;1),(1-\sigma+\varsigma+\eta,\rho;1),(-a,1;s)\end{matrix}\right]. \tag{9.82}$$

9.4 FURTHER OBSERVATIONS AND APPLICATIONS

The I-function contains several special cases as various special functions, for example, Inayat-Hussein $\overline{H}$-function, Fox H-function and Meijer's G-function. We recall the definition of the generalized H–function, that is the $\overline{H}$–function introduced by Inayat-Hussain in the form(see, for details, [13,14,21]):

$$\overline{H}(z) = \overline{H}_{p,q}^{m,n}[z] = \overline{H}_{p,q}^{m,n}\left[z \,\middle|\, \begin{matrix}(a_j,A_j;\varsigma_j)_{\overline{1,n}},\ldots,(a_j,A_j)_{\overline{n+1,p}}\\(b_j,B_j)_{\overline{1,m}},\ldots,(b_j,B_j;\varpi_j)_{\overline{m+1,q}}\end{matrix}\right]$$
$$= \frac{1}{2\pi i}\int_{\pounds}\chi(s)\,z^s\,ds \tag{9.83}$$

for all $z \neq 0$, where

$$\chi(s) = \frac{\prod_{j=1}^{m} \Gamma(b_j - B_j s) \prod_{j=1}^{n} \{\Gamma(1 - a_j + A_j s)\}^{\varsigma_j}}{\prod_{j=n+1}^{p} \Gamma(a_j - A_j s) \prod_{j=m+1}^{q} \{\Gamma(1 - b_j + B_j s)\}^{\varpi_j}} \tag{9.84}$$

which contains fractional powers of some of the gamma functions. Here, and in what follows, a_j $(j = 1, \ldots, p)$ and b_j $(j = 1, \ldots, q)$ are the complex parameters,

$$A_j > 0\ (j = 1, \ldots, p) \text{ and } B_j > 0\ (j = 1, \ldots, q),$$

the exponents

$$\varsigma_j\ (j = 1, \ldots, n) \text{ and } \varpi_j\ (j = m+1, \ldots, q)$$

can take non-integer values and $\pounds = \pounds_{(ic;\infty)}$ is a suitable contour of the Mellin-Barnes type starting at the point $c - i\infty$ and terminating at the point $c + i\infty (c \in \mathbb{R})$ with the usual indentations to separate one set of poles from other set of poles.

It has been established by Buschman and Srivastava [3, p. 4708] that the sufficient conditions for the absolute convergence of the contour integral in Eq. (9.83) is given by

$$\wedge = \sum_{j=1}^{m} B_j + \sum_{j=1}^{n} |\varsigma_j| A_j - \sum_{j=m+1}^{q} |\varpi_j| B_j - \sum_{j=n+1}^{p} A_j > 0.$$

This condition provides exponential decay of the integrand in Eq. (9.83), and region of absolute convergence of the contour integral (9.83) is given by

$$|\arg(z)| < \frac{\pi}{2} \wedge.$$

In previous section investigation, we established various fractional calculus results for the extended Hurwitz-Lerch Zeta $\Phi_{\lambda,\mu;\nu}^{(\rho,\tau;\kappa)}(t,s,a)$ by using the tools of *Saigo's*, *Riemann-Liouville* and *Erdélyi-Kober* fractional operators. From an application point of view, all the results obtained in previous sections can be written in terms of $\overline{H}$–function as particular cases. Here we mention results for extended Hurwitz-Lerch Zeta $\Phi_{\lambda,\mu;\nu}^{(\rho,\tau;\kappa)}(t,s,a)$.

Example 9.1. *Let* $\varsigma, \varpi, \eta, \sigma, \lambda, \mu, s \in \mathbb{C}$ *and* $a, \nu \in \mathbb{C} \setminus \mathbb{Z}_0^-$ *with* $\rho, \rho, \tau, \kappa \in \mathbb{R}^+$, *such that* $\Re(\varsigma) > 0$ *and* $\Re(\sigma) > \max[0, \Re(\varpi - \eta)]$. *Then for* $\Re(a) > 0$ *and* $\Re(s) > 0$, *following Saigo's fractional integral* $\mathscr{I}_{0+}^{\varsigma,\varpi,\eta}$ *of* $\Phi_{\lambda,\mu;\nu}^{(\rho,\tau;\kappa)}(t,s,a)$ *holds true:*

$$\left(\mathscr{I}_{0+}^{\varsigma,\varpi,\eta}\left\{t^{\sigma-1}\Phi_{\lambda,\mu;\nu}^{(\rho,\tau;\kappa)}(t^{\rho},s,a)\right\}\right)(x) = x^{\sigma-\varpi-1}\frac{\Gamma(\nu)}{\Gamma(\lambda)\Gamma(\mu)}$$
$$\times \overline{H}_{5,5}^{1,5}\left[-x^{\rho}\left|\begin{matrix}(1-\sigma,\rho;1),(1-\sigma+\varpi-\eta,\rho;1),(1-\lambda,\rho;1),(1-\mu,\tau;1),(1-a,1;s)\\(0,1),(1-\nu,\kappa;1),(-a,1;s),(1-\sigma+\varpi,\rho;1),(1-\sigma-\varsigma-\eta,\rho;1)\end{matrix}\right.\right]. \tag{9.85}$$

Example 9.2. *Let* $\varsigma, \varpi, \eta, \sigma, \lambda, \mu, s \in \mathbb{C}$ *and* $a, \nu \in \mathbb{C}\setminus\mathbb{Z}_0^-$ *with* $\rho, \rho, \tau, \kappa \in \mathbb{R}^+$, *such that* $\Re(\varsigma) > 0$ *and* $\Re(\sigma) < 1+\min[\Re(\varpi), \Re(\eta)]$. *Then for* $\Re(a) > 0$ *and* $\Re(s) > 0$, *following Saigo's fractional integral* $\mathscr{I}_{-}^{\varsigma,\varpi,\eta}$ *of* $\Phi_{\lambda,\mu;\nu}^{(\rho,\tau;\kappa)}(t,s,a)$ *holds true:*

$$\left(\mathscr{I}_{-}^{\varsigma,\varpi,\eta}\left\{t^{\sigma-1}\Phi_{\lambda,\mu;\nu}^{(\rho,\tau;\kappa)}(t^{\rho},s,a)\right\}\right)(x) = x^{\sigma-\varpi-1}\frac{\Gamma(\nu)}{\Gamma(\lambda)\Gamma(\mu)}$$
$$\times \overline{H}_{5,5}^{3,3}\left[-x^{\rho}\left|\begin{matrix}(1-\sigma,\rho),(1-\sigma+\varsigma+\varpi+\eta,\rho),(1-\lambda,\rho;1),(1-\mu,\tau;1),(1-a,1;s)\\(0,1),(1-\nu,\kappa;1),(-a,1;s),(1-\sigma+\varpi,\rho),(1-\sigma+\eta,\rho)\end{matrix}\right.\right]. \tag{9.86}$$

Example 9.3. *Let* $\varsigma, \varpi, \eta, \sigma, \lambda, \mu, s \in \mathbb{C}$ *and* $a, \nu \in \mathbb{C}\setminus\mathbb{Z}_0^-$ *with* $\rho, \rho, \tau, \kappa \in \mathbb{R}^+$, *such that* $\Re(\varsigma) \geq 0$ *and* $\Re(\sigma) > -\min[0, \Re(\varsigma+\varpi+\eta)]$. *Then for* $\Re(a) > 0$ *and* $\Re(s) > 0$, *following Saigo's fractional derivative* $\mathscr{D}_{0+}^{\varsigma,\varpi,\eta}$ *of* $\Phi_{\lambda,\mu;\nu}^{(\rho,\tau;\kappa)}(t,s,a)$ *holds true:*

$$\left(\mathscr{D}_{0+}^{\varsigma,\varpi,\eta}\left\{t^{\sigma-1}\Phi_{\lambda,\mu;\nu}^{(\rho,\tau;\kappa)}(t^{\rho},s,a)\right\}\right)(x) = x^{\sigma+\varpi-1}\frac{\Gamma(\nu)}{\Gamma(\lambda)\Gamma(\mu)}$$
$$\times \overline{H}_{5,5}^{1,5}\left[-x^{\rho}\left|\begin{matrix}(1-\sigma,\rho;1),(1-\sigma-\varsigma-\varpi-\eta,\rho;1),,(1-\lambda,\rho;1),(1-\mu,\tau;1),(1-a,1;s)\\(0,1),(1-\nu,\kappa;1),(-a,1;s),(1-\sigma-\varpi,\rho;1),(1-\sigma-\eta,\rho;1)\end{matrix}\right.\right]. \tag{9.87}$$

Example 9.4. *Let* $\varsigma, \varpi, \eta, \sigma, \lambda, \mu, s \in \mathbb{C}$ *and* $a, \nu \in \mathbb{C}\setminus\mathbb{Z}_0^-$ *with* $\rho, \rho, \tau, \kappa \in \mathbb{R}^+$, *such that* $\Re(\varsigma) \geq 0$ *and* $\Re(\sigma) < 1+\min[\Re(-\varpi-n), \Re(\varsigma+\eta)]$, $n = [\Re(\varsigma)]+1$. *Then for* $\Re(a) > 0$ *and* $\Re(s) > 0$, *following Saigo's fractional derivative* $\mathscr{D}_{-}^{\varsigma,\varpi,\eta}$ *of* $\Phi_{\lambda,\mu;\nu}^{(\rho,\tau;\kappa)}(t,s,a)$ *holds true:*

$$\left(\mathscr{D}_{-}^{\varsigma,\varpi,\eta}\left\{t^{\sigma-1}\Phi_{\lambda,\mu;\nu}^{(\rho,\tau;\kappa)}(t^{\rho},s,a)\right\}\right)(x) = x^{\sigma+\varpi-1}\frac{\Gamma(\nu)}{\Gamma(\lambda)\Gamma(\mu)}$$
$$\times \overline{H}_{5,5}^{3,3}\left[-x^{\rho}\left|\begin{matrix}(1-\lambda,\rho;1),(1-\mu,\tau;1),(1-a,1;s),(1-\sigma,\rho),(1-\sigma+\eta-\varpi,\rho)\\(0,1),(1-\nu,\kappa;1),(-a,1;s),(1-\sigma-\varpi,\rho),(1-\sigma+\varsigma+\eta,\rho)\end{matrix}\right.\right]. \tag{9.88}$$

Now we deduce the fractional integral formulas for the classical Riemann-Liouville and Erdélyi-Kober by letting $\varpi = -\varsigma$ and $\varpi = 0$ respectively, which are asserted by Corollaries 52 to 59 given below.

Corollary 52. *Let* $\varsigma, \varpi, \eta, \sigma, \lambda, \mu, s \in \mathbb{C}$ *and* $a, \nu \in \mathbb{C}\setminus\mathbb{Z}_0^-$ *with* $\rho, \rho, \tau, \kappa \in \mathbb{R}^+$, *such that* $\Re(\varsigma) > 0$ *and* $\Re(\sigma) > 0$. *Then for* $\Re(a) > 0$ *and* $\Re(s) > 0$, *following Riemann-Liouville fractional integral* $\mathscr{I}_{0+}^{\varsigma}$ *of* $\Phi_{\lambda,\mu;\nu}^{(\rho,\tau;\kappa)}(t,s,a)$ *holds true:*

$$\left(\mathscr{I}_{0+}^{\varsigma}\left\{t^{\sigma-1}\Phi_{\lambda,\mu;\nu}^{(\rho,\tau;\kappa)}(t^{\rho},s,a)\right\}\right)(x)$$
$$= x^{\sigma+\varsigma-1}\frac{\Gamma(\nu)}{\Gamma(\lambda)\Gamma(\mu)}\overline{H}_{4,4}^{1,4}\left[-x^{\sigma}\left|\begin{matrix}(1-\sigma,\rho;1),(1-\lambda,\rho;1),(1-\mu,\tau;1),(1-a,1;s)\\(0,1),(1-\nu,\kappa;1),(-a,1;s),(1-\sigma-\varsigma,\rho;1)\end{matrix}\right.\right]. \tag{9.89}$$

Corollary 53. *Let $\varsigma, \varpi, \eta, \sigma, \lambda, \mu, s \in \mathbb{C}$ and $a, \nu \in \mathbb{C}\setminus\mathbb{Z}_0^-$ with $\rho, \rho, \tau, \kappa \in \mathbb{R}^+$, such that $\Re(\varsigma) > 0$ and $\Re(\sigma) > -\Re(\eta)$. Then for $\Re(a) > 0$ and $\Re(s) > 0$, following Erdélyi-Kober fractional integral $\mathscr{I}_{\eta,\varsigma}^{+}$ of $\Phi_{\lambda,\mu;\nu}^{(\rho,\tau;\kappa)}(t,s,a)$ holds true:*

$$\left(\mathscr{I}_{\eta,\varsigma}^{+}\left\{t^{\sigma-1}\Phi_{\lambda,\mu;\nu}^{(\rho,\tau;\kappa)}(t^{\rho},s,a)\right\}\right)(x)$$
$$= x^{\sigma-1}\frac{\Gamma(\nu)}{\Gamma(\lambda)\Gamma(\mu)}\overline{H}_{4,4}^{1,4}\left[-x^{\rho}\left|\begin{array}{c}(1-\sigma-\eta,\rho;1),(1-\lambda,\rho;1),(1-\mu,\tau;1),(1-a,1;s)\\(0,1),(1-\nu,\kappa;1),(-a,1;s),(1-\sigma-\varsigma-\eta,\rho;1)\end{array}\right.\right]. \tag{9.90}$$

Corollary 54. *Let $\varsigma, \varpi, \eta, \sigma, \lambda, \mu, s \in \mathbb{C}$ and $a, \nu \in \mathbb{C}\setminus\mathbb{Z}_0^-$ with $\rho, \rho, \tau, \kappa \in \mathbb{R}^+$, such that $0 < \Re(\varsigma) < 1-\Re(\sigma)$. Then for $\Re(a) > 0$ and $\Re(s) > 0$, following Riemann-Liouville fractional integral $\mathscr{I}_{-}^{\varsigma}$ of $\Phi_{\lambda,\mu;\nu}^{(\rho,\tau;\kappa)}(t,s,a)$ holds true:*

$$\left(\mathscr{I}_{-}^{\varsigma}\left\{t^{\sigma-1}\Phi_{\lambda,\mu;\nu}^{(\rho,\tau;\kappa)}(t^{\rho},s,a)\right\}\right)(x)$$
$$= x^{\sigma+\varsigma-1}\frac{\Gamma(\nu)}{\Gamma(\lambda)\Gamma(\mu)}\overline{H}_{4,4}^{2,3}\left[-x^{\rho}\left|\begin{array}{c}(1-\lambda,\rho;1),(1-\mu,\tau;1),(1-a,1;s),(1-\sigma,\rho)\\(0,1),(1-\sigma-\varsigma,\rho),(1-\nu,\kappa;1),(-a,1;s)\end{array}\right.\right]. \tag{9.91}$$

Corollary 55. *Let $\varsigma, \varpi, \eta, \sigma, \lambda, \mu, s \in \mathbb{C}$ and $a, \nu \in \mathbb{C}\setminus\mathbb{Z}_0^-$ with $\rho, \rho, \tau, \kappa \in \mathbb{R}^+$, such that $\Re(\varsigma) > 0$ and $\Re(\sigma) < 1+\Re(\eta)$. Then for $\Re(a) > 0$ and $\Re(s) > 0$, following Erdélyi-Kober fractional integral $K_{\eta,\varsigma}^{-}$ of $\Phi_{\lambda,\mu;\nu}^{(\rho,\tau;\kappa)}(t,s,a)$ holds true:*

$$\left(K_{\eta,\varsigma}^{-}\left\{t^{\sigma-1}\Phi_{\lambda,\mu;\nu}^{(\rho,\tau;\kappa)}(t^{\rho},s,a)\right\}\right)(x)$$
$$= x^{\sigma-1}\frac{\Gamma(\nu)}{\Gamma(\lambda)\Gamma(\mu)}\overline{H}_{4,4}^{2,3}\left[-x^{\rho}\left|\begin{array}{c}(1-\lambda,\rho;1),(1-\mu,\tau;1),(1-a,1;s),(1-\sigma+\varsigma+\eta,\rho)\\(0,1),(1-\sigma+\eta,\rho),(1-\nu,\kappa;1),(-a,1;s)\end{array}\right.\right]. \tag{9.92}$$

Corollary 56. *Let $\varsigma, \varpi, \eta, \sigma, \lambda, \mu, s \in \mathbb{C}$ and $a, \nu \in \mathbb{C}\setminus\mathbb{Z}_0^-$ with $\rho, \rho, \tau, \kappa \in \mathbb{R}^+$, such that $\Re(\varsigma) \geq 0$ and $\Re(\sigma) > 0$. Then for $\Re(a) > 0$ and $\Re(s) > 0$, following Riemann-Liouville fractional differentiation $\mathscr{D}_{0+}^{\varsigma}$ of $\Phi_{\lambda,\mu;\nu}^{(\rho,\tau;\kappa)}(t,s,a)$ holds true:*

$$\left(\mathscr{D}_{0+}^{\varsigma}\left\{t^{\sigma-1}\Phi_{\lambda,\mu;\nu}^{(\rho,\tau;\kappa)}(t^{\rho},s,a)\right\}\right)(x)$$
$$= x^{\sigma+\varsigma-1}\frac{\Gamma(\nu)}{\Gamma(\lambda)\Gamma(\mu)}\overline{H}_{4,4}^{1,4}\left[-x^{\rho}\left|\begin{array}{c}(1-\sigma,\rho;1),(1-\lambda,\rho;1),(1-\mu,\tau;1),(1-a,1;s)\\(0,1),(1-\nu,\kappa;1),(-a,1;s),(1-\sigma+\varsigma,\rho;1)\end{array}\right.\right]. \tag{9.93}$$

Corollary 57. *Let $\varsigma, \varpi, \eta, \sigma, \lambda, \mu, s \in \mathbb{C}$ and $a, \nu \in \mathbb{C}\setminus\mathbb{Z}_0^-$ with $\rho, \rho, \tau, \kappa \in \mathbb{R}^+$, such that $\Re(\varsigma) \geq 0$ and $\Re(\sigma) > -\Re(\varsigma+\eta)$. Then for $\Re(a) > 0$ and $\Re(s) > 0$, following Erdélyi-Kober fractional derivative $\mathscr{D}_{\eta,\varsigma}^{+}$ of $\Phi_{\lambda,\mu;\nu}^{(\rho,\tau;\kappa)}(t,s,a)$ holds true:*

$$\left(\mathscr{D}_{\eta,\varsigma}^{+}\left\{t^{\sigma-1}\Phi_{\lambda,\mu;\nu}^{(\rho,\tau;\kappa)}(t^{\rho},s,a)\right\}\right)(x)$$
$$=x^{\sigma-1}\frac{\Gamma(\nu)}{\Gamma(\lambda)\Gamma(\mu)}\overline{H}_{4,4}^{1,4}\left[-x^{\rho}\left|\begin{matrix}(1-\sigma-\varsigma-\eta,\rho;1),(1-\lambda,\rho;1),(1-\mu,\tau;1),(1-a,1;s)\\(0,1),(1-\nu,\kappa;1),(-a,1;s),(1-\sigma-\eta,\rho;1)\end{matrix}\right.\right]. \tag{9.94}$$

Corollary 58. *Let $\varsigma, \varpi, \eta, \sigma, \lambda, \mu, s \in \mathbb{C}$ and $a, \nu \in \mathbb{C}\setminus\mathbb{Z}_0^-$ with $\rho, \rho, \tau, \kappa \in \mathbb{R}^+$, such that $\Re(\varsigma) \geq 0$ and $\Re(\sigma) < \Re(\varsigma) - [\Re(\varsigma)]$. Then for $\Re(a) > 0$ and $\Re(s) > 0$, following Riemann-Liouville fractional differentiation $\mathscr{D}_{-}^{\varsigma}$ of $\Phi_{\lambda,\mu;\nu}^{(\rho,\tau;\kappa)}(t,s,a)$ holds true:*

$$\left(\mathscr{D}_{-}^{\varsigma}\left\{t^{\sigma-1}\Phi_{\lambda,\mu;\nu}^{(\rho,\tau;\kappa)}(t^{\rho},s,a)\right\}\right)(x)$$
$$=x^{\sigma-\varsigma-1}\frac{\Gamma(\nu)}{\Gamma(\lambda)\Gamma(\mu)}\overline{H}_{4,4}^{2,3}\left[-x^{\rho}\left|\begin{matrix}(1-\lambda,\rho;1),(1-\mu,\tau;1),(1-a,1;s),(1-\sigma,\rho)\\(0,1),(1-\sigma+\varsigma,\rho),(1-\nu,\kappa;1),(-a,1;s)\end{matrix}\right.\right]. \tag{9.95}$$

Corollary 59. *Let $\varsigma, \varpi, \eta, \sigma, \lambda, \mu, s \in \mathbb{C}$ and $a, \nu \in \mathbb{C}\setminus\mathbb{Z}_0^-$ with $\rho, \rho, \tau, \kappa \in \mathbb{R}^+$, such that $\Re(\varsigma) \geq 0$ and $\Re(\sigma) < \Re(\varsigma+\eta) - [\Re(\varsigma)]$. Then for $\Re(a) > 0$ and $\Re(s) > 0$, following Erdélyi-Kober fractional differentiation $\mathscr{D}_{\eta,\varsigma}^{-}$ of $\Phi_{\lambda,\mu;\nu}^{(\rho,\tau;\kappa)}(t,s,a)$ holds true:*

$$\left(\mathscr{D}_{\eta,\varsigma}^{-}\left\{t^{\sigma-1}\Phi_{\lambda,\mu;\nu}^{(\rho,\tau;\kappa)}(t^{\rho},s,a)\right\}\right)(x)$$
$$=x^{\sigma-1}\frac{\Gamma(\nu)}{\Gamma(\lambda)\Gamma(\mu)}\overline{H}_{4,4}^{2,3}\left[-x^{\rho}\left|\begin{matrix}(1-\lambda,\rho;1),(1-\mu,\tau;1),(1-a,1;s),(1-\sigma+\eta,\rho)\\(0,1),(1-\sigma+\varsigma+\eta,\rho),(1-\nu,\kappa;1),(-a,1;s)\end{matrix}\right.\right]. \tag{9.96}$$

Here, we have mentioned only results for Saigo's fractional operators, Riemann-Liouville(R-L) fractional operators and Erdélyi-Kober(E-K) fractional operators for the extended Hurwitz-Lerch Zeta $\Phi_{\lambda,\mu;\nu}^{(\rho,\tau;\kappa)}(t,s,a)$ and remaining are left as an exercise for the interested readers.

9.5 CONCLUDING REMARKS

Our present investigation is motivated essentially by many potential avenues of applications of various generalized fractional integrations and differentiations formulas for known special functions. In this chapter, we have systematically investigated the generalized Saigo's hypergeometric fractional calculus operators to establish a number of key results for the families of extended Hurwitz-Lerch Zeta function defined by many authors, including (for example) involving Srivastava et al. [46], Garg et al. [11], Lin and Srivastava [16], and Goyal and Laddha [12]. The results are obtained in terms of the I–function introduced by Rathie [28]. Corresponding assertions as special case results are obtained for particular choices of parameters for the classical Riemann-Liouville and Erdélyi-Kober fractional integral and differential operators. Furthermore, we also observe that all the results derived in paper can also be represented in terms of generalized H–function, that is $\overline{H}$-function introduced by Inayat-Hussain [13,14]. Mathematical modeling of real-world problems usually leads to fractional differential equations and various other problems involving special functions in mathematical physics and their extensions and generalizations in one or more variables [8,23,27,47]. The results obtained in this chapter are also useful in such diverse and widely used fields of engineering and sciences such as electromagnetism, viscoelasticity, fluid dynamics, electrochemistry, biological population modeling, optics, and signal processing.

References

1. R. P. Agarwal, A. Kılıcman, R. K. Parmar and A. K. Rathie, Certain generalized fractional calculus formulas and integral transforms involving (p,q)–Mathieu-type series, *Adv. Differ. Equ.* **221** (2019), 1–11 , https://doi.org/10.1186/s13662-019-2142-0.
2. E. W. Barnes, The asymptotic expansion of integral functions defined by Taylor series, *Philos. Trans. Roy. Soc. London Ser. A* **206** (1906), 249–297.
3. R. G. Buschman and H. M. Srivastava, The H–function associated with a certain class of Feynman integrals, *J. Phys. A: Math. Gen.* **23** (1990), 4707–4710.
4. J. Choi, D. S. Jang and H. M. Srivastava, A generalization of the Hurwitz-Lerch Zeta function, *Integ. Transforms Spec. Funct.* **19** (2008), 65–79.
5. J. Choi and R. K. Parmar, An extension of the generalized Hurwitz-Lerch zeta function of two variables, *Filomat*, **31**(1) (2017), 91–96.
6. J. Choi and R. K. Parmar, Fractional integration and differentiation of the (p,q)–extended Bessel function, *Bull. Korean Math. Soc.*, **55**(2) (2018), 599–610.
7. J. Choi and R. K. Parmar, Fractional calculus of the (p,q)–extended Struve function, *Far. East J. Math. Sci.*, **103**(2) (2018), 541–559.
8. J. Choi, R. K. Parmar and P. Chopra, Extended Mittag-Leffler function and associated fractional calculus operators, *Georgian Math. J.*, **27(2)** (2020), 199–209.
9. J. Choi, R. K. Parmar and R. K. Raina, Extension of generalized Hurwitz Lerch Zeta function and associated properties, *Kyungpook Math. J.*, **57**(3) (2017), 393–400.
10. A. Erdélyi, W. Magnus, F. Oberhettinger and F. G. Tricomi, *Higher Transcendental Functions*, Vol. **I**. McGraw-Hill Book Company: New York, Toronto and London, 1953.
11. M. Garg, K. Jain and S. L. Kalla, A further study of general Hurwitz-Lerch zeta function, *Algebras Groups Geom.* **25** (2008), 311–319.
12. S. P. Goyal and R. K. Laddha, On the generalized Zeta function and the generalized Lambert function, *Ganita Sandesh* **11** (1997), 99–108.
13. A. A. Inayat-Hussain, New properties of hypergeometric series derivable from Feynman integrals. I: Transformation and reduction formulae, *J. Phys. A Math. Gen.* **20** (1987), 4109–4117.
14. A. A. Inayat-Hussain, New properties of hypergeometric series derivable from Feynman integrals. II: A generalization of the H-function, *J. Phys. A Math. Gen.* **20** (1987), 4119–4128.
15. D. Jankov, T. K. Pogany and R. K. Saxena, An extended general Hurwitz–Lerch Zeta function as a Mathieu (a,λ)-series, *Appl. Margematics Lett.*, **24** (2011), 1473–1476.
16. S. D. Lin and H. M. Srivastava, Some families of the Hurwitz-Lerch Zeta functions and associated fractional derivative and other integral representations, *Appl. Math. Comput.*, **154** (2004), 725–733.

17. A. A. Kilbas, H. M. Srivastava and J. J. Trujillo, *Theory and Applications of Fractional Differential Equations*, North-Holland Mathematical Studies, Vol. **204**. Elsevier (North-Holland) Science Publishers: Amsterdam, London and New York, 2006.
18. R. K. Parmar and R. K. Saxena, The incomplete generalized τ-hypergeometric and second τ-Appell functions, *J. Korean Math. Soc.*, **53**(2) (2016),363–379.
19. S. G. Samko, A. A. Kilbas and O. I. Marichev, *Fractional Integrals and Derivatives: Theory and Applications, Translated from the Russian: Integrals and Derivatives of Fractional Order and Some of Their Applications ("Nauka i Tekhnika", Minsk, 1987)*. Gordon and Breach Science Publishers: Reading, UK, 1993.
20. M. J. Luo, R. K. Parmar and R. K. Raina, On extended Hurwitz–Lerch zeta function, *J. Math. Anal. Appl.* **448** (2017), 1281–1304.
21. A. M. Mathai, R. K. Saxena and H. J. Haubold, *The H-Functions: Theory and Applications*. Springer: New York, 2010.
22. M. Saigo, A remark on integral operators involvig the Gauss hypergeometric functions, *Math. Rep. Kyushu Univ.* **11** (1977/78), 135–143.
23. , R. K. Parmar, A class of extended Mittag-Leffler functions and their properties related to integral transforms and fractional calculus, *Mathematics* **3(4)** (2015), 1069–1082.
24. R. K Parmar, J. Choi and S. D. Purohit, Further generalization of the extended Hurwitz-Lerch Zeta functions, *Bol. Soc. Paranaense de Matemática* **37**(1) (2019), 177–190.
25. R. K. Parmar and R. K. Saxena, Incomplete extended Hurwitz-Lerch Zeta functions and associated properties, *Commun. Korean Math. Soc.*, **32** (2017), 287–304.
26. R. K. Parmar and R. K. Raina, On a certain extension of the Hurwitz-Lerch Zeta function, *Ann. West Univ. Timisoara-Math.*, **52(2)** (2014), 157–170.
27. T. K. Pogány and R. K. Parmar, On *p*–extended Mathieu series, *Rad. Hrvat. Akad. Znan. Umjet. Mat. Znan.* **22** (2018), 107–117.
28. A. K. Rathie, A new generalization of generalized hypergeometric functions, *Le Matematiche* **52(2)** (1997), 297–310.
29. H. Singh and A. M. Wazwaz, Computational method for reaction diffusion-model arising in a spherical catalyst, *Int. J. Appl. Comput. Math.* **7(3)** (2021), 65.
30. H. Singh, Analysis for fractional dynamics of Ebola virus model, *Chaos Solitons Fractals*, **138** (2020), 109992,
31. H. Singh, Solving a class of local and nonlocal elliptic boundary value problems arising in heat transfer, *Heat Transfer*, **51** (2021), 1524–1542.
32. H. Singh, An efficient computational method for non-linear fractional Lienard equation arising in oscillating circuits. In: H. Singh, D. Kumar and D. Baleanu (Eds.), *Methods of Mathematical Modelling: Fractional Differential Equations*, pp. 39–50. CRC Press, Taylor & Francis Group: Boca Raton, FL, 2019.
33. H. Singh, Chebyshev spectral method for solving a class of local and nonlocal elliptic boundary value problems, *Int. J. Nonlinear Sci. Numer. Simul.*, (2021), 000010151520200235.
34. H. Singh and H. M. Srivastava, Numerical investigation of the fractional-order Liénard and Duffing equations arising in oscillating circuit theory, *Front. Phys.* **8** (2020), 120.
35. H. Singh, H. M. Srivastava and J. J. Nieto, *Handbook of Fractional Calculus for Engineering and Science*. CRC Press, Taylor & Francis Group: Boca Raton, FL, 2022.
36. H. Singh, D. Kumar and D. Baleanu, *Methods of Mathematical Modelling: Fractional Differential Equations*. CRC Press, Taylor & Francis: Boca Raton, FL, 2019.

37. H. Singh, H. M. Srivastava and D. Baleanu, *Methods of Mathematical Modelling: Infectious Disease*, Elsevier Science: Amsterdam, Netherlands, 2022.
38. H. M. Srivastava, A new family of the λ-generalized Hurwitz-Lerch Zeta functions with applications, *Appl. Math. Inf. Sci.* **8**(4) (2014), 1485–1500.
39. H. M. Srivastava and J. Choi, *Series Associated with the Zeta and Related Functions*. Kluwer Acedemic Publishers: Dordrecht, Boston and London, 2001.
40. H. M. Srivastava and J. Choi, *Zeta and q-Zeta Functions and Associated Series and Integrals*. Elsevier Science Publishers: Amsterdam, London and New York, 2012.
41. H. M.Srivastava, D. Jankov, T. K. Pogany and R. K. Saxena, Two-sided inequalities for the extended Hurwitz –Lerch Zeta function, *Compu. Math. Appl.*, **62** (2011), 516–522.
42. H. M. Srivastava and P. W. Karlsson, *Multiple Gaussian Hypergeometric Series*. Halsted Press (Ellis Horwood Limited, Chichester). John Wiley & Sons: New York, Chichester, Brisbane and Toronto, 1985.
43. H. M. Srivastava, M.-J. Luo and R. K. Raina, New results involving a class of generalized Hurwitz-Lerch Zeta functions and their applications, *Turkish J. Anal. Number Theory* **1**(1) (2013), 26–35.
44. H. M. Srivastava and H. L. Manocha, *A Treatise on Generating Functions*. Halsted Press (Ellis Horwood Limited, Chichester), John Wiley & Sons: New York, Chichester, Brisbane and Toronto, 1984.
45. H. M. Srivastava and R. K. Saxena, Operators of fractional integration and their applications, *Appl. Mathe. Compu.*, **118** (2001), 1–52.
46. H. M. Srivastava, R. K. Saxena, T. K. Pogány and R. Saxena, Integral and computational representations of the extended Hurwitz-Lerch Zeta function, *Integr. Transforms Spec. Funct.* **22**(7) (2011), 487–506.
47. D. L. Suthar, R. K. Parmar, S. D. Purohit, Fractional calculus with complex order and generalized hypergeometric functions, *Nonlinear Sci. Lett. A* **8(2)** (2017), 156–161.
48. V. M. Tripathi, H. M. Srivastava, H. Singh, C. Swarup and S. Aggarwal, Mathematical analysis of non-isothermal reaction-diffusion models arising in spherical catalyst and spherical biocatalyst, *Appl. Sci.* **11(21)** (2021) 10423.

10 Compact Difference Schemes for Solving the Equation of Fractional Oscillator Motion with Viscoelastic Damping

A. M. Elsayed
Zagazig University
Moscow State University of Civil Engineering

T. S. Aleroev
Moscow State University of Civil Engineering

CONTENTS

DOI: 10.1201/9781003368069-10

10.1 INTRODUCTION

Fractional partial differential equations (FPDEs) are utilized in a wide variety of scientific and engineering disciplines, including molecular spectrum, quantum science, chain-breaking of polymer materials, viscoelastic mechanics, electroanalytical chemistry, signal image processing, and anomalous ion diffusion in nerve cells [1–7]. In addition, fractional order PDEs were used to model the filtration and flow of a fluid in a porous fractal medium. Equations with fractional order derivatives and integrals appear in addition to classical ones when fractional derivatives (FDs) are used to model real physical processes or environments. Due to the material's dynamic and viscoelastic behavior [8–10], researchers have concentrated their efforts on fractional order physical models [11,12]. As a result, the fractional order model is frequently used to simulate the frequency distribution of structural damping mechanisms [13,14]. Process electrical and physical properties in relation to the order of the fractional operator. It's intriguing that these operators have an inherent multiscale existence. As a result, time-fractional operators enable memory effects (i.e., a response system depends on its past history), whereas space-fractional operators enable nonlocal and scale effects. Fractional analysis is frequently used in many scientific fields, such as friction, fluid dynamics, nonlinear biological processes, solid-state mechanics, field theory, and control theory [15–17]. Were the first to employ this technique in fractional differential equation calculations in mechanics. The sequential approach, according to [18,19], could be used to test such equations. The papers that came after this one concentrated on fractional mechanics using Lagrangian and Hamiltonian methods. When fractional constitutive relations are used to solve vibration problems in continuous structures (such as beams, bars, etc.), fractional differential equations emerge, which are similar to the equation of a forced, harmonic, damped oscillator [20–24]. These models will be referred to as forced fractional oscillators.

Enelund and Josefson [25] used the finite element method to investigate fractionally damped viscoelastic components. Concrete structures are utilized in the design of buildings and other facilities as they are strong, dependable, and robust. In addition moment, the concrete structure's surface is vulnerable to serious negative effects. In light of this, a composite with improved operating properties is currently being created on the basis of a polymer concrete blend, which has greater tolerance for humidity, chemicals, low temperatures, and toughness in comparison to concrete. The summary of viscoelastic solids by Craiem [7] is straightforward to generalize a fractional Zener model. As a consequence of the Maxwell model [26] eventually producing a plastic flow in response to an ongoing force, knowing how a scheme diffuses energy in free oscillations can be related to well-known viscoelastic behavior characteristics. This in specific might provide insight into the role of biological materials [27,28] in dissipating generated electricity by their capacity to reduce stress.

In simulation [29,30], polymer concrete can be represented as a collection of solid filler granules confined in a viscoelastic medium. The transverse movement under the control of the force of gravity or the external force of a filler granule [31,32] is described by the fractional oscillator formula. As a result, the FDEs are used in place of the second second-order differential equations when concrete is replaced with polymer concrete [33]. The application of fractional calculus to create more accurate mathematical representations of a variety of practical problems is given special consideration. In earlier works [34–37], many academics have discussed the theoretical development and application of fractional calculus. The numerical techniques for FDEs are necessary because the exact solutions to FDEs may be challenging to obtain. The most common method for numerically resolving partial differential problems is the finite difference technique [38,39], which is one of them. A fluid's flow and filtration through a highly permeable fractal medium were simulated using fractional order with PDEs. For polymeric materials that instantaneously form and exhibit some degree of frequency dependence, such as some Newtonian fluid motions [30,40] or certain molecular theories [31], fractional models are particularly helpful. In practice, fractional models are used to more accurately describe the viscoelastic properties of materials like rubber and concrete [4,32]. Through dynamic analysis and research into mathematical models of viscoelastic materials, the effectiveness of viscoelastic dampers is analyzed [41,42]. In real-world situations, viscoelastic dampers are frequently added to equipment or building structures to reduce vibration reactions.

First, we note that the FDs in space could be utilized to simulate irregular diffusion or scattering, and FDs in time can be utilized to stimulate certain mechanisms with memory. Special attention is required to be given to solve the equation of a fractional oscillator motion with viscoelastic damping in the range $D=\{0<r<\mathfrak{L},0<\theta<\mathfrak{T}\}$

$$\frac{\partial^2 v(r,\theta)}{\partial\theta^2}=\frac{\partial^2 v(r,\theta)}{\partial r^2}+bD_\theta^\gamma v(r,\theta)+cD_r^\alpha v(r,\theta)+f(r,\theta), \tag{10.1}$$

satisfying initial conditions

$$\begin{aligned} v(r,0) &= \varphi(r), \\ v_\theta(r,0) &= \psi(r), \end{aligned} \tag{10.2}$$

and boundary conditions

$$v(0,\theta)=v(\mathfrak{L},\theta)=0. \tag{10.3}$$

A granule's displacement through the x-axis at a given time is represented by $v(r,\theta)$, while $b,c-$ are arbitrary negative constants and $f(r,\theta)$ is an external forcing function, where $\gamma\in(1,2)$, D_r^α and D_θ^γ denote the spatial Liouville derivative of order α and temporal Caputo derivative with order γ respectively, The R–L fractional operator of $w\in C_{-1}^m$ and $\alpha>0$, for every $m\in\mathbb{N}\cup\{0\}$, and γ-Caputo derivative D_t^γ of $h\in C_{-1}^m$ with order $\gamma>0$ are defined, respectively, as [20]

$$D_r^\alpha w(\theta)=\frac{1}{\Gamma(m-\gamma)}\frac{d^m}{dr^m}\int_0^r(r-\zeta)^{m-\alpha-1}w(\zeta)d\zeta,\quad m-1<\alpha<m,m\in\mathbb{N}. \tag{10.4}$$

and

$$D_\theta^\gamma h(\theta) = \frac{1}{\Gamma(m-\gamma)} \int_0^\theta (\theta-\xi)^{m-\gamma-1} h^{(m)}(\xi) d\xi, \quad m-1<\gamma<m, m \in \mathbb{N}. \quad (10.5)$$

The model (10.1) is taken from [17,33,43–45], where FD operators are frequently used to describe vibration models. It is well known that the motion of vibrations with elastic and viscoelastic components can be accurately described by FD equations [15]. The results of [34] show how the results of problem-solving can be used to simulate changes in the deformation-strength characteristics of polymer concrete when subjected to gravity force. Researchers looked at polymer concrete made with polyester resin samples (chloride-1, Diane,1-dichloro-2, diacyl, and 2-diethylene). FDs are frequently used in structural dynamics to characterize the dissipative forces as well as the viscoelastic properties of complex materials [35].

In this work, our main objective is to formulate a numerical strategy for Eq. (10.1) and carry out a numerical comparison of the suggested approach. A variety of physical and mechanical engineering problems can be analyzed using analytical methods, which also take less time than the numerical approach. They have the advantage of illuminating the fundamentals of mechanical engineering problems and their physical implications. Researchers discovered that it is incredibly difficult to find precise solutions to PDEs. Numerous numerical methods, such as the homotopy perturbation methods, Adomian decomposition method, spectral method, finite difference scheme, Galerkin method, and finite element method [46–54], have been researched and developed to obtain numerical solutions to FPDEs. To produce an approximation solution, Huang et al. [39] took into account 2-D fractional-time super diffusion problems and developed a conservative linearized method with two Crank-Nicolson difference schemes.

Furthermore, the suggested scheme was shown that is solvable, and convergent with order $\mathcal{O}(\tau + h_x^2 + h_y^2 + \tau^\alpha)$ with τ time step and h grid size in the L_2 norm. Guo et al. [46] developed a novel Galerkin finite difference/spectral system to solve 3-D space-time fractional reaction-diffusion-wave equation of dispersed order in 2020, and numerical examples were used to demonstrate the stability and convergence of the suggested technique. In Ref. [47], Li developed a linearized fractional difference/finite element approximation for the fractional telegraph equation, and the proposed technique has been shown to be unconditionally stable by using the mathematical induction. Elsayed [43] developed Crank-Nicolson difference technique to deal with the fractional derivative vibration equation in 2020 and studied truncation error and stability. Lyu [55] suggested a finite difference technique to solve the time-fractional BBM model by adopting a weighted approximation and basing it on the $L2-1_\sigma$ algorithm. To construct a compact difference method for diffusion-wave equations of fractional order, the equivalent integro-differential equations and product trapezoidal law were used by Chen and Li [56]. In 2016, Wang et al. [5] studied finite difference methods for both temporal and spatial FDs for differential equations. They also present a precondition to improve the effectiveness of the schemes implementation in this case.

10.2 SOME APPLICATIONS OF FRACTIONAL OSCILLATOR MOTION WITH VISCOELASTIC DAMPING IN ENGINEERING

The conventional method for describing the rheological properties of viscoelastic dampers uses mechanical models with springs and flaps. Mechanical models made up of a number of securely connected springs and dashboards are necessary for a thorough explanation of viscoelastic shock absorbers. The dynamic analysis of structures with dampers is greatly complicated by this method because of the enormous number of motion equations that must be solved. The dynamic behavior of a single damper is described by a set of differential equations. In addition, a time-consuming nonlinear regression approach is advised, as described, for example, in order to determine the parameters of the aforementioned models. Even though fractional calculus has been around for more than 300 years and has been used in many branches of science and technology [16], the community that supports it still has a crucial job to do. The two most common inquiries we receive from scientists and engineers outside of our community when we discuss fractional calculus are about its applications and how scientists can use it in their respective fields. Many fractional calculus researchers working in theoretical fields, however, are also unfamiliar with the practical applications. It is necessary to provide a brief overview of the effective uses of fractional calculus in science and technology.

There is a wealth of information on other fractional order systems that describe different physical processes in Refs. [12,57,58]. However, the addition of a fractional derivative to the inertial terms of the motion equations is neither desirable nor justified for applications in mechanical engineering and civil engineering. FDs are helpful in engineering practice for describing dissipative forces in structural dynamics or the viscoelastic properties of contemporary materials [7,15,59]. In Ref. [60], where the solution is presented using the Laplace transform method, exact solutions for linear fractional oscillators based on various fractional calculus defining equations can be found. The analysis of an oscillator's damped vibrations frequently uses numerical solutions. When thinking about dynamic issues with viscoelastic damped structures, Bagley and Torvik [30] were among the first researchers to apply numerical solutions.

Viscoelastic effects are taken into account in a variety of mechanical and engineering systems, including aircraft, building structures, biomedical materials, and arch bridges [12], to describe the motion of the system. As is well known, viscoelasticity results in the development of a viscoelastic system, which includes energy dissipation resulting from elasticity as well as memory properties, on which the viscoelastic stress depends on both the present and all previous states of deformation. Typically, an integer derivative is used to describe how energy is lost as a result of damping. However, many works [1,3,21] introduced the substitution of the attenuation of the whole order of the derivative of the fractional order for the construction of models. For a good prediction of power phenomena, the fractional viscoelastic model is confirmed by Rossikhin [61]. The fractional theory of linear viscoelasticity was subsequently gradually enhanced by Mainardi [62].

As follows, the article is briefly summarized. We present the first scheme of the oscillator motion with viscoelastic damping; also, the scheme's stability and convergence are proved in Section 10.3. In Section 10.4, the compact difference of the second scheme is constructed and analyzed; in addition, it is rigorously proven that the proposed method is convergent and unconditionally stable. To validate our theoretical findings, numerical experiments are performed in Section 10.5.

10.3 CONSTRUCTION AND ANALYSIS OF SCHEME 1 FOR RIEMANN–LIOUVILLE WITH $0 < \alpha < 1$

A Crank-Nicolson difference method for the space-time fractional oscillation motion equation is constructed. In this scheme, the classical- central difference approximation and a $3-\gamma$ order formula are used for the 2nd order derivative and the derivative of caputo in temporal direction, respectively. Meanwhile, to deal with the spatial discretizations, the fractional difference approximation and the 2nd order difference quotient are applied. The unconditional stability and the convergence of the proposed scheme are discussed. The convergence order of the scheme is $\mathcal{O}(\tau^{3-\gamma}+h^2)$, where $(3-\gamma)$ is the order accuracy in time and h is the accuracy in space, respectively. We describe several definitions and lemmas, which are being used in the section.

Lemma 10.1. *(see [49]) For $n \geq 1$ and $\theta_k = k\tau, 0 \leq k \leq n$, we have*

$$0 \leq \sum_{k=1}^{n} \int_{\theta_{k-1}}^{\theta_k} \left\{ (\theta_n-\theta)^{2-\alpha} - \left[\frac{\theta-\theta_{k-1}}{\tau}(\theta_n-\theta_k)^{2-\alpha} + \frac{\theta_k-\theta}{\tau}(\theta_n-\theta_{k-1})^{2-\alpha} \right] \right\} d\theta$$
$$\leq \left[\frac{2-\alpha}{12} + \frac{2^{3-\alpha}}{3-\alpha} - (1+2^{1-\alpha}) \right] \tau^{3-\alpha}.$$

Lemma 10.2. *(see [49]) Suppose $g(\theta) \in C^2[0,\theta_n]$, then for $1<\gamma<2$, and $\theta_n = n\tau; n \in \mathbb{N}, \tau$ is a step size, the following inequality holds*

$$\left| \int_0^{\theta_n} \frac{g'(\theta)d\theta}{(\theta_n-\theta)^{\gamma-1}} - \frac{1}{\tau}\left[a_0 g(\theta_n) - \sum_{k=1}^{n-1}(a_{n-k-1}-a_{n-k})g(\theta_k) - a_{n-1}g(\theta_0) \right] \right|$$
$$0 \leq \frac{1}{2-\gamma}\left[\frac{2-\gamma}{12} + \frac{2^{3-\gamma}}{3-\gamma} - (1+2^{1-\gamma}) \right] \max_{0\leq\theta\leq\theta_n} |g''(\theta)|\tau^{3-\gamma},$$

where $a_k = \frac{\tau^{2-\gamma}}{2-\gamma}\left[(k+1)^{2-\gamma} - k^{2-\gamma}\right]$, for the integer $k \geq 0$.

So, we can easily find the following outcome based on the above lemmas.

Lemma 10.3. *(see [50]) For $1<\gamma<2$, and $v(\theta) \in C^3[0,\theta_n]$, then it holds*

$${}^{C}D_{\theta}^{\gamma}v(\theta_n) = \frac{1}{\tau\Gamma(2-\gamma)}\left[a_0\delta_{\widehat{\theta}}v(\theta_n) - \sum_{k=1}^{n-1}(a_{n-k-1}-a_{n-k})\delta_{\widehat{\theta}}v(\theta_k) - a_{n-1}v'(0) \right]$$
$$+\mathcal{O}(\tau^{3-\gamma}),$$

where $\delta_{\widehat{\theta}}v(\theta_n) = \frac{v(\theta_{n+1})-v(\theta_{n-1})}{2\tau}$,

Lemma 10.4. *(see [51]) For* $g(r,\theta) \in L^1(R)$, *in order to discretize the space-fractional R–L operator, we use the shifted Grünwald-difference formula , for* $0 < \alpha < 1$ *it holds*

$$({}^{Rl}D_r^{\alpha}g)(r_i,\theta) = h^{-\alpha}\sum_{j=0}^{i+1}\omega_j^{(\alpha)}g(r_{i-j+1},\theta) + \mathscr{O}(h^2),$$

where $r_i = ih$ *with step size* h, *and* $\omega_j^{(\alpha)} = (-1)^j\binom{\alpha}{j}$ *for* $j \geq 0$.

10.3.1 CONSTRUCTION OF SCHEME 1

We considered two fractional operators in this chapter. It should be noted that we use the numerical techniques offered in Refs. [49,53] directly to cope with the Caputo derivative; we acknowledge their priority of $3-\gamma$ th-order algorithm for Caputo derivative, but we introduce the algorithm for $\gamma \in (1,2)$ employing a slightly different technique in this chapter. In addition, we used a fractional R-L operator for $\alpha \in (0,1)$, which discretizes by a shifted Grünwald-difference formula, resulting in a C-R-L difference equation, requiring a huge computational performance once to solve Eq. (10.1).

In this chapter, first, we estimate the derivative of Caputo in temporal mesh size and a second-order derivative using a novel $3-\gamma$ th-order algorithm and the standard central difference method. We also use the fractional difference approximation and the 2nd difference quotient for dealing with spatial discretizations. Furthermore, a suggested system has been demonstrated to be convergent and stable, with a convergence order of $\mathscr{O}(h^2 + \tau^{3-\gamma})$, where h is the accuracy in space and $(3-\gamma)$ is the order accuracy in time, respectively.

To discretize Eqs. (10.1) and (10.2), introducing the step size $\tau = \frac{\mathfrak{T}}{\mathbb{N}}$ with non-zero integer numbers $\mathbb{N}$, and $\theta_n = n\tau; n = 0,1,...,\mathbb{N}$, also define a grid function time $\Omega_\tau = \{\theta_n | n \geq 0\}$. For a spatial discretization, let $h = \frac{\mathfrak{L}}{\mathbb{M}}$ and $r_i = ih; 0 \leq i \leq \mathbb{M}$, where $\mathbb{M}$ is a non-zero integer number, also defines a grid function space $\Omega_h = \{r_i| \quad 0 \leq i \leq \mathbb{M}\}$. Suppose on $\Omega_h \times \Omega_\tau$, their exist a grid functions $V = \{v_i^n| \quad 0 \leq i \leq \mathbb{M}, n \geq 0\}$, such that

$$v_i^{n+\frac{1}{2}} = \frac{1}{2}\left[v_i^{n+1} + v_i^n\right], \qquad \delta_\theta v_i^{n+\frac{1}{2}} = \frac{1}{\tau}\left[v_i^{n+1} - v_i^n\right],$$

$$\delta_{\hat{\theta}} v_i^n = \frac{1}{2\tau}\left[v_i^{n+1} - v_i^{n-1}\right], \qquad \delta_\theta^2 v_i^n = \frac{1}{\tau}\left[\delta_\theta v_i^{n+\frac{1}{2}} - \delta_\theta v_i^{n-\frac{1}{2}}\right],$$

$$\delta_r v_{i+\frac{1}{2}}^n = \frac{1}{h}\left[v_{i+1}^n - v_i^n\right], \qquad \delta_r^2 v_i^n = \frac{1}{h}\left[\delta_r v_{i+\frac{1}{2}}^n - \delta_r v_{i-\frac{1}{2}}^n\right],$$

where $v_i^{n+\frac{1}{2}}$ is the average of v at the points (r_i,θ_n) and (r_i,θ_{n+1}), $\delta_r v_{i+\frac{1}{2}}^n, \delta_\theta v_i^{n+\frac{1}{2}}$ is the 1nd order difference quotient of v and $\delta_r^2 v_i^n$ is the 2nd order-difference quotient at $(r_{i+1},\theta_n), (r_i,\theta_n)$, and (r_{i-1},θ_n).

Now, assuming Eq. (10.1) at the point (r_i,θ_n), we have

$$\frac{\partial^2 v(r_i,\theta_n)}{\partial\theta^2} = bD_\theta^\gamma v(r_i,\theta_n) + \frac{\partial^2 v(r_i,\theta_n)}{\partial r^2} + cD_r^\alpha v(r_i,\theta_n) + f(r_i,\theta_n), \tag{10.6}$$

assume that $v(r,\theta)\in C_{r,\theta}^{5,3}([0,L]\times[0,T])$, let $v_i^n = v(r_i,\theta_n)$ and $f_i^n = f(r_i,\theta_n)$, and applying Lemmas 10.2 and 10.4 and the difference quotient formula to Eq. (10.6), then

$$\delta_\theta^2 v_i^n = \frac{b}{\tau\Gamma(2-\gamma)}\left[a_0\delta_{\hat\theta} v_i^n - \sum_{k=1}^{n-1}(a_{n-k-1}-a_{n-k})\delta_{\hat\theta} v_i^k - a_{n-1}v_\theta(i,0)\right] + \delta_r^2 v_i^n$$
$$+ c\delta_r^\alpha v_i^n + f_i^n + \mathcal{O}(\tau^{3-\gamma}+h^2), \tag{10.7}$$

as $v_\theta(i,0)=\psi_i(r)$, then the above equation can written as

$$\delta_\theta^2 v_i^n = \frac{b}{\tau\Gamma(2-\gamma)}\left[a_0\delta_{\hat\theta} v_i^n - \sum_{k=1}^{n-1}(a_{n-k-1}-a_{n-k})\delta_{\hat\theta} v_i^k - a_{n-1}\psi_i\right] + \delta_r^2 v_i^n$$
$$+ \frac{c}{h^\alpha}\sum_{j=0}^{i+1}\omega_j^{(\alpha)} v_{i-j+1}^n + f_i^n + \mathcal{O}(\tau^{3-\gamma}+h^2). \tag{10.8}$$

Neglecting the truncation error term $\mathcal{O}(\tau^{3-\gamma}+h^2)$ from Eq. (10.8) and replacing v_i^n with its numerical solution V_i^n, obtaining the following scheme for Eq. (10.8)

$$\delta_\theta^2 V_i^n = \frac{b}{\tau\Gamma(2-\gamma)}\left[a_0\delta_{\hat\theta} V_i^n - \sum_{k=1}^{n-1}(a_{n-k-1}-a_{n-k})\delta_{\hat\theta} V_i^k - a_{n-1}\psi_i\right] + \delta_r^2 V_i^n$$
$$+ c\delta_r^\alpha V_i^n + f_i^n, \tag{10.9}$$

10.3.2 ANALYSIS OF SCHEME 1

Let us discuss the convergence and stability of the proposed method (10.9) throughout this section. Let us define a grid function $V=\{v_i^n|\quad 0\le i\le \mathbb{M}, n\ge 0;$ $v_0 = v_{\mathbb{M}} = 0\}$ is a grid function $\Omega_h\times\Omega_\tau$; for a grid functions $v,\mathrm{g}\in V$, we define the norms and the inner product as:

$$\|v^n\|_\infty = \max_{0\le i\le \mathbb{M}}|v_i^n|, \qquad |\delta_r v^n| = \sqrt{h\sum_{i=1}^{\mathbb{M}}(\delta_r v_{i-\frac{1}{2}}^n)^2}.$$

$$\langle v^n,\mathrm{g}^n\rangle = h\sum_{i=1}^{\mathbb{M}-1} v_i\mathrm{g}_i, \qquad \|v^n\|^2 = \langle \upsilon, v\rangle,$$

Lemma 10.5. *(see [51]) For $v,\mathrm{g}\in V$, there exists a linear difference operator to the operator δ_r^α that is denoted by $\delta_r^{\alpha/2}$, where*

$$\langle\delta_r^\alpha v,\mathrm{g}\rangle = \langle\delta_r^{\alpha/2}v,\delta_r^{\alpha/2}\mathrm{g}\rangle, \qquad \langle\delta_r^2 v,\mathrm{g}\rangle = -\langle\delta_r v,\delta_r \mathrm{g}\rangle,$$

and

$$(\delta_r^2 v, g) = -h\sum_{i=1}^{\mathbb{M}}(\delta_r v_{i-\frac{1}{2}})(\delta_r g_{i-\frac{1}{2}}), \qquad \|v^n\|_\infty \le \frac{\sqrt{L}}{2}|\delta_r v^n|,$$

also we find

$$\|v^n\|_\infty \le \mathbb{M}\|\delta_r^{\alpha/2} v^n\|, \qquad \|v^n\|_\infty \le \mathbb{M}\|\delta_\theta v^n|,$$

where $\mathbb{M}$ is a constant.

Lemma 10.6. *(see [49]) For every $F = \{F_1, F_2, F_3, \ldots\}$ and a constant r, we have*

$$\sum_{n=1}^{\mathbb{N}}\left[a_0 F_n - \sum_{k=1}^{n-1}(a_{n-k-1} - a_{n-k})F_k - a_{n-1}r\right] \ge \frac{\theta_{\mathbb{N}}^{1-\gamma}}{2}\tau\sum_{n=1}^{N} F_n^2 - \frac{\theta_{\mathbb{N}}^{2-\gamma}}{2(2-\gamma)}r^2; N = 1,2,\ldots,$$

where a_k is defined above in Lemma 10.2.

Theorem 10.1. *The numerical solution of the scheme* (10.9) *is stable, which holds*

$$\|V^{m+1}\|_\infty^2 \le \mathbb{M}[|\delta_r V^1|^2 + |\delta_r V^0|^2 + c\tau\|\delta_r^{\alpha/2} V^1\|^2 + c\tau\|\delta_r^{\alpha/2} V^0\|^2 + \frac{(b+1)\tau\theta_m^{1-\gamma}}{\Gamma(2-\gamma)} - \frac{b\theta_m^{2-\gamma}}{\Gamma(3-\gamma)}h\sum_{i=1}^{\mathbb{M}-1}\psi_i^2 + \tau\theta_m^{\gamma-1}\Gamma(2-\gamma)\sum_{n=1}^{m}(f^n)^2]. \tag{10.10}$$

Proof. Multiplying Eq. (10.10) by $h\tau\delta_{\hat\theta} V_i^n$ and summing over $1 \le i \le \mathbb{M}-1$; $1 \le n \le m$, we find

$$h\tau\sum_{i=1}^{\mathbb{M}-1}\sum_{n=1}^{m}(\delta_\theta^2 V_i^n)(\delta_{\hat\theta} V_i^n) = \frac{b}{\tau}\frac{h\tau}{\Gamma(2-\gamma)}\sum_{i=1}^{\mathbb{M}-1}\left\{\sum_{n=1}^{m}\left[a_0\delta_{\hat\theta} V_i^n - \sum_{k=1}^{n-1}(a_{n-k-1} - a_{n-k})\delta_{\hat\theta} V_i^k - a_{n-1}\psi_i\right]\delta_{\hat\theta} V_i^n\right\} + \tau\sum_{n=1}^{m}\left[h\sum_{i=1}^{\mathbb{M}-1}(\delta_r^2 V_i^n)(\delta_{\hat\theta} V_i^n)\right] + \tau c\sum_{n=1}^{m}\left[h\sum_{i=1}^{\mathbb{M}-1}(\delta_r^\alpha V_i^n)(\delta_{\hat\theta} v_i^n)\right] + h\sum_{i=1}^{\mathbb{M}-1}\left[\tau\sum_{n=1}^{m}(\delta_{\hat\theta} V_i^n)f_i^n\right]. \tag{10.11}$$

It is possible to turn the left-hand side

$$
\begin{aligned}
&h\tau\sum_{i=1}^{\mathbb{M}-1}\sum_{n=1}^{m}(\delta_{\theta}^{2}V_i^n)(\delta_{\widehat{\theta}}V_i^n)\\
&\quad=\tau\sum_{n=1}^{m}(\delta_{\theta}^{2}V^n,\delta_{\widehat{\theta}}V^n)=\tau\sum_{n=1}^{m}\left(\frac{1}{\tau}(\delta_{\theta}V^{n+\frac{1}{2}}-\delta_{\theta}V^{n-\frac{1}{2}}),\frac{1}{2}(\delta_{\theta}V^{n+\frac{1}{2}}+\delta_{\theta}V^{n-\frac{1}{2}})\right)\\
&\quad=\frac{1}{2}\sum_{n=1}^{m}\left(\|\delta_{\theta}V^{n+\frac{1}{2}}\|^2-\|\delta_{\theta}V^{n-\frac{1}{2}}\|^2\right)=\frac{1}{2}\left(\|\delta_{\theta}V^{m+\frac{1}{2}}\|^2-\|\delta_{\theta}V^{m-\frac{1}{2}}\|^2\right),
\end{aligned}
\tag{10.12}
$$

using Lemma 10.6, for the first term of the right part from Eq. (10.11)

$$
\begin{aligned}
&\frac{1}{\tau}\frac{h\tau}{\Gamma(2-\gamma)}\sum_{i=1}^{\mathbb{M}-1}\left\{\sum_{n=1}^{m}\left[a_0\delta_{\widehat{\theta}}V_i^n-\sum_{k=1}^{n-1}(a_{n-k-1}-a_{n-k})\delta_{\widehat{\theta}}V_i^k-a_{n-1}\psi_i\right]\delta_{\widehat{\theta}}V_i^n\right\}\\
&\quad\geq\frac{h}{\Gamma(2-\gamma)}\sum_{i=1}^{\mathbb{M}-1}\left[\frac{\tau t_m^{1-\gamma}}{2}\sum_{n=1}^{m}(\delta_{\widehat{\theta}}V_i^n)^2-\frac{\theta_m^{2-\gamma}}{2(2-\gamma)}\psi_i^2\right],\\
&\quad\geq\frac{\tau\theta_m^{1-\gamma}}{2\Gamma(2-\gamma)}\sum_{n=1}^{m}\|\delta_{\widehat{\theta}}V^n\|^2-\frac{\theta_m^{2-\gamma}}{2\Gamma(3-\gamma)}h\sum_{i=1}^{\mathbb{M}-1}\psi_i^2.
\end{aligned}
\tag{10.13}
$$

Applying Lemma 10.5 consequently,

$$
\begin{aligned}
&\tau\sum_{n=1}^{m}h\sum_{i=1}^{\mathbb{M}-1}(\delta_r^2V_i^n)(\delta_{\widehat{\theta}}V_i^n)=\tau\sum_{n=1}^{m}(\delta_r^2V^n,\delta_{\widehat{\theta}}V^n)=-\tau\sum_{n=1}^{m}h\sum_{i=1}^{\mathbb{M}}\delta_{\widehat{\theta}}(\delta_rV^n_{i-\frac{1}{2}})^2\\
&\quad=-\tau\sum_{n=1}^{m}\frac{h\sum_{i=1}^{\mathbb{M}}(\delta_rV^{n+1}_{i-\frac{1}{2}})^2-h\sum_{i=1}^{\mathbb{M}}(\delta_rV^{n-1}_{i-\frac{1}{2}})^2}{2\tau}=-\frac{1}{2}\sum_{n=1}^{m}(|\delta_rV^{n+1}|^2-|\delta_rV^{n-1}|^2)\\
&\quad=-\frac{1}{2}\left(|\delta_rV^{m+1}|^2+|\delta_rV^{m}|^2-|\delta_rV^{1}|^2-|\delta_rV^{0}|^2\right),
\end{aligned}
\tag{10.14}
$$

and

$$
\begin{aligned}
&\tau c\sum_{n=1}^{m}h\sum_{i=1}^{\mathbb{M}-1}(\delta_r^{\alpha}V_i^n)(\delta_{\widehat{\theta}}V_i^n)=\tau c\sum_{n=1}^{m}(\delta_r^{\alpha}V^n,\delta_{\widehat{\theta}}V^n)=\tau c\sum_{n=1}^{m}(\delta_r^{\alpha/2}V^n,\delta_{\widehat{\theta}}\delta_r^{\alpha/2}V^n)\\
&\quad=\frac{c}{2}\sum_{n=1}^{m}\left[\|\delta_r^{\alpha/2}V^{n+1}\|^2-\|\delta_r^{\alpha/2}V^{n-1}\|^2\right]\\
&\quad=-\frac{c}{2}\left(\|\delta_r^{\alpha/2}V^{m+1}\|^2+\|\delta_r^{\alpha/2}V^{m}\|^2-\|\delta_r^{\alpha/2}V^{1}\|^2-\|\delta_r^{\alpha/2}V^{0}\|^2\right).
\end{aligned}
\tag{10.15}
$$

Moreover,

$$h\sum_{i=1}^{\mathrm{M}-1}[\tau\sum_{n=1}^{m}(\delta_{\widehat{\theta}}V_i^n)f_i^n]\leq h\sum_{i=1}^{\mathrm{M}-1}\tau\sum_{n=1}^{m}\left[\frac{\theta_m^{1-\gamma}}{2\Gamma(2-\gamma)}(\delta_{\widehat{\theta}}V_i^n)^2+\frac{\theta_m^{\gamma-1}}{2}\Gamma(2-\gamma)(f_i^n)^2\right]$$
$$\leq\frac{\tau\theta_m^{1-\gamma}}{2\Gamma(2-\gamma)}\sum_{n=1}^{m}\|\delta_{\widehat{\theta}}V^n\|^2+\frac{\tau\theta_m^{\gamma-1}}{2}\Gamma(2-\gamma)\sum_{n=1}^{m}(f^n)^2. \tag{10.16}$$

Substituting from Eqs. (10.12)–(10.16) into Eq. (10.11), we find

$$\|\delta_\theta V^{m+\frac{1}{2}}\|^2+|\delta_r V^{m+1}|^2+|\delta_r V^m|^2+c\|\delta_r^{\alpha/2}V^{m+1}\|^2+c\|\delta_r^{\alpha/2}V^m\|^2$$
$$\leq\|\delta_\theta V^{m-\frac{1}{2}}\|^2+|\delta_r V^1|^2+|\delta_r v^0|^2+c\|\delta_r^{\alpha/2}v^1\|^2+c\|\delta_r^{\alpha/2}v^0\|^2$$
$$+\frac{(b+1)\tau\theta_m^{1-\gamma}}{\Gamma(2-\gamma)}\sum_{n=1}^{m}\|\delta_{\widehat{\theta}}v^n\|^2-\frac{b\theta_m^{2-\gamma}}{\Gamma(3-\gamma)}h\sum_{i=1}^{\mathrm{M}-1}\psi_i^2+\tau\theta_m^{\gamma-1}\Gamma(2-\gamma)\sum_{n=1}^{m}(f^n)^2. \tag{10.17}$$

Applying Lemma 10.5, we find

$$\|V^{m+1}\|_\infty^2\leq M[|\delta_r V^1|^2+|\delta_r V^0|^2+c\|\delta_r^{\alpha/2}V^1\|^2+c\|\delta_r^{\alpha/2}V^0\|^2$$
$$+\frac{(b+1)\tau\theta_m^{1-\gamma}}{\Gamma(2-\gamma)}\sum_{n=1}^{m}\|\delta_{\widehat{\theta}}V^n\|_\infty^2$$
$$-\frac{b\theta_m^{2-\gamma}}{\Gamma(3-\gamma)}h\sum_{i=1}^{\mathrm{M}-1}\psi_i^2+\tau\theta_m^{\gamma-1}\Gamma(2-\gamma)\sum_{n=1}^{m}(f^n)^2]. \tag{10.18}$$

Applying Gronwall inequality, Eq. (10.18) becomes

$$\|V^{m+1}\|_\infty^2\leq M[|\delta_r V^1|^2+|\delta_r V^0|^2+c\|\delta_r^{\alpha/2}V^1\|^2+c\|\delta_r^{\alpha/2}V^0\|^2+\frac{(b+1)\tau\theta_m^{1-\gamma}}{\Gamma(2-\gamma)}$$
$$-\frac{bt_m^{2-\gamma}}{\Gamma(3-\gamma)}h\sum_{i=1}^{\mathrm{M}-1}\psi_i^2+\tau\theta_m^{\gamma-1}\Gamma(2-\gamma)\sum_{n=1}^{m}(f^n)^2]. \tag{10.19}$$

This makes the proof complete. □

Theorem 10.2. *Let* $v(r,\theta)\in C_{r,\theta}^{5,3}([0,L]\times[0,T])$, *and suppose* $v(r,\theta)$ *is the exact solution of Eq.* (10.1) *and* $V(r,\theta)$ *is a numerical solution of scheme* (10.9), *which is defined as* $\{V_i^n|0\leq i\leq \mathrm{M},1\leq n\leq \mathrm{N}\}$. *So for* $n\tau\leq T$, *holds*

$$\|V^n-v^n\|\leq C(\tau^{3-\gamma}+h^2). \tag{10.20}$$

Proof. Subtracting Eq. (10.9) from Eq. (10.7) and denoting $e_i^n=v_i^n-V_i^n$, we have

$$\delta_\theta^2 e_i^n=\frac{b}{\tau\Gamma(2-\gamma)}\left[a_0\delta_{\widehat{\theta}}e_i^n-\sum_{k=1}^{n-1}(a_{n-k-1}-a_{n-k})\delta_{\widehat{\theta}}e_i^k\right]+\delta_r^2 e_i^n+c\delta_r^\alpha e_i^n$$
$$+\mathscr{O}(\tau^{3-\gamma}+h^2). \tag{10.21}$$

As similar to proofing of Theorem 10.1, multiplying Eq. (10.21) by $h(e_i^{n+1}+e_i^n)$ and summing over $1 \leq i \leq \mathbb{M}-1$, $1 \leq n \leq m$, we find

$$\begin{aligned}\|\delta_\theta e^{m+\frac{1}{2}}\|^2+|\delta_r e^{m+1}|^2+|\delta_r e^m|^2+c\|\delta_r^{\alpha/2}e^{m+1}\|^2+c\|\delta_r^{\alpha/2}e^m\|^2 \\ \leq \|\delta_\theta e^{m-\frac{1}{2}}\|^2+|\delta_r e^1|^2+|\delta_r e^0|^2+c\|\delta_r^{\alpha/2}e^1\|^2+c\|\delta_r^{\alpha/2}e^0\|^2 \\ +\frac{(b+1)\tau\theta_m^{1-\gamma}}{\Gamma(2-\gamma)}\sum_{n=1}^{m}\|\delta_{\widehat{\theta}}e^n\|^2+2\Gamma(2-\gamma)T^\gamma(C(\tau^{3-\gamma}+h^2)^2.\end{aligned} \quad (10.22)$$

As $e_i^0=0$ for $0 \leq i \leq \mathbb{M}$, applying Lemma 10.5, Gronwalls inequality, then we obtain

$$\|e^{m+1}\|_\infty^2 \leq C(\tau^{3-\gamma}+h^2)^2.$$

This proof is complete. □

10.4 CONSTRUCTION AND ANALYSIS OF SCHEME 2 FOR $1<\alpha<2$

Our fundamental objective in this work is to form a numerical strategy for Eq. (10.1) and perform the comparing numerical examination for the suggested method. Using the integral operator, we first convert Eq. (10.1) into the equivalent partial integro-differential equations to reduce the requirement for smoothness in time. Second, to deal with temporal direction, the Crank-Nicolson technique is used. The midpoint formula is then used to discretize the 1nd order derivative, and the weighted-shifted convolution quadrature formula is used to approximate the first-order integral. For spatial approximations, the classical-central difference formula and a compact difference formula are used.

Let us assume an equivalent form of Eq. (10.1), that is by integrate both sides of Eq. (10.1) under the first-order integral operator ${}_0J_\theta$; where ${}_0J_\theta v(.,\theta)=\int_0^\theta v(.,q)dq$. get

$$v_\theta(r,\theta)=bD_\theta^{\gamma-1}v(r,\theta)+cJ_\theta D_r^\alpha u(r,\theta)+J_\theta\frac{\partial^2 v(r,\theta)}{\partial r^2}+\mathbb{F}, \quad (10.23)$$

where $0<\beta=\gamma-1<1$, and $\mathbb{F}=J_\theta f(r,\theta)$.

Lemma 10.7. *Suppose $v(\cdot,\theta)\in C^2([0,T])$ then it satisfy*

$${}_0J_\theta v\left(\cdot,\theta_{n+1/2}\right)=\frac{1}{2}\left[{}_0J_\theta v\left(\cdot,\theta_{n+1}\right)+{}_0J_\theta v\left(\cdot,\theta_n\right)\right]+O\left(\tau^2\right).$$

and

$${}_0^CD_\theta^\gamma v\left(\cdot,\theta_{n+1/2}\right)=\frac{1}{2}\left({}_0^CD_\theta^\gamma v\left(\cdot,\theta_{n+1}\right)+{}_0^CD_\theta^\gamma v\left(\cdot,\theta_n\right)\right)+O\left(\tau^2\right).$$

Lemma 10.8. *Let $v(\cdot,t)\in C^2([0,\mathfrak{T}]); v(\cdot,0)=v_\theta(\cdot,0)=0$, the generating function weight's $\{\varpi_k\}$ of the function $\left(3/2-2z+z^2/2\right)^{-1}$,defined as :*

$$|{}_0J_{\theta_{n+1}}v(\cdot,\theta)-\tau\sum_{k=0}^{n+1}\varpi_{n+1-k}v\left(\cdot,\theta_k\right)|\leq O\left(\tau^2\right)$$

To discretize Eq. (10.23), introducing the step size $\tau = \frac{\mathfrak{T}}{\mathbb{N}}$ with $\mathbb{N}$ positive integer, and $\theta_n = n\tau; n = 0, 1, ..., \mathbb{N}$, also define a grid function time $\Omega_\tau = \{\theta_n | n \geq 0\}$. For a spatial discretization, let $h = \frac{\mathfrak{L}}{\mathbb{M}}$ and $r_i = ih; 0 \leq i \leq \mathbb{M}$, where $\mathbb{M}$ is a non-zero integer number, also define a grid function space $\Omega_h = \{r_i | \quad 0 \leq i \leq \mathbb{M}\}$. Suppose on $\Omega_h \times \Omega_\tau$, their exist a grid functions $V = \{v_i^n | \quad 0 \leq i \leq \mathbb{M}, n \geq 0\}$, such that for any $w, g \in V$, we define the following norms, semi-norm $\|.\|_{\check{H}}$ and the inner product, as follows

$$v_i^{n+\frac{1}{2}} = \frac{1}{2}\left[v_i^{n+1} + v_i^n\right], \qquad \delta_\theta v_i^{n+\frac{1}{2}} = \frac{1}{\upsilon}\left[v_i^{n+1} - v_i^n\right],$$

$$\langle v^n, g^n \rangle = h\sum_{i=1}^{\mathbb{M}-1} v_i g_i, \qquad \|v^n\|^2 = \langle v, v \rangle,$$

$$\|v^n\|_\infty = \max_{0 \leq i \leq \mathbb{M}} |v_i^n|, \qquad \langle \delta_r^2 v, g \rangle = -\langle \delta_r v, \delta_r g \rangle,$$

$$\langle \delta_r v, \delta_r g \rangle_{\check{H}} = \langle \delta_r v, \delta_r g \rangle - \frac{h^2}{12}\langle \delta_r^2 v, \delta_r^2 g \rangle, \qquad \|\delta_r v\|_{\check{H}} = \sqrt{\langle \delta_r v, \delta_r v \rangle_{\check{H}}}.$$

Also, we utilize the discretization [49,63] for the spatial derivatives supplied by

$$D_r^\alpha \mathrm{f}(r_i) = \frac{1}{\Gamma(4-\alpha)h^\alpha}\sum_{s=0}^{i+1} q_s^\alpha \mathrm{f}(r_{i-s+1}) + \mathcal{O}(h^2), \qquad \delta_r^2 v_i^n = \frac{v_{i+1}^n - 2v_i^n + v_{i-1}^n}{h^2},$$

where $\mathrm{f} \in \mathscr{C}^4(R)$ and defined by

$$q_s^\alpha = \begin{cases} 1, & s = 0 \\ 2^{3-\alpha} - 4, & s = 1 \\ 3^{3-\alpha} - 4.2^{3-\alpha} + 6, & s = 2 \\ (s+1)^{3-\alpha} - 4s^{3-\alpha} + 6(s-1)^{3-\alpha} - 4(s-2)^{3-\alpha} + (s-3)^{3-\alpha}, & s \geq 3, \end{cases} \tag{10.24}$$

and

$$\mathscr{H} v_i = \begin{cases} (1 + \frac{h^2}{12}\delta_r^2)v_i = \frac{1}{12}(v_{i-1} + 10v_i + v_{i+1}), & 1 \leq i \leq \mathbb{M}-1 \\ v_i, & i = 0, \mathbb{M}. \end{cases}$$

To raise the accuracy in $\frac{\partial^2 v}{\partial r^2}$, we use following lemma.

Lemma 10.9. *[64] If $\mu(\theta) \in \mathscr{C}^6[r_{i-1}, r_{i+1}], 1 \leq i \leq \mathbb{M}-1$, so it holds that*

$$\frac{1}{12}[\mu''(r_{i-1}) + 10\mu''(r_i) + \mu''(r_{i+1})] = \frac{1}{\mu^2}[\mu(r_{i-1}) - 2\mu(r_i) + \mu(r_{i+1})] + \mathcal{O}(\mu^4).$$

Lemma 10.10. *(Grownall's inequality [65]) Suppose that v_n and ϑ_n are nonnegative sequences, and $\{\phi_n\}$ is a sequence which satisfies*

$$\phi_0 \leq \hbar_0, \quad \phi_n \leq \hbar_0 + \sum_{s=0}^{n-1}\vartheta_s + \sum_{s=0}^{n-1}\mathrm{v}_s\phi_s, \quad \hbar_0 \geq 0; n \geq 1,$$

so, a sequence $\{\phi_n\}$ *fulfills*

$$\phi_n \leq \left(\hbar_0 + \sum_{s=0}^{n-1} \vartheta_s\right) \exp\left(\sum_{s=0}^{n-1} \nu_s\right); \quad n \geq 1.$$

Lemma 10.11. *[54] Let* $\left\{\varpi_j^{\alpha}\right\}_{j=0}^{\infty}$ *and* $\left\{\omega_j^{\alpha}\right\}_{j=0}^{\infty}$ *which are defined as in Lemma 10.4 and Lemma 10.2. Then, for any* $n \in N$ *and* $v \in \Omega_h$*, the following inequality hold*

$$\sum_{m=0}^{n} \sum_{j=0}^{m} \omega_j \langle \mathscr{H} v^{m-j}, v^m \rangle \geq 0,$$
$$\sum_{m=0}^{n} \sum_{j=0}^{m} \varpi_j^{\alpha} \langle \mathscr{H} v^{m-j}, v^m \rangle \geq 0,$$

Lemma 10.12. *[49] For any* $\alpha \in (1,2)$*, the sequence* $q_s^{(\alpha)}$ *which is defined in Eq. (10.24) fulfills the next characteristics:*

$$q_1^{(\alpha)} < 0,\ q_0^{(\alpha)} \geq q_3^{(\alpha)} ... \geq 0,\ q_0^{(\alpha)} + q_2^{(\alpha)} \geq 0,\ q_2^{(\alpha)} \begin{cases} > 0, & \alpha \in (\alpha_0, 2) \\ \leq 0, & \alpha \in (1, \alpha_0), \end{cases} \sum_{s=0}^{\infty} q_s^{(\alpha)} = 0,,$$

and $\alpha_0 \approx 1.5545$ *is the root of the Eq.* $3^{3-\alpha} - 4.2^{3-\alpha} + 6; \alpha \in (1,2)$.

Therefore, a weighted Crank-Nicolson method for Eq. 10.23 at the point $(r_i, \theta_{n+\frac{1}{2}})$ and with zero-initial values, thus the Caputo derivative can be equivalent to Riemann derivative using Lemma 10.4 for $0 < \beta < 1$, thus

$$\begin{aligned} &\frac{v_i^{n+1} - v_i^n}{\tau} - \frac{b}{2}\tau^{-\beta}\left(\sum_{k=0}^{n+1} \omega_k^{(\beta)} v_i^{n+1-k} + \sum_{k=0}^{n} \omega_k^{(\beta)} v_i^{n-k}\right) \\ &= \frac{\tau}{2}\left(\sum_{k=0}^{n+1} \varpi_k \delta_r^2 v_i^{n+1-k} + \sum_{k=0}^{n} \varpi_k \delta_r^2 v_i^{n-k}\right) + \frac{c}{2}\tau\left(\sum_{k=0}^{n+1} \varpi_k \delta_r^{\alpha} v_i^{n+1-k} + \sum_{k=0}^{n} \varpi_k \delta_r^{\alpha} v_i^{n-k}\right) \\ &+ \frac{\tau}{2}(\mathbb{F}_i^{n+1} + \mathbb{F}_i^n) + \mathscr{O}(\tau^2 + h^2). \end{aligned} \tag{10.25}$$

where $1 \leq i \leq \mathbb{M} - 1, 0 \leq n \leq \mathbb{N} - 1, v_i^n$ is a numerical solution of $v(x_i, t_n), \mathsf{F}_i^n = \mathsf{F}(x_i, t_n)$.

10.4.1 CONSTRUCTION OF SCHEME 2

By rearranging the Eq. (10.25), then yields

$$v_i^{n+1} - v_i^n - \frac{b}{2}\tau^{1-\beta}\left(\sum_{k=0}^{n+1}\omega_k^{(\beta)}v_i^{n+1-k} + \sum_{k=0}^{n}\omega_k^{(\beta)}v_i^{n-k}\right)$$

$$= \frac{\tau^2}{2}\left(\sum_{k=0}^{n+1}\varpi_k\delta_r^2 v_i^{n+1-k} + \sum_{k=0}^{n}\varpi_k\delta_r^2 v_i^{n-k}\right) + \frac{c}{2}\tau^2\left(\sum_{k=0}^{n+1}\varpi_k\delta_r^\alpha v_i^{n+1-k} + \sum_{k=0}^{n}\varpi_k\delta_r^\alpha v_i^{n-k}\right)$$

$$+ \frac{\tau^2}{2}(\mathbb{F}_i^{n+1} + \mathbb{F}_i^n) + \mathscr{O}(\tau^3 + \tau h^2). \tag{10.26}$$

Denoting $\rho = \tau^{1-\beta}/2$ and $\eta = \frac{\tau^2}{2}$, so we suggest the following Crank–Nicolson method, which is based on Lemma 10.9.

$$\mathscr{H}(v_i^{n+1} - v_i^n) - b\rho\mathscr{H}\left(\sum_{k=0}^{n+1}\omega_k^{(\beta)}v_i^{n+1-k} + \sum_{k=0}^{n}\omega_k^{(\beta)}v_i^{n-k}\right)$$

$$= \eta\left(\sum_{k=0}^{n+1}\varpi_k\delta_r^2 v_i^{n+1-k} + \sum_{k=0}^{n}\varpi_k\delta_r^2 v_i^{n-k}\right) + c\eta\left(\sum_{k=0}^{n+1}\varpi_k\delta_r^\alpha v_i^{n+1-k} + \sum_{k=0}^{n}\varpi_k\delta_r^\alpha v_i^{n-k}\right)$$

$$+ \eta\mathscr{H}(\mathbb{F}_i^{n+1} + \mathbb{F}_i^n) + \tau\rho_i^{n+1}. \tag{10.27}$$

where, $\rho_i^{n+1} = \mathscr{O}(\tau^2 + h^2)$. Ignoring the truncation error term in Eq. (10.27) and replacing v_i^n by its numerical solution V_i^n, so for Eq. (10.27) we obtain

$$\mathscr{H}(V_i^{n+1} - V_i^n) - b\rho\mathscr{H}\left(\sum_{k=0}^{n+1}\omega_k^{(\beta)}V_i^{n+1-k} + \sum_{k=0}^{n}\omega_k^{(\beta)}V_i^{n-k}\right)$$

$$= \eta\left(\sum_{k=0}^{n+1}\varpi_k\delta_r^2 V_i^{n+1-k} + \sum_{k=0}^{n}\varpi_k\delta_r^2 V_i^{n-k}\right) + c\eta\left(\sum_{k=0}^{n+1}\varpi_k\delta_r^\alpha V_i^{n+1-k} + \sum_{k=0}^{n}\varpi_k\delta_r^\alpha V_i^{n-k}\right)$$

$$+ \eta\mathscr{H}(\mathbb{F}_i^{n+1} + \mathbb{F}_i^n), \tag{10.28}$$

with

$$\begin{aligned} &V_0^n = V_M^n = 0, && 1 \le n \le \mathrm{N}, \\ &V_i^0 = 0, \quad V_i^n = 0, && 0 \le i \le \mathrm{M}. \end{aligned} \tag{10.29}$$

10.4.2 ANALYSIS OF SCHEME 2

Theorem 10.3. *Let* $v(r,\theta) \in \mathscr{C}_{r,\theta}^{6,3}([0,\mathfrak{L}] \times [0,\mathfrak{T}])$, *be the exact solution of the Eqs.* (10.1)–(10.3) *and* $V(r,\theta)$ *is a numerical solution of scheme* (10.27)–(10.28), *which defined as* $\{v_i^n | 0 \le i \le \mathrm{M}, 0 \le j \le \mathrm{N}\}$. *So for* $n\tau \le \mathfrak{T}$, *it satisfy that*

$$\|V^j - v^j\| \le \tilde{c}(\tau^2 + h^2).$$

Proof. Subtracting Eq. (10.28) from Eq. (10.27) and denote the error $\mathscr{E}_i^j = v_i^j - V_i^j$, then we have

$$\mathscr{H}(\mathscr{E}_i^{j+1} - \mathscr{E}_i^j) = b\rho\mathscr{H}\sum_{s=0}^{j}\omega_s^{(\beta)}(\mathscr{E}_i^{j+1-s} + \mathscr{E}_i^{j-s}) + \eta\sum_{s=0}^{j}\varpi_s\delta_r^2(\mathscr{E}_i^{j+1-s} + \mathscr{E}_i^{j-s})$$
$$+ c\eta\sum_{s=0}^{j}\varpi_s\delta_r^{\alpha}(\mathscr{E}_i^{j+1-s} + \mathscr{E}_i^{j-s}) + \tau\rho_i^{j+1}, \tag{10.30}$$

where, $\mathscr{E}_i^0 = 0, \quad 0 \le i \le \mathbb{M}$.

We can readily rewrite Eq. (10.30) in a matrix form:

$$\overline{C}(\mathscr{E}^{j+1} - \mathscr{E}^j) = b\rho\overline{C}\sum_{s=0}^{j}\omega_s^{(\beta)}(\mathscr{E}^{j+1-s} + \mathscr{E}^{j-s}) + \eta\sum_{s=0}^{j}\varpi_s\overline{B}(\mathscr{E}^{j+1-s} + \mathscr{E}^{j-s})$$
$$+ c\eta\sum_{s=0}^{j}\varpi_s\overline{A}(\mathscr{E}^{j+1-s} + \mathscr{E}^{j-s}) + \tau\overline{\rho^{j+1}}, \tag{10.31}$$

with $\overline{A} = c/\Gamma(4-\alpha)h^{\alpha}E$, and $\|\overline{\rho^{j+1}}\| \le (\tau^2 + h^2)$. By multiplying Eq. (10.31) with the identity matrix I of size N, we get

$$C(\mathscr{E}^{j+1} - \mathscr{E}^j) = b\rho C\sum_{s=0}^{j}\omega_s^{(\beta)}(\mathscr{E}^{j+1-s} + \mathscr{E}^{j-s}) + \eta\sum_{s=0}^{j}\varpi_s B(\mathscr{E}^{j+1-s} + \mathscr{E}^{j-s})$$
$$+ c\eta\sum_{s=0}^{j}\varpi_s A(\mathscr{E}^{j+1-s} + \mathscr{E}^{j-s}) + \tau\rho^{j+1}, \tag{10.32}$$

where,

$$E = \begin{pmatrix} q_1^{(\alpha)} & q_0^{(\alpha)} & 0 & \cdots & 0 \\ q_2^{(\alpha)} & q_1^{(\alpha)} & q_0^{(\alpha)} & \cdots & \\ \vdots & q_2^{(\alpha)} & q_1^{(\alpha)} & \ddots & \vdots \\ \vdots & \cdots & \ddots & \ddots & q_0^{(\alpha)} \\ q_N^{(\alpha)} & q_{N-1}^{(\alpha)} & \cdots & q_2^{(\alpha)} & q_1^{(\alpha)} \end{pmatrix}, \qquad C = \frac{1}{12}\begin{pmatrix} 10 & 2 & & \cdots & 0 \\ 1 & 10 & 1 & \cdots & \\ \vdots & \ddots & \ddots & \ddots & \vdots \\ & & 1 & 10 & 1 \\ & & \cdots & 2 & 10 \end{pmatrix},$$

$$B = \begin{pmatrix} 2 & -2 & & \cdots & 0 \\ -1 & 2 & -1 & \cdots & \\ \vdots & \ddots & \ddots & \ddots & \vdots \\ & & -1 & 2 & -1 \\ & & \cdots & -2 & 2 \end{pmatrix}.$$

Multiplying Eq. (10.32) by $h(\mathscr{E}^{j+1}+\mathscr{E}^{j})^T$, so obtain

$$h(\mathscr{E}^{j+1}+\mathscr{E}^{j})^T C(\mathscr{E}^{j+1}-\mathscr{E}^{j}) = b\rho\sum_{s=0}^{j}\omega_s^{(\beta)}(\mathscr{E}^{j+1}+\mathscr{E}^{j})^T C(\mathscr{E}^{j+1-s}+\mathscr{E}^{j-s})$$
$$+\eta\sum_{s=0}^{j}\varpi_s(\mathscr{E}^{j+1}+\mathscr{E}^{j})^T B(\mathscr{E}^{j+1-s}+\mathscr{E}^{j-s})$$
$$+c\eta\sum_{s=0}^{j}\varpi_s(\mathscr{E}^{j+1}+\mathscr{E}^{j})^T A(\mathscr{E}^{j+1-s}+\mathscr{E}^{j-s})+\tau h(\mathscr{E}^{j+1}+\mathscr{E}^{j})^T\rho^{j+1}, \qquad (10.33)$$

By the Gershgorin theorem, Lemmas 10.11 and 10.12, we could investigate into A and B are negative definite matrices, follow

$$(\mathscr{E}^{j+1}+\mathscr{E}^{j})^T A(\mathscr{E}^{j+1-s}+\mathscr{E}^{j-s})<0, \quad (\mathscr{E}^{j+1}+\mathscr{E}^{j})^T B(\mathscr{E}^{j+1-s}+\mathscr{E}^{j-s})<0,$$

then summing over j from 0 to $n-1$, it is deduced

$$h(\mathscr{E}^{j+1}+\mathscr{E}^{j})^T C(\mathscr{E}^{j+1}-\mathscr{E}^{j}) = h\left[(\mathscr{E}^{j+1})^T C\mathscr{E}^{j+1}-(\mathscr{E}^{j})^T C\mathscr{E}^{j}\right],$$
$$h(\mathscr{E}^{n})^T C\mathscr{E}^{n} \ge \|\mathscr{E}^{n}\|^2,$$

$$\|\mathscr{E}^{n}\|^2 \le \tau\sum_{j=0}^{n-1}\langle\rho^{j+1},(\mathscr{E}^{j+1}+\mathscr{E}^{j})\rangle \le \frac{1}{2}\|\mathscr{E}^{n}\|^2+\frac{\tau}{2}\|\mathscr{E}^{n-1}\|^2+\frac{\tau^2}{2}\|\rho^{n}\|^2+\frac{\tau}{2}\|\rho^{n}\|^2+$$
$$+\frac{\tau}{2}\sum_{j=1}^{n-2}\|\mathscr{E}^{j}\|^2+\frac{\tau}{2}\sum_{j=1}^{n-1}\|\mathscr{E}^{j}\|^2+\tau\sum_{j=0}^{n-2}\|\rho^{j+1}\|^2 \le \frac{1}{2}\|\mathscr{E}^{n}\|^2+\tau\sum_{j=1}^{n-1}\|\mathscr{E}^{j}\|^2$$
$$+\tau\sum_{j=0}^{n-1}\|\rho^{j+1}\|^2, \qquad (10.34)$$

which gives,

$$\|\mathscr{E}^{n}\|^2 \le 2\tau\sum_{j=1}^{n-1}\|\mathscr{E}^{j}\|^2+2\tau\sum_{j=0}^{n-1}\|\rho^{j+1}\|^2 \le 2\tau\sum_{j=1}^{n-1}\|\mathscr{E}^{j}\|^2+C(\tau^2+h^2)^2.$$

□

Theorem 10.4. *Consider V_i^n is the numerical solution of scheme* (10.28)–(10.29) *which is stable and hold*

$$\|V^K\|_\infty^2 \le 2\|V^0\|_{\check{H}}^2+2\tau\sum_{n=0}^{K-1}\|V^n\|^2+\tau\sum_{n=0}^{K-1}\|\tau^\theta\lambda_{n+1}(\delta_x^\alpha+\delta_x^2)V^0+2\mathscr{H}\psi+2\mathscr{H}\mathrm{f}^{\frac{1}{2}}\|^2.$$

Proof. Multiplying Eq. (10.28) by $h\mathscr{H}(V_i^{n+1}+V_i^n)$ and getting sum over $1 \leq i \leq \mathbb{M}-1$,

$$\langle \mathscr{H}(V_i^{n+1}-V_i^n), \mathscr{H}(V_i^{n+1}+V_i^n)\rangle = b\rho \sum_{k=0}^{n} \omega_k^{(\beta)} \langle \mathscr{H}(V^{n+1-k}+V^{n-k}),$$
$$\mathscr{H}(V^{n+1}+V^n)\rangle = \eta \sum_{k=0}^{n} \varpi_k \langle \delta_r^2 (V^{n+1-k}+V^{n-k}),$$
$$\mathscr{H}(V^{n+1}+V^n)\rangle + c\eta \sum_{k=0}^{n} \varpi_k \langle \delta_r^{\alpha} (V^{n+1-k}+V^{n-k}),$$
$$\mathscr{H}(V^{n+1}+V^n)\rangle + \eta \langle \mathscr{H}(\mathbb{F}^{n+1}+\mathbb{F}^n), \mathscr{H}(V^{n+1}+V^n)\rangle + b\rho\omega_{n+1}^{(\beta)} \langle \mathscr{H}V^o,$$
$$\mathscr{H}(V^{n+1}+V^n)\rangle + \eta\varpi_{n+1}\langle \delta_r^2 V^0, \mathscr{H}(V^{n+1}+V^n)\rangle + c\eta\varpi_{n+1}\langle \delta_r^{\alpha}(V^0+V^{n-k}),$$
$$\mathscr{H}(V^{n+1}+V^n)\rangle. \tag{10.35}$$

Further, calculations give that

$$\|\mathscr{H}\mathscr{W}^{n+1}\|^2 - \|\mathscr{H}\mathscr{W}^n\|^2 = b\rho \sum_{k=0}^{n} \omega_k^{(\beta)} \langle \mathscr{H}(V^{n+1-k}+V^{n-k}), \mathscr{H}(V^{n+1}+V^n)\rangle$$
$$= \eta \sum_{k=0}^{n} \varpi_k \langle \delta_r^2 (V^{n+1-k}+V^{n-k}), \mathscr{H}(V^{n+1}+V^n)\rangle + c\eta \sum_{k=0}^{n} \varpi_k \langle \delta_r^{\alpha}(V^{n+1-k}+V^{n-k}),$$
$$\mathscr{H}(V^{n+1}+V^n)\rangle + \eta\langle \mathscr{H}(\mathbb{F}^{n+1}+\mathbb{F}^n), \mathscr{H}(V^{n+1}+V^n)\rangle + b\rho\omega_{n+1}^{(\beta)}\langle \mathscr{H}V^o,$$
$$\mathscr{H}(V^{n+1}+V^n)\rangle + \eta\varpi_{n+1}\langle \delta_r^2 V^0, \mathscr{H}(V^{n+1}+V^n)\rangle + c\eta\varpi_{n+1}\langle \delta_r^{\alpha}(V^0+V^{n-k}),$$
$$\mathscr{H}(V^{n+1}+V^n)\rangle, \tag{10.36}$$

After applying Cauchy-Schwarz inequality, and getting the sum Eq. (10.36) over n from 0 to $K-1$,

$$\begin{aligned}
\|\mathscr{H}\mathscr{W}^K\|^2 - \|\mathscr{H}\mathscr{W}^0\|^2 \leq & \sum_{n=0}^{K-1} b\rho \sum_{k=0}^{n} \omega_k^{(\beta)} \langle \mathscr{H}(V^{n+1-k}+V^{n-k}), \mathscr{H}(V^{n+1}+V^n)\rangle \\
& = \eta \sum_{n=0}^{K-1}\sum_{k=0}^{n} \varpi_k \langle \delta_r^2 (V^{n+1-k}+V^{n-k}), \mathscr{H}(V^{n+1}+V^n)\rangle \\
& + c\eta \sum_{n=0}^{K-1}\sum_{k=0}^{n} \varpi_k \langle \delta_r^{\alpha} (V^{n+1-k}+V^{n-k}), \mathscr{H}(V^{n+1}+V^n)\rangle \\
& + \eta \sum_{n=0}^{K-1} \langle \mathscr{H}(\mathbb{F}^{n+1}+\mathbb{F}^n), \mathscr{H}(V^{n+1}+V^n)\rangle \\
& + b\rho \sum_{n=0}^{K-1} \omega_{n+1}^{(\beta)} \langle \mathscr{H}V^o, \mathscr{H}(V^{n+1}+V^n)\rangle \\
& + \eta \sum_{n=0}^{K-1} \varpi_{n+1} \langle \delta_r^2 V^0, \mathscr{H}(V^{n+1}+V^n)\rangle \\
& + c\eta \sum_{n=0}^{K-1} \varpi_{n+1} \langle \delta_r^{\alpha}(V^0+V^{n-k}), \mathscr{H}(V^{n+1}+V^n)\rangle.
\end{aligned} \tag{10.37}$$

According to Lemmas 10.11, and here note that b, c are negative constants, we inferred that the first three terms on the whole right side of Eq. (10.37) are negative, this follow

$$\|\mathscr{H}V^K\|^2 - \|\mathscr{H}V^0\|^2 \leq \eta \sum_{n=0}^{K-1} \langle \mathscr{H}(\mathbb{F}^{n+1} + \mathbb{F}^n), \mathscr{H}(V^{n+1} + V^n)\rangle + b\rho \sum_{n=0}^{K-1} \omega_{n+1}^{(\beta)} \langle \mathscr{H}V^o,$$

$$\mathscr{H}(V^{n+1} + V^n)\rangle + \eta \sum_{n=0}^{K-1} \varpi_{n+1} \langle \delta_r^2 V^0, \mathscr{H}(V^{n+1} + V^n)\rangle + c\eta \sum_{n=0}^{K-1} \varpi_{n+1} \langle \delta_r^\alpha (V^0 + V^{n-k}),$$

$$\mathscr{H}(V^{n+1} + V^n)\rangle. \tag{10.38}$$

or

$$\|\mathscr{H}\mathscr{W}^K\|^2 - \|\mathscr{H}\mathscr{W}^0\|^2 \leq \eta \sum_{n=0}^{K-1} \|\mathscr{H}(\mathbb{F}^{n+1} + \mathbb{F}^n)\| \|\mathscr{H}(V^{n+1} + V^n)\|$$

$$+ b\rho \sum_{n=0}^{K-1} \omega_{n+1}^{(\beta)} \|\mathscr{H}V^o\| \|\mathscr{H}(V^{n+1} + V^n)\| + \eta \sum_{n=0}^{K-1} \varpi_{n+1} \|\delta_r^2 V^0\| \|\mathscr{H}(V^{n+1} + V^n)\|$$

$$+ c\eta \sum_{n=0}^{K-1} \varpi_{n+1} \|\delta_r^\alpha V^0\| \|\mathscr{H}(V^{n+1} + V^n)\|. \tag{10.39}$$

By applying Young's inequality,

$$\|\mathscr{H}V^K\|^2 \leq \|\mathscr{H}V^0\|^2 + \frac{\eta}{2} \sum_{n=0}^{K-1} \|\mathscr{H}\mathsf{F}^{n+1/2}\|^2 + \frac{\eta}{2} \sum_{n=0}^{K-1} \|\mathscr{H}(V^{n+1} + V^n)\|$$

$$+ \frac{b\rho}{2} \sum_{n=0}^{K-1} \omega_{n+1}^{(\beta)} \|\mathscr{H}V^o\|^2 + \frac{b\rho}{2} \sum_{n=0}^{K-1} \|\mathscr{H}(V^{n+1} + V^n)\|^2$$

$$+ \frac{\eta}{2} \sum_{n=0}^{K-1} \varpi_{n+1} \|\delta_r^2 V^0\|^2 + \frac{\eta}{2} \sum_{n=0}^{K-1} \|\mathscr{H}(V^{n+1} \mid V^n)\|^2$$

$$+ c\frac{\eta}{2} \sum_{n=0}^{K-1} \varpi_{n+1} \|\delta_r^\alpha V^0\|^2 + \frac{\eta}{2} \sum_{n=0}^{K-1} \|\mathscr{H}(V^{n+1} + V^n)\|^2. \tag{10.40}$$

Then, we get

$$\|V^K\|_\infty^2 \leq 2\|V^0\|_{\check{H}}^2 + 2\eta \sum_{n=0}^{K-1} \|\mathscr{H}\mathsf{F}^{n+1/2}\|^2 + 2b\eta \sum_{n=0}^{K-1} \omega_{n+1}^{(\beta)} \|V^0\|_{\check{H}}^2$$

$$+ 2\eta \sum_{n=0}^{K-1} \varpi_{n+1} \|\delta_r^2 V^0\|^2 + 2c\eta \sum_{n=0}^{K-1} \varpi_{n+1} \|\delta_r^\alpha V^0\|^2. \tag{10.41}$$

This implies,

$$\|V^K\|_\infty^2 \leq e^{2T} \left(2(1 + b\eta\omega_{n+1}^{(\beta)}) \|V^0\|_{\check{H}}^2 + 2\eta \sum_{n=0}^{K-1} \|\varpi_{n+1}(\delta_r^\alpha + \delta_r^2)V^0 + \mathscr{H}\mathsf{F}^{n+1/2}\|^2 \right).$$

□

10.5 NUMERICAL EXAMPLE

In this section, we introduce numerical examples for demonstrating the computational performance and theoretical findings of our proposed methods.

Example 10.1. *Consider the following 1-D time-space fractional oscillator motion*

$$\frac{\partial^2 v(r,\theta)}{\partial\theta^2} = \frac{\partial^2 v(r,\theta)}{\partial r^2} + bD_\theta^\gamma v(r,\theta) + cD_r^\alpha v(r,\theta) + f(r,\theta) \quad 0<r<\mathfrak{L}, \tag{10.42}$$
$$v(r,0)=0, \quad v_\theta(r,0)=0,$$
$$v(0,\theta)=v(\mathfrak{L},\theta)=0, \quad 0<\theta<\mathfrak{T},$$

where $c=-0.5, b=-1.8, \mathfrak{L}=1$, *and* $\mathfrak{T}=1$. *The exact solution is given as* $v(r,\theta)=(\theta^{2+\gamma})r^2(1-r)^2$. *The function* $f(r,\theta)$ *is*

$$\begin{aligned} f(r,\theta) =&(2+\gamma)(1+\gamma)r^2(1-r)^2\theta^\gamma + 1.8\frac{\Gamma(3+\gamma)}{\Gamma(3)}r^2(1-r)^2\theta^2 \\ &-(12r^2-12r+2)(\theta^{2+\gamma}) \\ &+0.5(\theta^{2+\gamma})\left(\frac{\Gamma(5)}{\Gamma(5-\alpha)}r^{4-\alpha} - 2\frac{\Gamma(4)}{\Gamma(4-\alpha)}r^{3-\alpha} + \frac{\Gamma(3)}{\Gamma(3-\alpha)}r^{2-\alpha}\right). \end{aligned} \tag{10.43}$$

First, we note that the exact solution $v(r,\theta)$ *of Eq.* (10.42) *fulfill all the smoothness conditions needed by the schemes* (10.9) *and* (10.28). *In Figure 10.1 , the approximate and the exact solution of scheme* (10.9) *are shown for* $\gamma=1.5$ *and* $\alpha=0.5$, *respectively. Also in Figure 10.2, let us take step size* $\tau=h=1/40$ *to plot the curves at* $\mathfrak{T}=1$ *for comparing the numerical solution of scheme 1 at* $\alpha=0.5$, *and numerical solution of scheme 2 at* $\alpha=1.5$, *with the exact solutions for* $\gamma=1.5$. *This assured us that the exact solutions accord well with our numerical results of the schemes. As an analysis of the error in a numerical solution, consider the* L_2*-norm*

$$\varepsilon(\tau,h)=\sqrt{h\sum_{i=1}^{M-1}|v_i^N-V_i^N|^2},$$

where we can roughly calculate the convergence rate order R_r *and* R_θ *by*

$$R_r \simeq log_2[\varepsilon(\tau,2h)/\varepsilon(\tau,h)], \quad \tau\longrightarrow 0$$
$$R_\theta \simeq log_2[\varepsilon(\tau,h)/\varepsilon(2\tau,h)], \quad h\longrightarrow 0$$

Furthermore, in Table 10.1, we set up different step sizes to compute the errors and numerical convergence of scheme 2 and the order of the Crank-Nicolson difference scheme 2 for different values of α *and* γ *at time* $\mathfrak{T}=1$. *Next, in Table 10.2, we set up different step sizes to compute the errors and numerical convergence of scheme 1 and also analyze how the error* $\varepsilon(\tau,h)$ *and convergence rate* R_x *of the Crank-Nicolson difference scheme change with* $\mathbb{M}$, *for different values of* α *and* γ. *It can be seen that the errors of the method reduce as the step size* h *and* τ *decrease.*

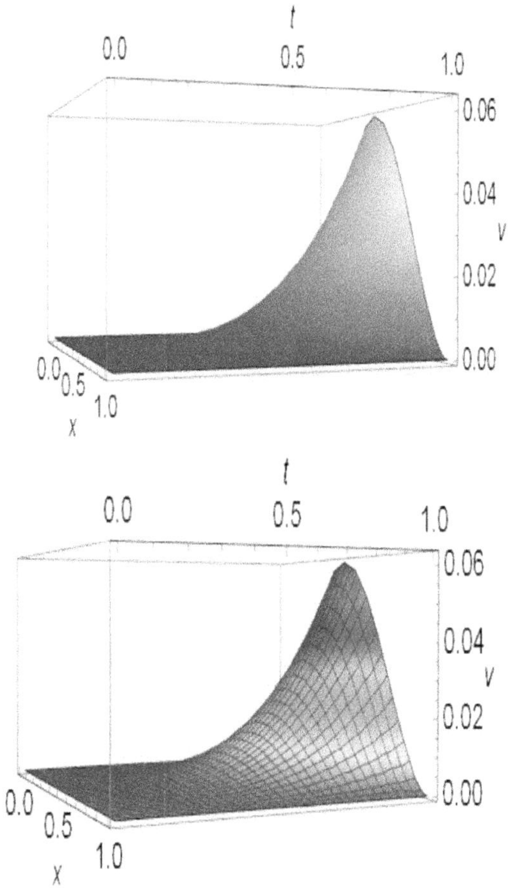

Figure 10.1 The numerical scheme 1 and the exact solution for $\alpha = 0.5$ and $\gamma = 1.5$ at $c = -0.5, b = -1.8$.

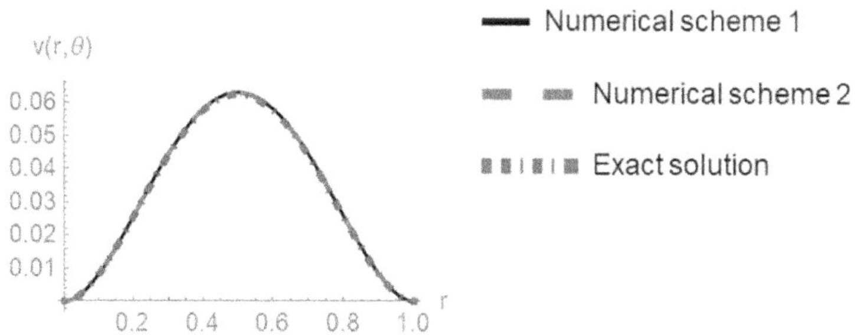

Figure 10.2 A comparison between the exact solution and the numerical schemes, when $t = 1$.

Table 10.1
The Error and Order of Crank-Nicolson Difference Scheme 2, for Different α and β

	$\alpha = 1.1, \beta = 1.9$		$\alpha = 1.2, \beta = 1.8$	
$h = \tau$	$\varepsilon(\tau, h)$	Order	$\varepsilon(\tau, h)$	Order
1/5	5.738×10^{-3}		4.954×10^{-3}	
1/10	1.474×10^{-3}	1.961	1.221×10^{-3}	2.020
1/20	3.518×10^{-4}	2.066	2.852×10^{-4}	2.098
1/40	8.382×10^{-5}	2.069	6.225×10^{-5}	2.196

Table 10.2
The Error and Order of Crank-Nicolson Difference Scheme 1, for Different α and β

	$\alpha = 0.1, \beta = 1.9$		$\alpha = 0.2, \beta = 1.8$	
$h = \tau$	$\varepsilon(\tau, h)$	Order	$\varepsilon(\tau, h)$	Order
1/5	4.893×10^{-3}		3.504×10^{-3}	
1/10	2.515×10^{-3}	0.959	1.663×10^{-3}	1.075
1/20	1.238×10^{-3}	1.023	7.581×10^{-4}	1.133
1/40	5.948×10^{-4}	1.0576	3.376×10^{-4}	1.167

10.6 CONCLUSIONS

In this chapter, the equation of a fractional oscillator motion with viscoelastic damping has been described and demonstrated. Using finite difference schemes, we deduce the numerical solution of the oscillation problem. The suggested Crank-Nicolson difference scheme demonstrated stability and convergence with an accuracy of second-order in space and $(3-\gamma)$ order accuracy in time for the first scheme 1, and converge with an accuracy of second-order in space and time for the first scheme 2. We found that the exact solution is in perfect agreement with our numerical solutions. Numerical simulations carried out to show the method's effectiveness revealed that it is precise, reliable, and stable. Both numerical schemes and theoretical analyses show that the suggested methods are efficient for solving the equation of a fractional oscillator motion with viscoelastic damping and other FPDEs. For future work, we will study the proposed problem for local and non-local boundary conditions with different analytical and numerical methods.

AUTHOR CONTRIBUTIONS

Conceptualization, T.A. and A.M.E.; Investigation, T.A. and A.M.E.; Methodology, A.M.E.; Resources, A.M.E.; Software, A.M.E.; Supervision, T.A.; Validation, T.A.; Visualization, T.A.; Writing – original draft, A.M.E.; Writing – review and editing, T.A. and A.M.E. All authors have read and agreed to the published version of the manuscript.

References

1. Labedzki P., Pawlikowski R. and Radowicz A., Axial vibrations of bars using fractional viscoelastic material models, *Vibrations in Physical Systems*, 29 (2018), p. 2018009.
2. Labedzki P., Pawlikowski R. and Radowicz A., Transverse vibration of a cantilever beam under base excitation using fractional rheological model, *AIP Conference Proceedings*, 2029(1) (2018), p. 020034.
3. Martin O., Nonlinear dynamic analysis of viscoelastic beams using a fractional rheological model, *Applied Mathematical Modelling*, 43 (2017), pp. 351–359.
4. Paola M. Di, Heuer R. and Pirrotta A., Fractional viscoelastic Euler-Bernoulli beam, *International Journal of Solids and Structures*, 50(2013), pp. 3505–3510.
5. Wang Z., Vong S. and Lei S. , Finite difference schemes for two-dimensional time-space fractional differential equations, *International Journal of Computer Mathematics* 93(3) (2016) , pp. 578–595.
6. Dai Z., Peng Y. , Mansy H. A., Sandler R. H. and Royston T. J. A model of lung parenchyma stress relaxation using fractional viscoelasticity, *Medical Engineering & Physics* 37 (2015), pp. 752–758.
7. Craiem D., Rojo Perez F. J. , Atienza Riera J. M., Guinea Tortuero G. V. and Armentano R. L., Fractional calculus applied to model arterial viscoelasticity, *Latin American Applied Research* 38 (2008) , pp. 141–145 .
8. Agrawal O. P., Formulation of Euler–Lagrange equations for fractional variational problems, *Journal of Mathematical Analysis and Applications* 272 (2002), pp. 368–379.
9. Riewe F., Nonconservative Lagrangian and Hamiltonian mechanics, *Physical Review E* 53 (1996), pp. 1890–1899.
10. Singh H. and Wazwaz A. M. , Computational method for reaction diffusion-model arising in a spherical catalyst. *International Journal of Applied and Computational Mathematics* 7 (2021), pp. 1–11.
11. Agrawal O. P., A general formulation and solution scheme for fractional optimal control problems, *Nonlinear Dynamics* 38 (2004), pp. 323–337.
12. Hilfer J. R. *Applications of Fractional Calculus in Physics*, World Scientific Publishing Company: Singapore (2000), pp. 87–130.
13. Tarasov V. E., Review of some promising fractional physical models, *International Journal of Modern Physics B* 27 (2013), p. 1330005.
14. Riewe F., Mechanics with fractional derivatives, *Physical Review E* 55 (1997), pp. 3581–3592.
15. Ingman D. and Suzdalnitsky J., Control of damping oscillations by fractional differential operator with time-dependent order. *Computer Methods in Applied Mechanics and Engineering* 193 (2004), pp. 5585–5595.
16. Shen K. L. and Soong T. T., Modeling of viscoelastic dampers for structural applications. *Journal of Engineering Mechanics* 121 (1995), pp. 694–701.

17. Aleroev T. S., Mahmoud, E. I. and Elsayed A. M., Mathematical model of the polymer concrete by fractional calculus with respect to a spatial variable. *IOP Conference Series: Materials Science and Engineering* 1129 (2021), pp. 12–31.
18. Aleroev T. S., The analysis of the polymer concrete characteristics by fractional calculus. *Journal of Physics: Conference Series*1425 (2020), p. 012112.
19. Harendra S. and Srivastava H. M., Numerical investigation of the fractional-order Liénard and duffing equations arising in oscillating circuit theory. *Frontiers in Physics*, 8 (2020), p. 120.
20. Podlubny I., *Fractional Differential Equations : An Introduction to Fractional Derivatives, Fractional Differential Equations, to Methods of Their Solution and Some of Their Applications*. San Diego : Academic Press (1999).
21. Shen K. L. and Soong T. T. , Modeling of viscoelastic dampers for structural applications, *Journal of Engineering Mechanics-ASCE* 121 (1995), pp. 694–701.
22. Mahmoud E. I. and Aleroev T. S., Boundary value problem of space-time fractional advection diffusion. *Mathematics* 10(17) (2022), p. 3160.
23. Ingman D. and Suzdalnitsky J., Iteration method for equation of viscoelastic motion with fractional differential operator of damping, *Computer Methods in Applied Mechanics and Engineering* 190 (2001), pp. 5027–5036.
24. El-Nabulsi R., Cosmology with a fractional action principle, *Romanian Reports on Physics* 59 (2007), pp. 763–771.
25. Enelund M. and Josefson B. L. , Time-domain finite element analysis of viscoelastic structures with fractional derivatives constitutive relations, *AIAA Journal* 35 (1997), pp. 1630–1637.
26. Taloni A., Chechkin A. and Klafter J., Generalized elastic model yields a fractional langevin equation description. *Physical Review Letters* 104(16) (2010), p. 160602.
27. Freed A. D. and Diethelm K., Fractional calculus in biomechanics: A 3D viscoelastic model using regularized fractional derivative kernels with application to the human calcaneal fat pad. *Biomechanics and Modeling in Mechanobiology* 5(4) (2006), pp. 203–215.
28. Tripathi V. M., Srivastava M. H., Singh H., Swarup C. and Aggarwal S., Mathematical analysis of non-isothermal reaction–diffusion models arising in spherical catalyst and spherical biocatalyst. *Applied Sciences* 11(21) (2021), p. 10423.
29. Kirianova L., Modeling of strength characteristics of polymer concrete via the wave equation with a fractional derivative, *Mathematics* 8(10) (2020), pp. 18–43.
30. Torvik P. J. and Bagley R. L., On the appearance of the fractional derivative in the behavior of real materials. *Journal of Applied Mechanics* 51 (1984), pp. 294–298.
31. Bagley R. L., A theoretical basis for the application of fractional calculus to viscoelasticity. *Journal of Rheology* 27 (1983), p. 201.
32. Di Paola M., Pirrotta A. and Valenza A., Visco-elastic behavior through fractional calculus: An easier method for best fitting experimental results. *Mechanics of Materials* 43 (2011), pp. 799–806.
33. Aleroev T. S. and Elsayed A. M., Analytical and approximate solution for solving the vibration string equation with a fractional derivative, *Mathematics* 8(7) (2020), p. 1154.
34. Ray S., Sahoo S. and Das S., Formulation and solutions of fractional continuously variable order mass spring damper systems controlled by viscoelastic and viscous-viscoelastic dampers, *Advances in Mechanical Engineering* 8 (2016), pp. 1–17.
35. Aleroev T., Erokhin S. and Kekharsaeva E., Modeling of deformation-strength characteristics of polymer concrete using fractional calculus, *IOP Conference Series: Materials Science and Engineering* 365(3) (2018), p. 032004.

36. Bhrawy A. H. and Zaky M. A., A method based on the Jacobi tau approximation for solving multi-term time–space fractional partial differential equations, *Journal of Computational Physics* 281 (2015), pp.876–895.
37. Xu M. and Tan W., Intermediate processes and critical phenomena: Theory, method and progress of fractional operators and their applications to modern mechanics, *Science in China Series G* 49(3) (2006), pp. 257–272.
38. Mahmoud E. and Orlov V., Numerical solution of two dimensional time-space fractional Fokker Planck equation with variable coefficients. *Mathematics* 9(11) (2021), p. 1260.
39. Huang J., Tang Y., Vázquez L. and Yang J., Two finite difference schemes for time fractional diffusion-wave equation, *Numerical Algorithms* 64 (2013), pp. 707–720.
40. Singh H., Analysis for fractional dynamics of Ebola virus model. *Chaos, Solitons and Fractals* 138 (2020), p. 109992.
41. Aleroev T., On a class of positive definite operators and their application in fractional calculus. *Axioms* 11(6) (2022), p. 272.
42. Singh H., An efficient computational method for non-linear fractional Lienard equation arising in oscillating circuits. In H. Singh, D. Kumar and D. Baleanu (Eds.),Methods of Mathematical Modelling. CRC Press: Boca Raton, FL, pp. 39–50 (2019).
43. Elsayed A. and Orlov V., Numerical scheme for solving time–space vibration string equation of fractional derivative, *Mathematics* 8 (2020), p. 1069.
44. Aleroev T. S., Elsayed A. M. and Mahmoud E. I., Solving one dimensional time-space fractional vibration string equation, *IOP Conference Series: Materials Science and Engineering* 1129 (2021), 012030.
45. Orlov V. N., Elsayed A. M. and Mahmoud, E. I., Two linearized schemes for one-dimensional time and space fractional differential equations. *Mathematics* 10(19) (2022), p. 3651.
46. Guo S., Mei L., Zhang Z., Li C., Li M. and Wang Y., The linearized finite difference/spectral-Galerkin scheme for three-dimensional distributed-order time-space fractional nonlinear reaction-diffusion-wave equation: Numerical simulations of Gordon-type solitons. *Computer Physics Communications* 252 (2020), pp. 107–144.
47. Zhengang Z. and Changpin L., Fractional difference/finite element approximations for the time–space fractional telegraph equation. *Applied Mathematics and Computation* 219 (2012), pp. 2975–2988.
48. Singh H., Solving a class of local and nonlocal elliptic boundary value problems arising in heat transfer. *Heat Transfer* 51(2) (2022), pp. 1524–1542.
49. Sun Z. and Wu X., A fully discrete difference scheme for a diffusion-wave system, *Applied Numerical Mathematics* 56(2) (2006), pp.193–209.
50. Chen J., Liu F., Anh V., Shen S., Liu Q. and Liao C., The analytical solution and numerical solution of the fractional diffusion-wave equation with damping. *Applied Mathematics and Computation* 219 (2012), pp. 1737–1748.
51. Li C. P. and Zeng F. Z., *Numerical Methods for Fractional Calculus*, 1st ed. Chapman and Hall/CRC: New York, p. 300 (2015).
52. Singh H., Chebyshev spectral method for solving a class of local and nonlocal elliptic boundary value problems. *International Journal of Nonlinear Sciences and Numerical Simulation* (2021), 000010151520200235.
53. Changpin L., Rifang W., Hengfei D. and Gianni P., High-order approximation to Caputo derivatives and Caputo-type advection-diffusion equations a Dedicated to Professor Francesco Mainardi on the occasion of his retirement. *Communications in Applied and Industrial Mathematics.* 299 (2015), pp. 159–175.

54. Wang Z. and Vong S., Compact difference schemes for the modified anomalous fractional sub-diffusion equation and the fractional diffusion-wave equation, *Journal of Computational Physics* 277 (2013), pp. 1–15.
55. Lyu P. and Vong S. A., Linearized second-order finite difference scheme for time fractional generalized BBM equation. *Applied Mathematics Letters* 78 (2018), pp. 16–23.
56. Chen A. and Li C., Numerical solution of fractional diffusion-wave equation, *Numerical Functional Analysis and Optimization* 37 (2016), pp. 19–39.
57. Sarwar S., Khaled M. F. and Muhammad A., Abundant wave solutions of conformable space-time fractional order Fokas wave model arising in physical sciences. *Alexandria Engineering Journal* 60(2) (2021), pp. 2687–2696.
58. Jin S., Jiaquan X., Jingguo Q. and Yiming C., A numerical method for simulating viscoelastic plates based on fractional order model. *Fractal and Fractional* 6 (2022), pp. 3–11.
59. Faal R. T., Sourki R.,Crawford B., Vaziri R. and Milanib A. S., Using fractional derivatives for improved viscoelastic modeling of textile composites. Part II: Fabric under different temperatures. *Composite Structures* 248 (2020), p. 112494.
60. Kazem S., Exact solution of some linear fractional differential equations by Laplace transform. *International Journal of Nonlinear Science* 16 (2013), pp. 3–11.
61. Rossikhin Y. and Shitikova M., New approach for the analysis of damped vibrations of fractional oscillators. *Shock and Vibration* 16 (2009), pp. 365–387.
62. Mainardi F., *Fractional Calculus and Waves in Linear Viscoelasticity: An Introduction to Mathematical Models*. World Scientific: Singapore (2010).
63. Sousa E. and Li C., A weighted finite difference method for the fractional diffusion equation based on the riemann–liouville derivative, *Applied Numerical Mathematics* 90 (2015), pp. 22–37.
64. Gao G. H. and Sun Z. Z., A compact finite difference scheme for the fractional subdiffusion equations, *Journal of Computational Physics* 230(3) (2011), pp. 586–595.
65. Quarteroni A. and Valli A., *Numerical Approximation of Partial Different Equations*, Springer: Berlin, Germany (1994).

11 Dynamics of the Dadras-Momeni System in the Frame of the Caputo-Fabrizio Fractional Derivative

Chandrali Baishya
Tumkur University

P. Veeresha
CHRIST (Deemed to be University)

CONTENTS

11.1 INTRODUCTION

The study of dynamical systems associated with complex nature, particularly chaos is the most captivating issue in all of academia. Numerous real applications such as power systems [1], circuit [2], medicine [3], biology [4], chemical reactors [5], and others employ chaotic behavior. In literature, the dynamical systems like Lorenz system [6], Newton–Leipnik system [7], Chen system [8] , Lü system [9], and exhibit interesting chaotic behaviors. For the past 10 years, chaos control is another area where remarkable development can be seen. The sliding mode controller [10],

DOI: 10.1201/9781003368069-11

adaptive sliding mode controller [11], backstepping technique [12], and linear feedback controllers [13] are some of the largest applied techniques that researchers have adopted to control chaos in dynamical systems with chaotic behavior.

Although the fractional derivative (FD) is as old as the classical order derivative, the notion has received noticeable attention and was considered significantly from 1967 when it was first modified as part of the Caputo investigation [14]. Riemann-Liouville and Caputo FD received special attention from the research community over the decades. Continuity and differentiability of an arbitrary function are not mandatory with Riemann-Liouville FD at the origin. In the model formulation, the Caputo FD has the benefits of allowing standard initial and boundary conditions [14]. Additionally, under the Caputo FD, the derivative of a constant is zero. Although these FDs have a lot of advantages, they cannot be used in every case. As the derivative of a constant is not zero in the Riemann-Liouville sense, it forms many differences in the assumptions of modeling, and the initial conditions also need to be generalized with fractional order. In addition, fractional derivation produces singularity at the origin whenever an arbitrary function is a constant at the origin. For instance, Mittag-Leffler and exponential functions. The Caputo derivative is defined only for the differentiable functions. But despite not having the first-order derivative, the functions might have FDs of all orders less than one in the Riemann-Liouville sense. The commonly used fractional differential operators have some limitations for simulating real-world problems due to the singularity of the power-law-based FDs. In this connection, Caputo and Fabrizio have nurtured a new operator with fractional order. They stressed that numerous physical events are non-singular and that employing singular operators which simulate non-singular events might lead to confusing conclusions. To resolve this problem, a fractional differential operator with exponential function as kernel known as Caputo-Fabrizio FD has been formulated. The exponential functions are eigenfunctions of the decay differential ordinary equations and also they are the generator of an evolution partial differential equations [15].

Recently, rigorous applications of fractional calculus have been noticed in various fields of science and engineering by using FDs, involving Riemann–Liouville, Caputo, Weyl, Riesz, Grünwald–Letnikov, Marchaud and Hifler, Caputo-Fabrizio, and Atangana-Baleanu operators. In this connection, the most vital topics include anomalous diffusion, power law, history-dependent process, vibration and control, continuous time random walk, Riesz potential, fractional predator-prey system, Brownian motion, fractals, nonlocal phenomena, biomedical engineering, porous media, fractional transforms, variational principles, soft matter mechanics, signal processing, geophysics, acoustic dissipation, robotics, system identification, percolation, telecommunications, relaxation, rheology, viscoelasticity, and chaos. Significant literature of the above-mentioned works can be found in Refs. [16–35]. On the other hand, in engineering fields, chaotic dynamics have been used in modeling of real-life systems, in performance enhancement of existing systems, especially in secure communications, antenna and radar systems, biomedical engineering, civil, mechanical, robotics, power systems, etc. As a result, fractional chaotic dynamics have immerged as an attractive area of research for many researchers [36–40].

It is observed that Lorenz system [41,42] and Chen system [43–45], and other fractional dynamical systems behave chaotic or hyperchaotic. Particularly, dynamical and control [36–40] chaotic systems have drawn a lot of interest from a variety of scientific sectors. The fractional-order control and synchronization of chaotic systems have been the most fascinating areas in recent years, and numerous academics have made significant contributions. To control the variety of fractional-order chaotic systems, several methods are employed and which can be observed in the literature. One of the most significant strategies that is popular as an efficient control strategy is the sliding mode control(SMC) approach [46–49]. The control rule to enforce the states of the model is the basic feature of SMC to switch from their initial states onto some preset sliding surface. Moreover, the surface should be designed to ensure the system holds the qualities like capability, tracking ability, disturbance rejection, and stability. In a Master-Slave framework, to synchronize fractional-order chaotic systems a controller based on active sliding mode theory was presented in Ref. [46]. For a nonlinear system, an intelligent resilient fractional surface SMC is investigated in Ref. [50]. The modified Duffing systems with synchronizations of two uncoupled chaotic systems associated with fractional order have been examined in Ref. [51]. To manage chaos to equilibria in the modified Van der Pol–Duffing model, the fractional Routh–Hurwitz conditions are employed [52]. In Ref. [53], the authors present a new hyperchaotic system and as a result the nonlinear state observer due to the pole insertion technique to synchronize. Between fractional-order chaotic systems, the function projective synchronization was investigated in Ref. [54]. In Ref. [55], the presence of mild solutions and optimal controls in the α norm for semi-linear fractional evolution equations have been studied. To establish the chaos synchronization for the hyperchaotic system with fractional order, three approaches are proposed in Ref. [56]. Applying classical control theory in a bounded domain, the authors have investigated a fractional diffusion equation [57]. With the aid of an adaptive adjustment mechanism and the generalized T-S fuzzy model, a simple and effective strategy for controlling chaotic systems with fractional order is nurtured in Ref. [58]. In Ref. [59], after evidencing the chaotic behavior of the modified coupled dynamic system, the authors applied the feedback control approach to regulate chaos in the system by defining the conditions that suppressed chaos to unstable equilibrium points.

In this chapter, we study the Dadras-Momeni (DM) chaotic attractor using the Caputo-Fabrizio FD. We have observed in the literature that the DM model exhibits interesting chaotic behavior for integer order derivatives. But, its behavior under fractional-order derivative has not been analyzed yet. The prime goal of this study is to conduct a critical analysis of the complex system with fractional order of commensurate and non-commensurate derivatives using numerical simulations. For this purpose, we have designed a SMC law to govern the chaos in the DM model of fractional order with or without external disturbances and uncertainties. With the assistance of numerical results, we can observe that one can maintain and stabilize system states on the sliding surface and the controller can restrain chaos in the system.

The rest of the present investigation is sequenced as: in Section 11.2, and fundamentals associated with Caputo-Fabrizio fractional order are discussed. The subsequent section is associated with the fractional-order DM system. In Section 11.4, the existence and uniqueness of the solution of the projected model are analyzed. In Section 11.5 the dynamics study of the FDM system is presented and then in the next segment, to globally and asymptotically stabilize the SMC is designed. In Section 11.7, the numerical study is conducted to validate the attained results, and the concluding remarks are drawn in the lost section.

11.2 PRELIMINARIES

Here, we have presented some results related to the Caputo-Fabrizio FDs that are applied in this work to prove the theoretical aspects. Let $H^1(a,b) = \{f : f \in L^2(a,b), f' \in L^2(a,b)\}$, where $L^2(a,b)$ is the space of square integrable functions on the interval (a,b) and $C[a,b]$ be the set of vector functions which operate on $H^1(a,b)$. The norm of $X(t) = (x_1(t), x_2(t), x_3(t)) \in [a,b]$ is given by $||X(t)|| = \sum_{i=1}^{3} Sup|x_i(t)|$.

Definition 1. *The Caputo FD of order α for the function $g(t)$ (where $g(t)$ is k times continuously differentiable and $g^{(k)}(t)$ integrable in $[t_0, T]$ is defined as [17]*

$$^{C}D_t^{\alpha} g(t) = \frac{1}{\Gamma(1-\alpha)} \int_{t_0}^{t} \frac{g'(\tau)}{(t-\tau)^{\alpha}} d\tau, \tag{11.1}$$

where $b > a$ and $g \in H^1(a,b)$ is a positive integer such that $0 < \alpha < 1$.

The kernel $(t-\tau)^{-\alpha}$ in Eq. (11.1) has singularity at $t = \tau$. In 2015, by changing the kernel $(t-\tau)^{-\alpha}$ with the function $exp\left(-\frac{\alpha(t-\tau)}{1-\alpha}\right)$ and $\frac{1}{\Gamma(1-\alpha)}$ with $\frac{B(\alpha)}{1-\alpha}$ Caputo and Fabrizio [15] proposed the following definition with exponential kernel

$$^{CF}D_t^{\alpha} f(t) = \frac{B(\alpha)}{(1-\alpha)} \int_a^t exp\left(-\frac{\alpha}{1-\alpha}(t-\tau)\right) f'(\tau) d\tau, \tag{11.2}$$

where $B(\alpha)$ is a normalization function (any smooth positive function) such that $B(0) = B(1) = 1$. According to the Caputo-Fabrizio FD, derivative of a constant is zero and the kernel does not have singularity for $t = \tau$. Later, by changing the kernel $(t-\tau)^{-\alpha}$ with the function $exp\left(-\frac{\alpha(t-\tau)}{1-\alpha}\right)$ and $\frac{1}{\Gamma(1-\alpha)}$ with $\frac{1}{\sqrt{2\pi(1-\alpha^2)}}$ in Eq. (11.1), Losada and Nieto [60] has derived a new Caputo-Fabrizio FD of order $0 < \alpha < 1$ as follows.

$$^{CF}D_t^{\alpha} f(t) = \frac{(2-\alpha)B(\alpha)}{2(1-\alpha)} \int_a^t exp\left(-\frac{\alpha}{1-\alpha}(t-\tau)\right) f'(\tau) d\tau. \tag{11.3}$$

Definition 2. *The Caputo-Fabrizio fractional integral operator expressed using the technique in Ref. [60] is as follows:*

$$I_t^{\alpha} f(t) = \frac{2(1-\alpha)}{2B(\alpha) - \alpha B(\alpha)} f(t) + \frac{2\alpha}{2B(\alpha) - \alpha B(\alpha)} \int_0^t f(\tau) d\tau, \; t \geq 0. \tag{11.4}$$

The following result can be found according to the above definition [60], we have

$$\frac{2(1-\alpha)}{(2-\alpha)B(\alpha)}+\frac{2\alpha}{(2-\alpha)B(\alpha)}=1.$$

The above expression results in a formula as given below

$$B(\alpha)=\frac{2}{2-\alpha},\ 0<\alpha<1.$$

Therefore, Nieto and Losada nurture the new Caputo derivative of order $0<\alpha<1$ and which is presented as

$$^{CF}D_t^{\alpha}f(t)=\frac{1}{1-\alpha}\int_0^t exp\left(-\frac{\alpha}{1-\alpha}(t-\tau)\right)f'(\tau)d\tau,\ t\geq 0, \tag{11.5}$$

where $f\in H^1(0,b)$. For more details, the readers are referred to [12, 13]. Moreover, The connected integral was proposed by Losada and Nieto [60].
For Eq. (11.2), the Laplace transform (LT) is given by

$$L[^{CF}D_t^{\alpha}f(t)]=\frac{B(\alpha)(sF(s)-f(0))}{s+\alpha(1-s)},\ s>0. \tag{11.6}$$

Here, $F(s)=L[f(t)]$ signifies the LT of $f(t)$.

11.3 MODEL FORMULATION

In this segment, we considered the chaotic system proposed by Dadras and Momeni [61], which is familiarly called DM system. The corresponding system is given by

$$\begin{aligned}\frac{du}{dt}&=v-\beta u+\gamma vw,\\ \frac{dv}{dt}&=\lambda v-uw+w,\\ \frac{dw}{dt}&=\delta uv-\sigma w.\end{aligned}$$

In this study, we have considered DM system in the frame of the Caputo-Fabrizio FD as follows:

$$\begin{aligned}{}^{CF}D_t^{\alpha_1}u&=v-\beta u+\gamma vw,\\ {}^{CF}D_t^{\alpha_2}v&=\lambda v-uw+w,\\ {}^{CF}D_t^{\alpha_3}w&=\delta uv-\sigma w,\end{aligned} \tag{11.7}$$

with $u(0)=x_0, v(0)=y_0, w(0)=z_0$.

11.4 EXISTENCE AND UNIQUENESS OF SOLUTIONS FOR THE PROJECTED SYSTEM

Here, we illustrated the existence and uniqueness using the fixed point theory. After employing CF fractional integral operator cited in Eqs. (11.3)–(11.7), we have

$$\begin{aligned} u(t)-u(0) &= {}^{CF}I_t^{\alpha_1}\left(v-\beta u+\gamma v w\right), \\ v(t)-v(0) &= {}^{CF}I_t^{\alpha_2}\left(\lambda v-uw+w\right), \\ w(t)-w(0) &= {}^{CF}I_t^{\alpha_3}\left(\delta uv-\sigma w\right). \end{aligned} \tag{11.8}$$

Now, we consider the following kernels for computational convenience

$$\begin{aligned} K_1(t,u) &= v-\beta u+\gamma v w, \\ K_2(t,v) &= \lambda v-uw+w, \\ K_3(t,w) &= \delta uv-\sigma w. \end{aligned} \tag{11.9}$$

We define the function $\Theta(\alpha)=\frac{2(1-\alpha)}{2B(\alpha)-\alpha B(\alpha)}$ and $\Omega(\alpha)=\frac{2\alpha}{2B(\alpha)-\alpha B(\alpha)}$. To prove the required results, we will assume that u, v, and w are non-negative bounded functions such that $||u(t)||\leq\rho_1$, $||v(t)||\leq\rho_2$, $||w(t)||\leq\rho_3$, where ρ_i, $i=1,2,3$ are positive constants. Let us consider $\xi_1=\rho_2+\beta+\gamma\rho_2\rho_3$, $\xi_2=\lambda+\rho_1\rho_3+\rho_3$, $\xi_3=\delta\rho_1\rho_2+\sigma$. After applying the definition of the FC fractional integral to Eq. (11.8), one can obtain

$$\begin{aligned} u(t)-u(0) &= \Theta(\alpha_1)K_1(t,u)-\Theta(\alpha_1)K_1(0,u(0))+\Omega(\alpha_1)\int_0^t K_1(\tau,u)d\tau, \\ v(t)-v(0) &= \Theta(\alpha_2)K_2(t,v)-\Theta(\alpha_2)K_2(0,v(0))+\Omega(\alpha_2)\int_0^t K_2(\tau,v)d\tau, \\ w(t)-w(0) &= \Theta(\alpha_3)K_3(t,w)-\Theta(\alpha_3)K_3(0,w(0))+\Omega(\alpha_3)\int_0^t K_3(\tau,w)d\tau. \end{aligned} \tag{11.10}$$

We consider the kernel K_1 and any two functions u and u_1. Then

$$||K_1(t,u)-K_1(t,u_1)||=\beta||u-u_1||. \tag{11.11}$$

Similarly, from the other two equations of Eq. (11.9) we get

$$||K_2(t,v)-K_2(t,v_1)||=\lambda||v-v_1||,\ ||K_3(t,w)-K_3(t,w_1)||=\sigma||u-u_1||. \tag{11.12}$$

In terms of the kernels, the state variable can be presented using Eq. (11.10), as follows

$$\begin{aligned} u(t) &= u(0)-\Theta(\alpha_1)K_1(0,u(0))+\Theta(\alpha_1)K_1(t,u)+\Omega(\alpha_1)\int_0^t K_1(\tau,u)d\tau, \\ v(t) &= v(0)--\Theta(\alpha_2)K_1(0,v(0))+\Theta(\alpha_2)K_2(t,v)+\Omega(\alpha_2)\int_0^t K_2(\tau,v)d\tau, \\ w(t) &= w(0)-\Theta(\alpha_3)K_1(0,w(0))+\Theta(\alpha_3)K_3(t,w)+\Omega(\alpha_3)\int_0^t K_3(\tau,w)d\tau. \end{aligned} \tag{11.13}$$

Using given initial conditions, the initial components of the above recursive formulas are obtained as

$$u_0(t) = u(0),\ v_0(t) = v(0),\ w_0(t) = w(0).$$

For the recursive formulas, the differences between the consecutive terms are given by

$$\begin{aligned}
\Psi_{1n}(t) = u_n(t) - u_{n-1}(t) = \Theta(\alpha_1)(K_1(t,u_{n-1}) - K_1(t,u_{n-2}))\\
+\Omega(\alpha_1)\int_0^t (K_1(\tau,u_{n-1}) - K_1(\tau,u_{n-2}))d\tau,\\
\Psi_{2n}(t) = v_n(t) - v_{n-1}(t) = \Theta(\alpha_2)(K_2(t,v_{n-1}) - K_2(t,v_{n-2}))\\
+\Omega(\alpha_2)\int_0^t (K_2(\tau,v_{n-1}) - K_2(\tau,v_{n-2}))d\tau,\\
\Psi_{3n}(t) = w_n(t) - w_{n-1}(t) = \Theta(\alpha_3)(K_3(t,w_{n-1}) - K_2(t,w_{n-2}))\\
+\Omega(\alpha_3)\int_0^t (K_3(\tau,w_{n-1}) - K_3(\tau,w_{n-2}))d\tau.
\end{aligned} \tag{11.14}$$

Note that

$$u_n(t) = \sum_{i=1}^{n} \Psi_{1i},\ v_n(t) = \sum_{i=1}^{n} \Psi_{2i},\ w_n(t) = \sum_{i=1}^{n} \Psi_{3i}. \tag{11.15}$$

For the differences Ψ_{1n}, Ψ_{2n}, Ψ_{3n}, the recursive inequalities are formulate as follows:

$$\begin{aligned}
&||\Psi_{1n}(t)|| = ||u_n(t) - u_{n-1}(t)||\\
&= ||\Theta(\alpha_1)(K_1(t,u_{n-1}) - K_1(t,u_{n-2})) + \Omega(\alpha_1)\int_0^t (K_1(\tau,u_{n-1}) - K_1(\tau,u_{n-2}))d\tau||.
\end{aligned} \tag{11.16}$$

Now, Eq. (11.16) simplifies after employing the triangle inequality as

$$\begin{aligned}
||u_n(t) - u_{n-1}(t)|| \le \Theta(\alpha_1)||(K_1(t,u_{n-1}) - K_1(t,u_{n-2}))||\\
+\Omega(\alpha_1)\int_0^t ||(K_1(\tau,u_{n-1}) - K_1(\tau,u_{n-2}))||d\tau.
\end{aligned}$$

Then, since the kernel K_1 satisfies (11.11), we have

$$||u_n(t) - u_{n-1}(t)|| \le \Theta(\alpha_1)\beta||u_{n-1} - u_{n-2}|| + \Omega(\alpha_1)\beta\int_0^t ||u_{n-1} - un - 2||d\tau.$$

Thus, we obtain

$$||\Psi_{1n}(t)|| \le \Theta(\alpha_1)\beta||\Psi_{1(n-1)}(t)|| + \Omega(\alpha_1)\beta\int_0^t ||\Psi_{1(n-1)}(\tau)||d\tau. \tag{11.17}$$

Similarly, we can establish the following

$$\begin{aligned}
||\Psi_{2n}(t)|| \le \Theta(\alpha_2)\lambda||\Psi_{2(n-1)}(t)|| + \Omega(\alpha_2)\lambda\int_0^t ||\Psi_{2(n-1)}(\tau)||d\tau,\\
||\Psi_{3n}(t)|| \le \Theta(\alpha_3)\sigma||\Psi_{3(n-1)}(t)|| + \Omega(\alpha_3)\sigma\int_0^t ||\Psi_{3(n-1)}(\tau)||d\tau.
\end{aligned} \tag{11.18}$$

Theorem 11.1. *If there exists a time* $t_\alpha > 0$ *such that the following inequalities admit* $\Theta(\alpha_1)\beta + \Omega(\alpha_1)\beta t_\alpha < 1$, $\Theta(\alpha_2)\lambda + \Omega(\alpha_2)\lambda t_\alpha < 1$, $\Theta(\alpha_3)\sigma + \Omega(\alpha_3)\sigma t_\alpha < 1$, *then solutions exists for Eq.* (11.7).

Proof. Clearly, we assumed $u(t)$, $v(t)$ and $w(t)$ are bounded and each of the kernels admits a Lipschitz condition. Now, using Eqs. (11.17)–(11.18) recursively, the following relations can be obtained.

$$\begin{aligned} ||\Psi_{1n}(t)|| &\leq ||u(0)||\,(\Theta(\alpha_1)\beta + \Omega(\alpha_1)\beta t)^n, \\ ||\Psi_{2n}(t)|| &\leq ||v(0)||\,(\Theta(\alpha_2)\lambda + \Omega(\alpha_2)\lambda t)^n, \\ ||\Psi_{3n}(t)|| &\leq ||w(0)||\,(\Theta(\alpha_3)\sigma + \Omega(\alpha_3)\sigma t)^n. \end{aligned} \tag{11.19}$$

The above system confirms the existence and smoothness of Eq. (11.15). To ensure the required condition, we prove the functions $u_n(t)$, $v_n(t)$, $w_n(t)$ converge to a the solutions of the system (11.7). Now, we construct $A_n(t), B_n(t), C_n(t)$ as the remainder terms after n iterations, i.e.,

$$\begin{aligned} A_n(t) &= u_n(t) - (u(t) - u(0)), \\ B_n(t) &= v_n(t) - (v(t) - v(0)), \\ C_n(t) &= w_n(t) - (w(t) - w(0)). \end{aligned} \tag{11.20}$$

Then, using the triangle inequality and Eq. (11.11), we have

$$\begin{aligned} ||A_n(t)|| &= ||\Theta(\alpha_1)(K_1(t,u) - K_1(t,u_{n-1})) + \Omega(\alpha_1)\int_0^t (K_1(\tau,u) - K_1(\tau,u_{n-1}))d\tau|| \\ &\leq \Theta(\alpha_1)||(K_1(t,u) - K_1(t,u_{n-1}))|| + \Omega(\alpha_1)\int_0^t ||(K_1(\tau,u) - K_1(\tau,u_{n-1}))||d\tau \\ &\leq \Theta(\alpha_1)\beta||u - u_{n-1}|| + \Omega(\alpha_1)\beta||u - u_{n-1}||. \end{aligned}$$

Similarly,

$$||A_n(t)|| \leq [(\Theta(\alpha_1) + \Omega(\alpha_1)t)\,\beta]^{n+1}\rho_1. \tag{11.21}$$

At t_α, we obtain

$$||A_n(t)|| \leq [(\Theta(\alpha_1) + \Omega(\alpha_1)t_\alpha)\,\beta]^{n+1}\rho_1. \tag{11.22}$$

Taking limit on Eq. (11.22) as $n \to \infty$ and then considering $\Theta(\alpha_1)\beta + \Omega(\alpha_1)\beta t_\alpha < 1$, we have

$$||A_n(t)|| \to 0.$$

Similarly, we have the following relation:

$$||B_n(t)|| \leq [(\Theta(\alpha_2) + \Omega(\alpha_2)t_\alpha)\,\lambda]^{n+1}\rho_2, \tag{11.23}$$

$$||C_n(t)|| \leq [(\Theta(\alpha_3) + \Omega(\alpha_3)t_\alpha)\,\sigma]^{n+1}\rho_3. \tag{11.24}$$

Similarly taking limits on Eqs. (11.23)–(11.24) as $n \to \infty$ and using conditions $\Theta(\alpha_2)\lambda + \Omega(\alpha_2)\lambda t_\alpha < 1$, $\Theta(\alpha_3)\sigma + \Omega(\alpha_3)\sigma t_\alpha < 1$, we get $||B_n(t)|| \to 0$ and $||C_n(t)|| \to 0$. Therefore, this proves the required condition. □

Theorem 11.2. *If the following conditions admit*

$$(1-\Theta(\alpha_1)\beta-\Omega(\alpha_1)\beta t)>0,\ (1-\Theta(\alpha_2)\lambda-\Omega(\alpha_2)\lambda t)>0,$$
$$(1-\Theta(\alpha_3)\sigma-\Omega(\alpha_3)\sigma t)>0, \tag{11.25}$$

then system (11.7) *has a unique solutions.*

Proof. Let us assume that in addition to $(u(t),v(t),w(t))$, $(\bar{u}(t),\bar{v}(t),\bar{w}(t))$ is also a solution of the system (11.7) satisfying the Theorem 11.1. Then

$$u(t)-u_1(t)=\Theta(\alpha_1)(K_1(t,u)-K_1(t,\bar{u}))+\Omega(\alpha_1)\int_0^t(K_1(\tau,u)-K_1(\tau,\bar{u}))d\tau. \tag{11.26}$$

Now, Eq. (11.26) simplifies after employing triangle inequality, as

$$||u(t)-\bar{u}(t)||\leq\Theta(\alpha_1)||K_1(t,u)-K_1(t,\bar{u})||+\Omega(\alpha_1)\int_0^t||K_1(\tau,u)-K_1(\tau,\bar{u})||d\tau. \tag{11.27}$$

Using Lipschitz condition for the kernel K_1, one can obtain

$$||u(t)-\bar{u}(t)||\leq\Theta(\alpha_1)\beta||u(t)-\bar{u}(t)||+\Omega(\alpha_1)\beta t||u(t)-\bar{u}(t)||. \tag{11.28}$$

From Eq. (11.28) we obtain

$$||u(t)-\bar{u}(t)||(1-\Theta(\alpha_1)\beta-\Omega(\alpha_1)\beta t)\leq 0. \tag{11.29}$$

From condition (11.25) we must have $||u(t)-\bar{u}(t)||=0$. This implies $u(t)=\bar{u}(t)$. Applying similar procedure we can conclude that $v(t)=\bar{v}$, $w(t)=\bar{w}(t)$. This proves the uniqueness of the solution of the system (11.7). □

11.5 DYNAMICS OF THE SYSTEM (11.7)

If $\alpha_1=\alpha_2=\alpha_3$, then system (11.7) is called as commensurate otherwise, it is incommensurate order system [62]. Now, at one of its equilibrium points $E=(u^*,v^*,w^*)$, the Jacobian matrix of system (11.7) is given by

$$J=\begin{pmatrix}-\beta & \gamma w+1 & \gamma v\\ -w & \lambda & 1-u\\ \delta v & \delta u & -\sigma\end{pmatrix}.$$

For an incommensurate fractional-order system, an equilibrium point E^* of the system is asymptotically stable if

$$|arg(\Lambda)|>\frac{\pi}{2M}, \tag{11.30}$$

satisfied for all roots Λ of the following equation

$$det(diag([\Lambda^{M\alpha_1},\Lambda^{M\alpha_2},\Lambda^{M\alpha_3}]))=0, \tag{11.31}$$

where M denotes the least common multiple (LCM) of the denominators $x_i's$ of $\alpha_i's$, $\alpha_i = \frac{y_i}{x_i}$, x_i and $y_i \in \mathbb{Z}^+$, $i = 1,2,3$. Then, Eq. (11.30) simplifies to

$$\frac{\pi}{2M} - \min_i\{|arg(\Lambda_i)|\} < 0. \tag{11.32}$$

Therefore, an equilibrium point will be asymptotically stable if its roots Λ_i admit Eq. (11.32). In fractional-order systems, the term $\frac{\pi}{2M} - min_i\{|arg(\Lambda_1)|\}$ is known as the instability measure for equilibrium points. This measure not a sufficient condition even though it is a necessary [62] for the presence of chaos in system. The equilibrium points of the system (11.7) can be obtained as [61]:

$$\begin{aligned}
&\mathscr{E}_0(0,0,0),\\
&\mathscr{E}_1\left(\frac{\delta+\sqrt{\mathscr{G}_1}}{2\delta}, \frac{\sigma}{\gamma}\left(\frac{-1+\sqrt{1+\mathscr{G}_2}}{\delta+\sqrt{\mathscr{G}_1}}\right), \frac{-1+\sqrt{1+\mathscr{G}_2}}{2\gamma}\right),\\
&\mathscr{E}_2\left(\frac{\delta+\sqrt{\mathscr{G}_1}}{2\delta}, \frac{\sigma}{\gamma}\left(\frac{-1-\sqrt{1+\mathscr{G}_2}}{\delta+\sqrt{\mathscr{G}_1}}\right), \frac{-1-\sqrt{1+\mathscr{G}_2}}{2\gamma}\right),\\
&\mathscr{E}_3\left(\frac{\delta-\sqrt{\mathscr{G}_1}}{2\delta}, \frac{\sigma}{\gamma}\left(\frac{-1+\sqrt{1+\mathscr{G}_2}}{\delta-\sqrt{\mathscr{G}_1}}\right), \frac{-1+\sqrt{1+\mathscr{G}_3}}{2\gamma}\right),\\
&\mathscr{E}_4\left(\frac{\delta-\sqrt{\mathscr{G}_1}}{2\delta}, \frac{\sigma}{\gamma}\left(\frac{-1-\sqrt{1+\mathscr{G}_2}}{\delta-\sqrt{\mathscr{G}_1}}\right), \frac{-1-\sqrt{1+\mathscr{G}_3}}{2\gamma}\right),
\end{aligned} \tag{11.33}$$

where $\mathscr{G}_1 = \delta^2 + 4\lambda\sigma\delta$, $\mathscr{G}_2 = \frac{2\beta\gamma}{\sigma}(\delta + 2\lambda\sigma + \sqrt{\mathscr{G}_1})$, and $\mathscr{G}_3 = \delta + 2\lambda\sigma - \sqrt{\mathscr{G}_1}$.

11.6 DESIGN OF SLIDING MODE CONTROLLER

The discontinuous control to force helps the system state trajectories to preset sliding surfaces and it is the main objective of the SMC law on which the corresponding phenomenon has anticipated consequences including stability. The SMC law is an efficient control algorithm with diverse features including the chaotic control of the system, insensitivity to external guarantees and disturbances, and parameter uncertainties. The controller is the unification of a variable structure control part (which imposes the trajectories to reach the sliding surface) and an equivalent control part (that labels the system character when the trajectories stay over the sliding surface).

To devise a controller with sliding mode, we need to follow a couple of steps. The sliding surface needs to be developed such that it signifies an anticipated system dynamics is the first step. As the second step, the switching control law needs to be proposed to make sure that the sliding mode exists at every point of the surface and any states outside the surface are aided to reach the surface in a finite time [48]. To control chaos, the control unit $q(t)$ is generalized to the second component of Eq. (11.7). The proposed fractional-order model is

$$\begin{aligned}
{}^{CF}D_t^{\alpha_1} u &= v - \beta u + \gamma v w,\\
{}^{CF}D_t^{\alpha_2} v &= \lambda v - uw + w + q(t),\\
{}^{CF}D_t^{\alpha_3} w &= \delta uv - \sigma w.
\end{aligned} \tag{11.34}$$

The sliding surface is defined to include a SMC scheme, as

$$\rho(t) = {}^{CF}D_t^{\alpha_2-1}v(t) + \int_0^t \phi(\tau)d\tau, \tag{11.35}$$

where $\phi(t)$ is a function given by

$$\phi(t) = \beta u + uw - \lambda v. \tag{11.36}$$

The invariance conditions of the surface must be held in the sliding mode, which are given as

$$\begin{aligned} &\rho(t) = 0, \\ &\frac{d}{dt}(\rho(t)) = 0. \end{aligned} \tag{11.37}$$

From Eq. (11.37), we get

$${}^{CF}D_t^{\alpha_2}v(t) = -\phi(t) = -\beta u - uw + \lambda v. \tag{11.38}$$

According to the system (11.34) and (11.38), the equivalent control law is derived as

$$\begin{aligned} q_{eq}(t) &= {}^{CF}D^{\alpha_2}v(t) - \lambda v + uw - w \\ &= -\beta u - uw + \lambda v - \lambda v + uw - w \\ &= -\beta u - w. \end{aligned} \tag{11.39}$$

To state into sliding surface, the reaching mode control scheme is design, and reaching law is considered as

$$q_r(t) = -K_r sign(\rho).$$

Here

$$sign(\rho) = \begin{cases} 1 & \rho > 0, \\ 0 & \rho = 0, \\ -1 & \rho < 0. \end{cases}$$

and K_r is the reach gain of the controller. Now, the switching control action are contracted by

$$q(t) = q_{eq}(t) + q_r(t) = -\beta u(t) - w(t) - K_r sign(\rho). \tag{11.40}$$

Theorem 11.3. *The system* (11.34) *with control law* (11.40) *is globally asymptotically stable.*

Proof. Let us choose the Lyapunov function as

$$X = \frac{1}{2}\rho^2. \tag{11.41}$$

Its derivative gives

$$\begin{aligned}\dot{X} &= \rho\dot{\rho} = \rho[{}^{CF}D_t^{\alpha_2}v(t) + \beta u + uw - \lambda v] \\ &= \rho[w + q(t) + \beta u] \\ &= \rho[w - \beta u - w - K_r sign(\rho) + \beta u] \\ &= -K_r|\rho| < 0.\end{aligned}$$

Therefore, the Lyapunov function that is $X > 0$, $\dot{X} < 0$, is satisfied the condition. Hence the closed loop system in the presence of the controller (11.40) is globally asymptotically stable. □

Theorem 11.4. *Suppose the system* (11.34) *is perturbed by external disturbances and uncertainties. Then the system has*

$$\begin{aligned}{}^{CF}D_t^{\alpha_1}u &= v - \beta u + \gamma vw, \\ {}^{CF}D_t^{\alpha_2}v &= \lambda v - uw + w + \Delta h(u,v,w) + \xi(t) + q(t), \\ {}^{CF}D_t^{\alpha_3}w &= \delta uv - \sigma w,\end{aligned} \tag{11.42}$$

where $\Delta h(u,v,w)$ and $\xi(t)$ are such that $|\Delta h(u,v,w)| < \eta_1$ and $|\xi(t)| < \eta_2$, η_1 and η_2 being positive constant. Then sliding mode controller is globally asymptotically stable if $\eta_1 + \eta_2 < K_r$.

Proof. Let us set the Lyapunov function as

$$X = \frac{1}{2}\rho^2.$$

We have

$$\begin{aligned}\dot{X} &= \rho\dot{\rho} = \rho[{}^{CF}D_t^{\alpha_2}v(t) + \beta u + uw - \lambda v] \\ &= \rho[w + \Delta h + \xi(t) + q(t) + \beta u] \\ &= \rho[w + \Delta h + \xi(t) - \beta u - w - K_r sign(\rho) + \beta u] \\ &= \rho[\Delta h + \xi(t) - K_r sign(\rho)] \\ &\leq (\eta_1 + \eta_2 - K_r)|\rho|.\end{aligned}$$

$\dot{X} < 0$ if $\eta_1 + \eta_2 < K_r$. □

11.7 NUMERICAL SIMULATIONS BY SINGLE STEP ADAMS-BASHFORTH-MOULTON METHOD

In recent years, many new analytical and numerical schemes have been nurtured for solving the diverse class of nonlinear fractional differential equations depicting real-world phenomena involving the Caputo-Fabrizio derivative. They include collocation [63,64], homotopy perturbation [65], homotopy analysis [66], He's variational iteration [67], Adomian's decomposition [68], predictor-corrector [69–73], and others. In this chapter, we have applied the single-step Adams–Bashforth–Moulton method to solve Caputo-Fabrizio FD equation (11.7). If $0 < \alpha < 1$, then the unique solution of the following initial value problem

$$ {}^{CF}D_t^{\alpha} f(t) = g(t, f(t)) = z(t),\ t \geq 0, \tag{11.43} $$

$$ f(0) = f_0, \tag{11.44} $$

The corresponding integral equation is

$$ f(x) = f(0) + \frac{1-\alpha}{B(\alpha)}(g(x) - g(0)) + \frac{\alpha}{B(\alpha)} \int_0^x z(t)dt. \tag{11.45} $$

The time interval $[0,t]$ discretize in steps of h and achieve $t_0 = 0, t_{k+1} = t_k + h,\ k = 0, 1, 2, ..., n\text{–}1,\ t_n = t$. The recursive formulas is contracted using Eq. (11.45), as

$$ f(x_{k+1}) = f(0) + \frac{1-\alpha}{B(\alpha)}(g(x_k) - g(0)) + \frac{\alpha}{B(\alpha)} \int_0^{x_{k+1}} \tilde{z}_{k+1}(t)dt. \tag{11.46} $$

The integrand is approximated as $\tilde{z}_{k+1}(t) = z(t_i)$ and we get the explicit predictor formula as

$$ f_{k+1}^{p} = f(0) + \frac{1-\alpha}{B(\alpha)}(g(x_k, f_k) - g(x_0, f_0)) + \frac{\alpha h}{B(\alpha)} \sum_{i=0}^{k} z(x_i). \tag{11.47} $$

If we approximate the integrand in Eq. (11.46) as $\tilde{z}_{k+1}(t) = \frac{t_{i+1}-t}{h} z(t_i) + \frac{t-t_i}{h} z(t_{i+1})$ we get the corrector formula as

$$ \begin{aligned} f_{k+1} = f(0) &+ \frac{1-\alpha}{B(\alpha)}(g(x_k, f_k) - g(x_0, f_0)) \\ &+ \frac{\alpha h}{2B(\alpha)} \left(\sum_{i=0}^{k} g(x_i, f(x_i)) + g(x_{k+1}, f^P(x_{k+1})) \right). \end{aligned} \tag{11.48} $$

Applying the predictor-corrector method defined in Eqs. (11.47) and (11.48), the numerical simulations for the system of Eq. (11.7) are performed. We have considered the values of the parameters as $\beta = 3$, $\gamma = 2.7$, $\lambda = 4.7$, $\delta = 2$ and $\sigma = 9$ to evaluate the behavior of the system (11.7) under the influence of Caputo-Fabrizio FD. With these parameter values, the points of equilibrium of the system (11.7) are:

$E_0(0,0,0)$, $E_1(5.12601, -2.40454, -2.73904)$, $E_2(-4.12601, 2.44714, -2.24376)$, $E_3(-4.12601, -2.0432, 1.87339)$, $E_4(5.12601, 2.0794, 2.36867)$. Eigen values of the Jacobian matrix are as follows:

At E_0: $(-9, 4.7, -3)$
E_1: $(-11.7857, 2.24287 + 6.85806i, 2.24287 - 6.85806i)$
E_2: $(-11.6814, 2.19068 + 6.1881i, 2.19068 - 6.1881i)$
E_3: $(-10.7669, 1.73347 + 6.00245i, 1.73347 - 6.00245i)$

E_4: $(-11.0247, 1.86234 + 6.68306i, 1.86234 - 6.68306i)$. We notice that all the points of equilibrium points are saddle points. By using the single-step Adams-Bashforth method the numerical simulations presented in Section 11.7.

Case I: Commenturate order
From Figures 11.1–11.4 we have considered commensurate fractional-order system with initial condition $(5, 0, -4)$. In Figures 11.1–11.3, for $\alpha = 1$, $\alpha = 0.99$, $\alpha = 0.97$ we observed two-scroll chaotic attractor. Whereas for $\alpha = 0.96$ in Figure 11.4 we notice four-scroll chaotic attractor. As the FD value decreases further, for $\alpha = 0.95$ we observe that the system experiences stable behavior (Figure 11.5). In Figures 11.6 and 11.7 we have shown the state of the system (11.34) under the controller (11.40) and sliding surface (11.35) for $\alpha = 1$ and $\alpha = 96$. This shows that the sliding controller helps the system reach stabilization.

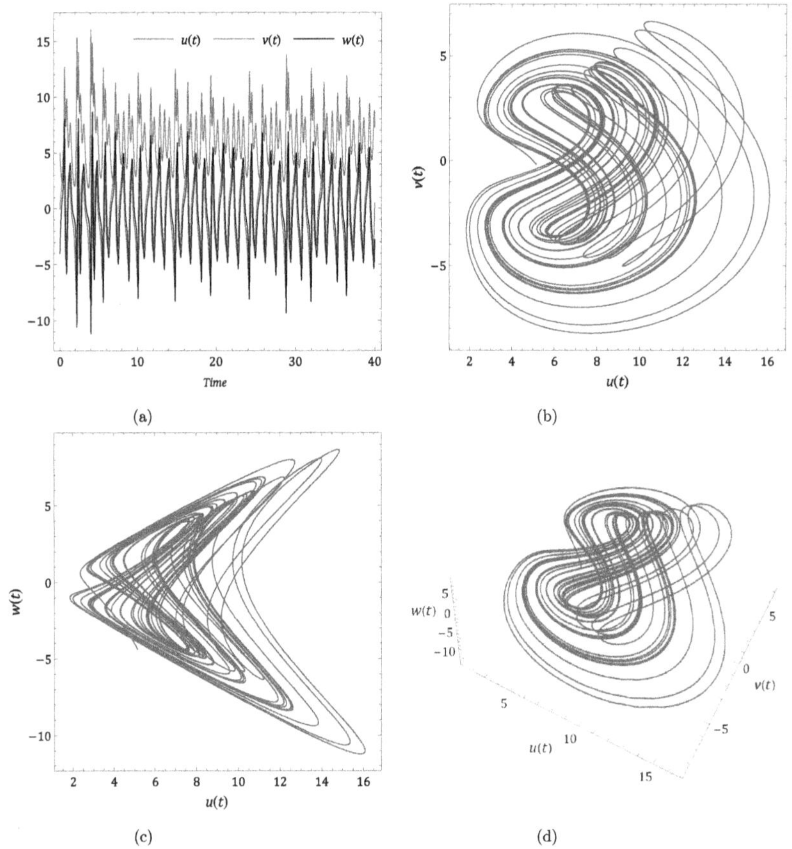

Figure 11.1 (a) Variation of u, v, and w with respect to time, (b) 2D projection of u and v, (c) 2D projection of u and w, (d) 3D projection of u, v, and w, in the system (11.7) for commensurate derivative for $\alpha = 1$.

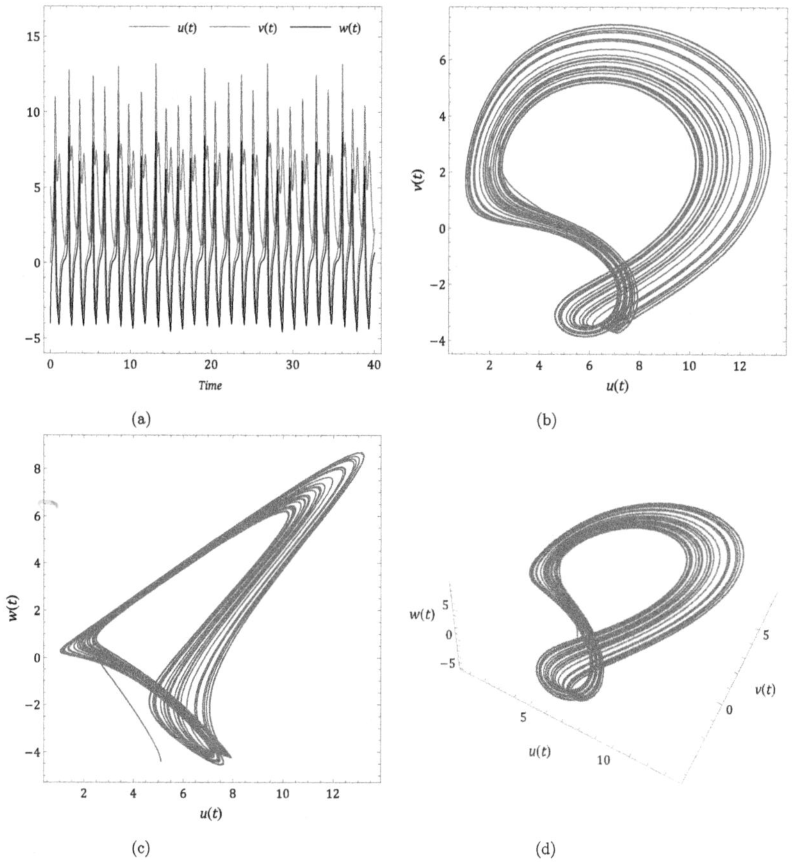

Figure 11.2 (a) Variation of u, v, and w with respect to time, (b) 2D projection of u and v, (c) 2D projection of u and w, (d) 3D projection of u, v, and w, in the system (11.7) for commensurate derivative for $\alpha = 0.99$

Case II: Non-commensurate order
In Figures 11.8 and 11.9 we have presented the behavior of the non-commensurate fractional-order system (11.7) with $\alpha_1 = 1$, $\alpha_2 = 0.96$, $\alpha_3 = 0.97$. The effect of the controller (11.40) and sliding surface (11.35) can be observed in Figure 11.9.

Case III: Commensurate and non-commensurate order with uncertainty and an external disturbance.
We have perturbed (11.7) by an uncertainty term $\Delta h(u,v,w) = 0.5sin(\pi u)cos(\pi v)$ $sin(2\pi w)$ and an external disturbance $\xi(t) = 0.1cos(2t)$ where $|\Delta h(u,v,w)| \leq \eta_1 =$ 0.5 and $\xi(t) \leq \eta_2 = 0.1$ and obtain the system (11.42). The perturbed system's response to the controller can be observed in Figures 11.10 and 11.11. It is observed that even in the presence of uncertainty and external disturbance, the chaos is controlled by the controller (11.40) and sliding surface (11.35) in the system (11.7).

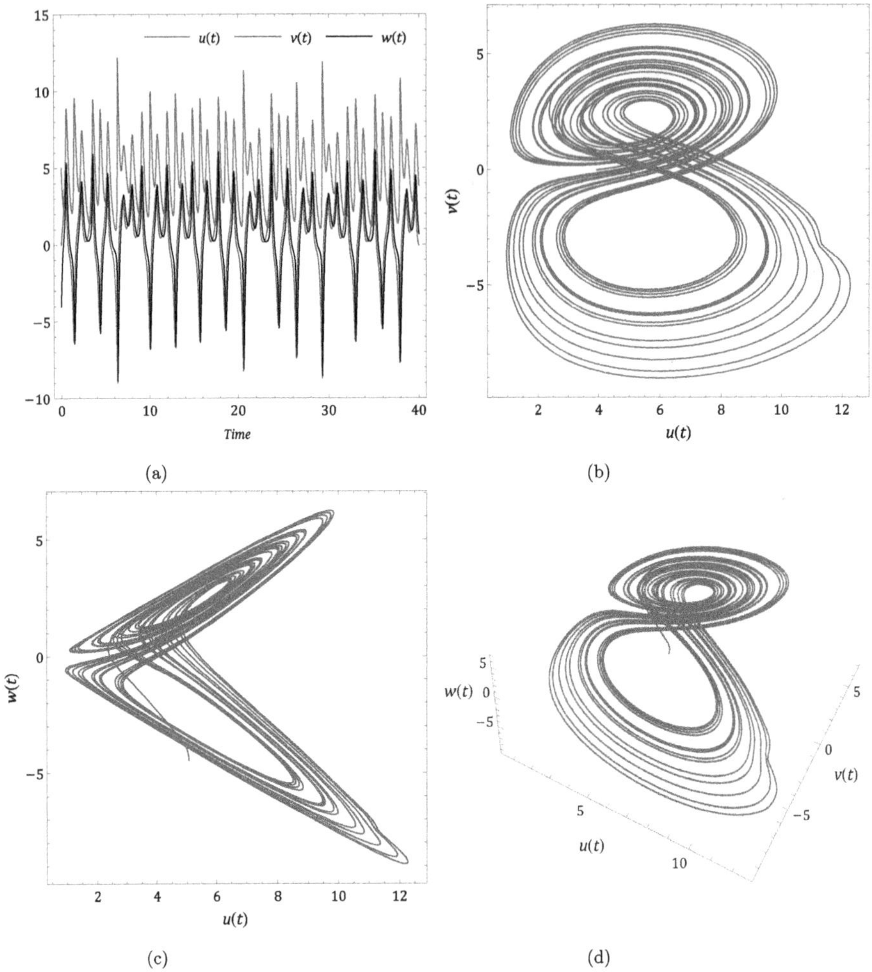

Figure 11.3 (a) Variation of u, v, and w with respect to time, (b) 2D projection of u and v, (c) 2D projection of u and w, (d) 3D projection of u, v, and w, in the system (11.7) for commensurate derivative for $\alpha = 0.97$.

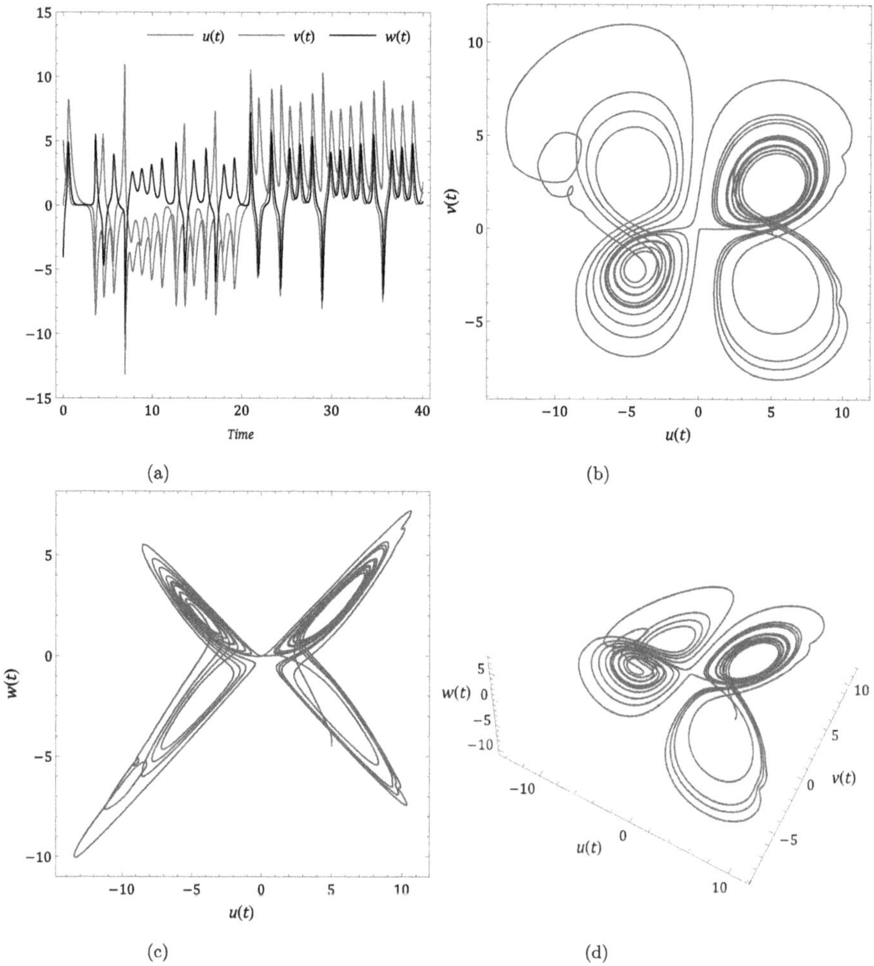

Figure 11.4 (a) Variation of u, v, and w with respect to time, (b) 2D projection of u and v, (c) 2D projection of u and w, (d) 3D projection of u, v, and w, in the system (11.7) for commensurate derivative for $\alpha = 0.96$.

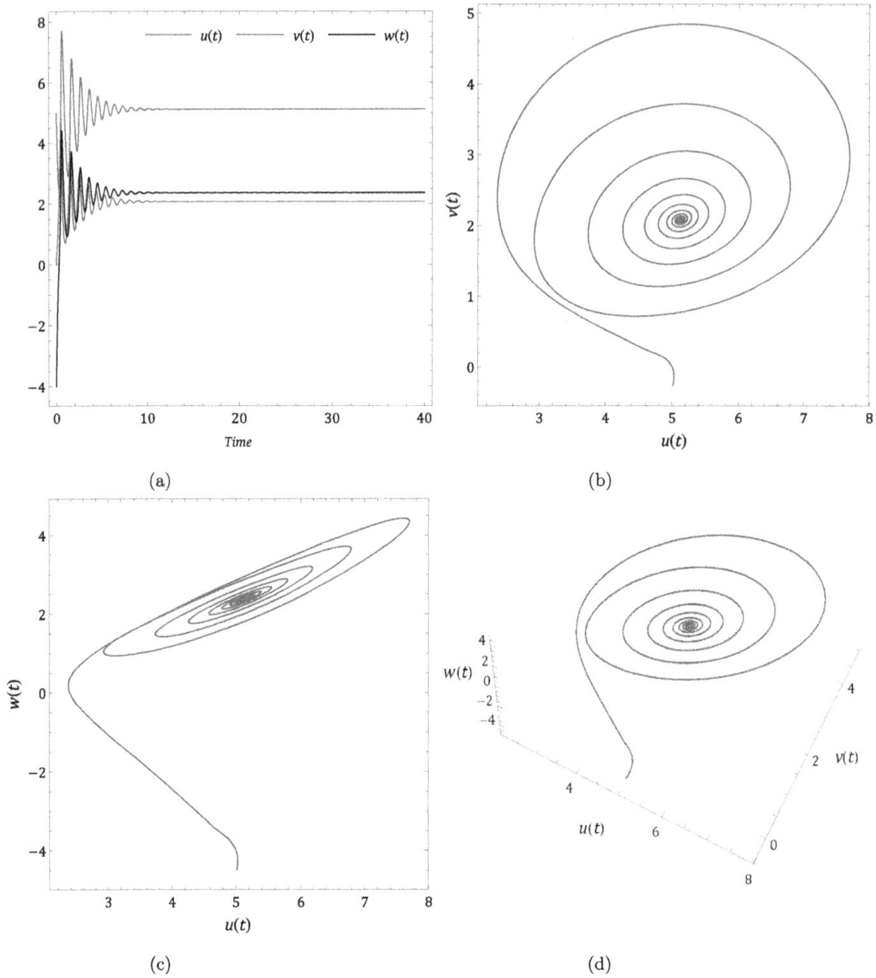

Figure 11.5 (a) Variation of u, v, and w with respect to time, (b) 2D projection of u and v, (c) 2D projection of u and w, (d) 3D projection of u, v, and w, in the system (11.7) for commensurate derivative for $\alpha = 0.95$.

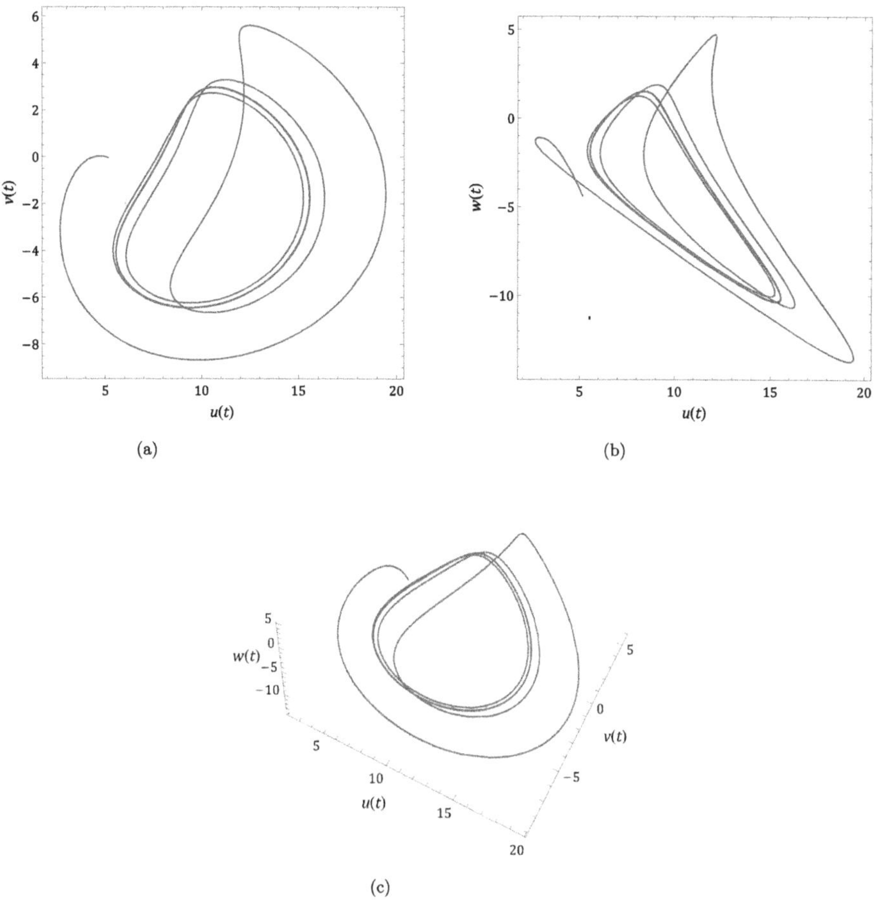

Figure 11.6 (a) 2D projection of u and v, (b) 2D projection of u and w, (c) 3D projection of u, v, and w, for the Controlled commenturate system (11.34) for $\alpha = 1$.

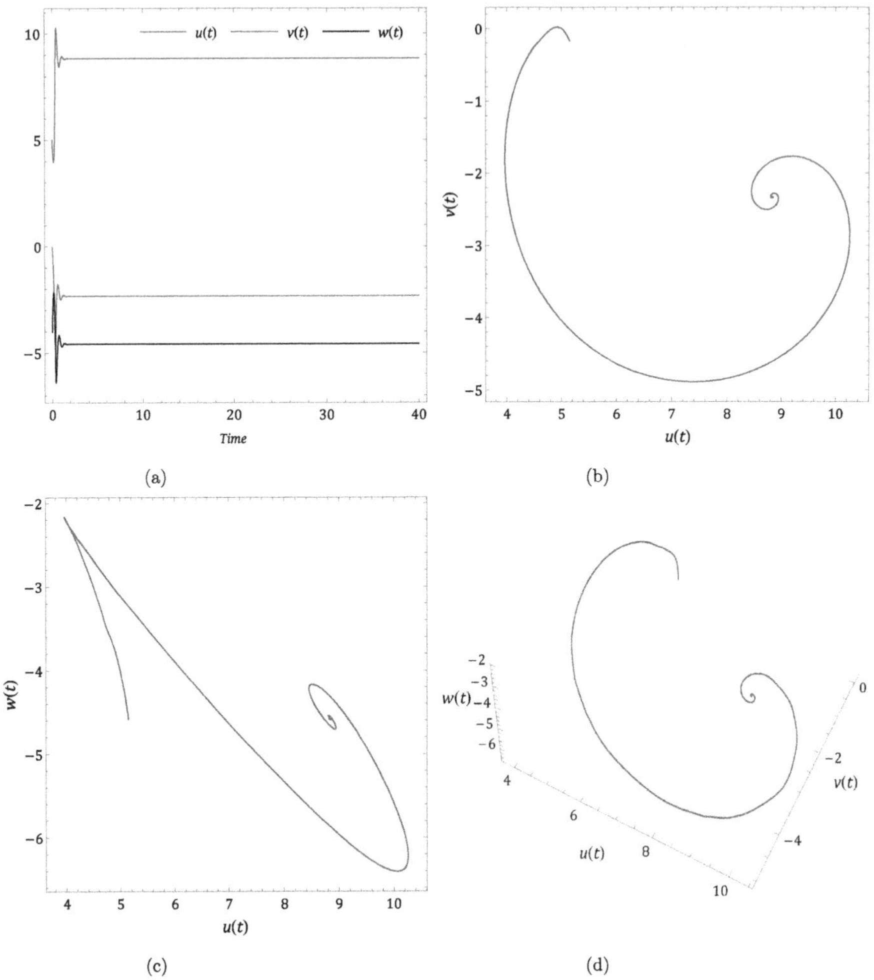

Figure 11.7 (a) Variation of u, v, and w with respect to time, (b) 2D projection of u and v, (c) 2D projection of u and w, (d) 3D projection of u, v, and w, for the controlled commenturate system (11.34) for $\alpha = 0.96$.

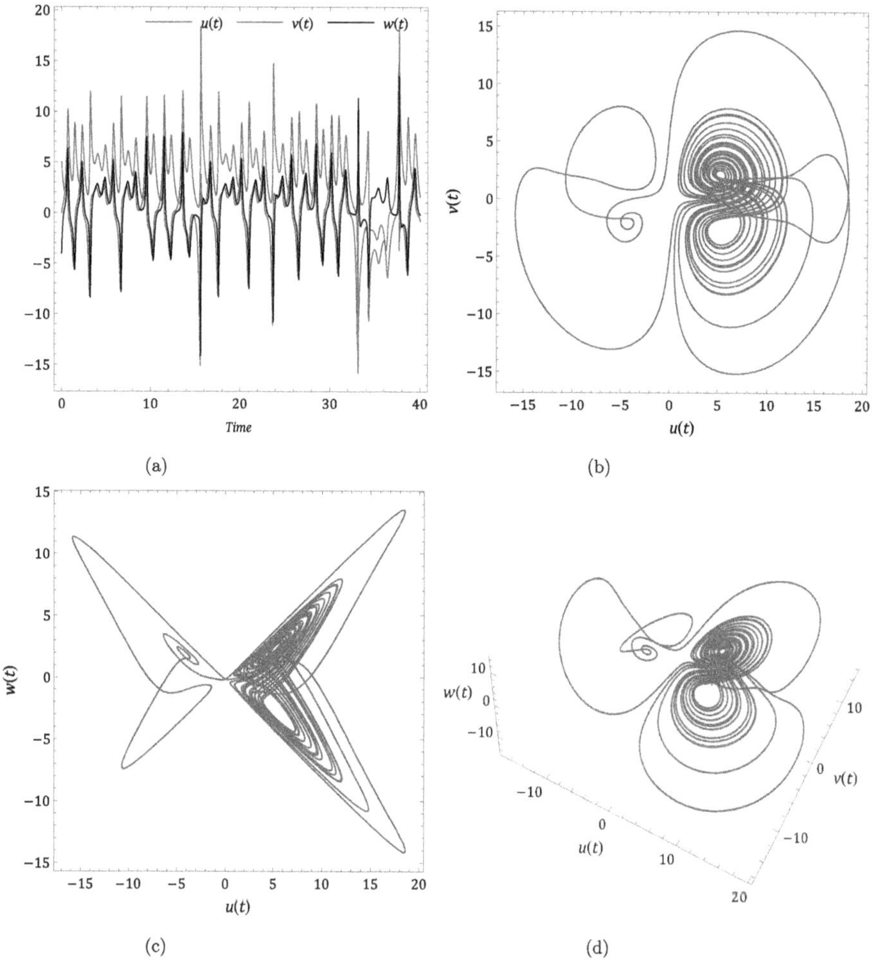

Figure 11.8 (a) Variation of u, v, and w with respect to time, (b) 2D projection of u and v, (c) 2D projection of u and w, (d) 3D projection of u, v, and w, for non-commenturate system (11.7) for $\alpha_1 = 1$, $\alpha_2 = 0.96$, $\alpha_3 = 0.97$ without control.

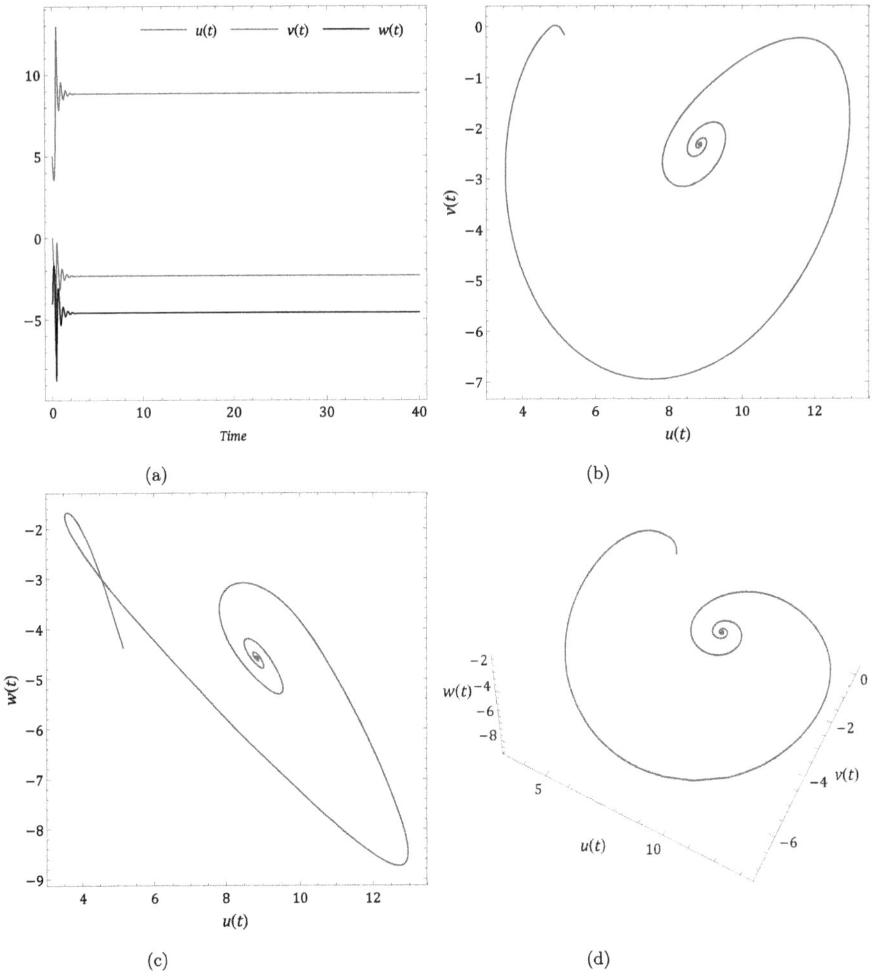

Figure 11.9 (a) Variation of u, v, and w with respect to time, (b) 2D projection of u and v, (c) 2D projection of u and w, (d) 3D projection of u, v, and w, for the non-commenturate system (11.34) for $\alpha_1 = 1$, $\alpha_2 = 0.96$, $\alpha_3 = 0.97$ with control.

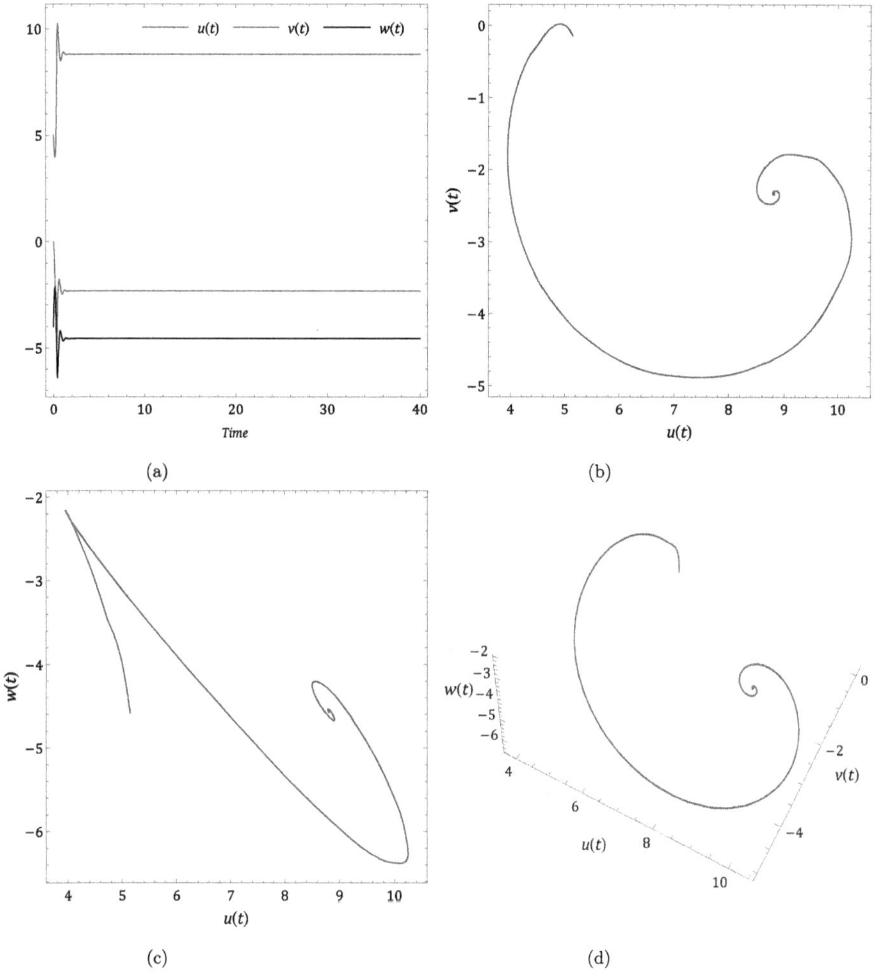

Figure 11.10 (a) Variation of u, v, and w with respect to time, (b) 2D projection of u and v, (c) 2D projection of u and w, (d) 3D projection of u, v, and w, for the controlled commenturate system (11.42) for $\alpha = 0.96$ with uncertainty and external disturbance.

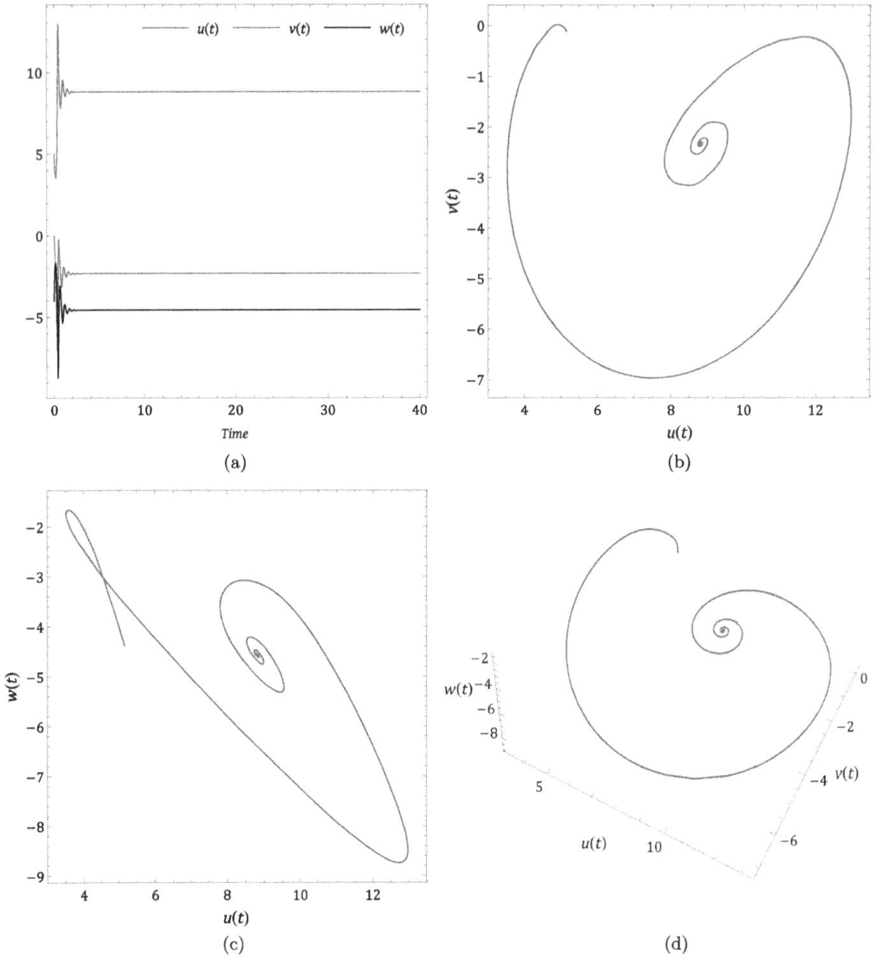

Figure 11.11 (a) Variation of u, v, and w with respect to time, (b) 2D projection of u and v, (c) 2D projection of u and w, (d) 3D projection of u, v, and w, for the controlled non-commenturate system (11.42) for $\alpha_1 = 1,\ \alpha_2 = 0.96,\ \alpha_3 = 0.97$ with uncertainty and external disturbance.

11.8 CONCLUSION

In this chapter, the analysis of DM fractional-order chaotic systems is effectively presented in the context of the Caputo-Fabrizio operator. We have established the existence and uniqueness of the solutions of the system. An SMC law has been developed using the Lyapunov stability theorem to regulate chaos in systems. We have observed that as the order reduces, the solution of the system exhibits a switch from chaotic behavior to a stable profile. Again, it is noticed that the states of the fractional-order system can be stabilized using the SMC approach. Even in the presence of uncertainty and external disruption, the SMC approach stabilizes the system. These observations are effectively validated by numerical simulations in this chapter using the single-step predictor-corrector method. Even though the DM system exhibits interesting chaotic behavior, this system is not yet linked to any physical phenomenon. This can be a future direction of study concerning this system. Moreover, one can observe the behavior of this system under the influence of different FDs and different control techniques.

References

1. A. M. Harb and N. Abdel-Jabbar, "Controlling hopf bifurcation and chaos in a small power system," *Chaos, Solitons & Fractals*, vol. 18, no. 5, pp. 1055–1063, 2003.
2. Y. Liu, "Circuit implementation and finite-time synchronization of the 4d rabinovich hyperchaotic system," *Nonlinear Dynamics*, vol. 61, no. 1, pp. 89–96, 2012.
3. J. E. Skinner, M. Molnar, T. Vybiral, and M. Mitra, "Application of chaos theory to biology and medicine," *Integrative Physiological and Behavioral Science: The Official Journal of the Pavlovian Society*, vol. 27, no. 1, pp. 39–53, 1992.
4. J. Ma, C. N. Wang, J. Tang, and Y. F. Xia, "Suppression of the spiral wave and turbulence in the excitability-modulated media," *International Journal of Theoretical Physics*, vol. 48, pp. 150–157, 2009.
5. P. Lamba and J. L. Hudson, "Experiments on bifurcations to chaos in a forced chemical reactor," *Chemical Engineering Science*, vol. 42, no. 1, pp. 1–8, 1987.
6. E. N. Lorenz, "Deterministic nonperiodic flow," *Journal of the Atmospheric Sciences*, vol. 20, no. 2, pp. 130–141, 1963.
7. R. Leipnik and T. A. Newton, "Double strange attractors in rigid body motion with linear feedback control," *Physics Letters A*, vol. 86, no. 2, pp. 63–67, 1981.
8. G. Chen and T. Ueta, "Yet another chaotic attractor," *International Journal of Bifurcation and Chaos*, vol. 09, no. 7, pp. 1465–1466, 1999.
9. J. Lü, G. Chen, and S. Zhang, "Dynamical analysis of a new chaotic attractor," *International Journal of Bifurcation and Chaos*, vol. 12, no. 5, pp. 1001–1015, 2002.
10. S. Dadras, H. R. Momeni, and V. J. Majd, "Sliding mode control for uncertain new chaotic dynamical system," *Chaos, Solitons & Fractals*, vol. 41, no. 4, pp. 1857–1862, 2009.
11. M. M. Roopaei, B. R. Sahraei, and T. C. Lin, "Adaptive sliding mode control in a novel class of chaotic systems," *Communications in Nonlinear Science and Numerical Simulation*, vol. 15, no. 12, pp. 4158–4170, 2010.
12. M. T. Yassen, "Chaos control of chaotic dynamical systems using backstepping design," *Chaos, Solitons & Fractals*, vol. 27, no. 2, pp. 537–548, 2006.
13. Q. Jia, "Chaos control and synchronization of the newtonleipnik chaotic system," *Chaos, Solitons & Fractals*, vol. 35, no. 4, pp. 814–824, 2008.

14. M. Caputo, "Linear models of dissipation whose q is almost frequency independent-II," *Geophysical Journal International*, vol. 13, pp. 529–539, 1967.
15. M. Caputo and M. Fabrizio, "A new definition of fractional derivative without singular kernel," *Progress in Fractional Differentiation and Applications*, vol. 1, no. 2, pp. 73–85, 2015.
16. A. A. Kilbas, H. M. Srivastava, and J. Trujillo, *Theory and Applications of Fractional Differential Equations*, North-Holland Mathematics Studies, vol. 204. Elsevier: Berlin, Germany, 2006.
17. I. Podlubny, *Fractional Differential Equations*. Academic Press: Cambridge, MA, 1999.
18. B. Ross, "A brief history and exposition of the fundamental theory of fractional calculus," in *Fractional Calculus and Its Applications: Proceedings of the International Conference*, University of New Haven, June 1974, pp. 1–36, Springer, 1975.
19. P. Veeresha, H. M. Baskonus, and W. Gao, "Strong interacting internal waves in rotating ocean: novel fractional approach," *Axioms*, vol. 10, no. 2, p. 123, 2021.
20. P. Veeresha and D. Baleanu, "A unifying computational framework for fractional gross–pitaevskii equations," *Physica Scripta*, vol. 96, no. 12, p. 125010, 2021.
21. C. Baishya, S. J. Achar, P. Veeresha, and D. G. Prakasha, "Dynamics of a fractional epidemiological model with disease infection in both the populations," *Chaos: An Interdisciplinary Journal of Nonlinear Science*, vol. 31, no. 4, p. 043130, 2021.
22. H. Singh and A.-M. Wazwaz, "Computational method for reaction diffusion-model arising in a spherical catalyst," *International Journal of Applied and Computational Mathematics*, vol. 7(3), p. 65, 2021.
23. H. Singh, "Analysis for fractional dynamics of ebola virus model," *Chaos, Solitons & Fractals*, vol. 138, p. 109992, 2020.
24. C. Baishya, "Dynamics of fractional holling type-ii predator-prey model with prey refuge and additional food to predator," *Journal of Applied Nonlinear Dynamics*, vol. 10, no. 2, pp. 315–328, 2020.
25. H. Singh, "Solving a class of local and nonlocal elliptic boundary value problems arising in heat transfer," *Heat Transfer*, vol. 51, no. 2, pp. 1524–1542, 2022.
26. V. Tripathi, H. Srivastava, H. Singh, C. Swarup, and S. Aggarwal, "Mathematical analysis of non-isothermal reaction–diffusion models arising in spherical catalyst and spherical biocatalyst," *Applied Sciences*, vol. 11, 2021.
27. H. Singh, "An efficient computational method for non-linear fractional lienard equation arising in oscillating circuits," In H. Singh, D. Kumar and D. Baleanu (Eds.), *Methods of Mathematical Modelling*, CRC Press, Taylor & Francis Group: Boca Raton, FL, pp. 39–50, 2019.
28. C. Baishya, "Dynamics of fractional stage structured predator prey model with prey refuge," *Indian Journal of Ecology*, vol. 47, no. 4, pp. 1118–1124, 2020.
29. H. Singh, "Chebyshev spectral method for solving a class of local and nonlocal elliptic boundary value problems," *International Journal of Nonlinear Sciences and Numerical Simulation*, 000010151520200235, 2021.
30. H. Singh and H. M. Srivastava, "Numerical investigation of the fractional-order liénard and duffing equations arising in oscillating circuit theory," *Frontiers in Physics*, vol. 8, p. 120, 2020.
31. C. Baishya and P. Veeresha, "Laguerre polynomial-based operational matrix of integration for solving fractional differential equations with non-singular kernel," *Proceedings of the Royal Society A*, vol. 477, pp. 20210438, 2021.

32. H. Singh, H. M. Srivastava, and J. J. Nieto, *Handbook of Fractional Calculus for Engineering and Science*. CRC Press, Taylor & Francis Group: Boca Raton, FL, 2022.
33. H. Singh, D. Kumar, and D. Baleanu, *Methods of Mathematical Modelling | Fractional Differential Equations*. CRC Press, Taylor & Francis Group: Boca Raton, FL, 2019.
34. C. Baishya, "An operational matrix based on the Independence polynomial of a complete bipartite graph for the Caputo fractional derivative," *SeMA Journal*, vol. 79, pp. 1–19, 2021.
35. H. Singh, H. Srivastava, and D. Baleanu, *Methods of Mathematical Modelling: Infectious Disease*. Elsevier Science: Berlin, Germany, 2022.
36. Y. Chen, H. Ahn, and D. Xue, "Robust controllability of interval fractional order linear time invariant systems," *Signal Process*, vol. 86(10), pp. 2794–2802, 2006.
37. M. S. Tavazoei and M. Haeri, "Chaos control via a simple fractional-order controller," *Physics Letters A*, vol. 372, pp. 798–807, 2008.
38. Y. Luo, Y. Chen, H. Ahn, and Y. Pi, "Fractional order robust control for cogging effect compensation in PMSM position servo systems: Stability analysis and experiments," *Control Engineering Practice*, vol. 18, no. 9, pp. 1022–1036, 2010.
39. C. Tricaud and Y. Chen, "An approximate method for numerically solving fractional order optimal control problems of general form," *Computers & Mathematics with Applications*, vol. 59, no. 5, pp. 1644–1655, 2010.
40. M. Shahiri, R. Ghaderi, A. Ranjbar, and S. Momani, "Chaotic fractional-order coullet system: Synchronization and control approach," *Communications in Nonlinear Science and Numerical Simulation*, vol. 15, pp. 665–674, 2010.
41. A. K. Alomari, M. S. M. Noorani, R. Nazar, and C. P. Li, "Homotopy analysis method for solving fractional lorenz system," *Communications in Nonlinear Science and Numerical Simulations*, vol. 15, pp. 1864–1872, 2010.
42. W. M. Ahmad, R. El-Khazali, and Y. Al-Assaf, "Stabilization of generalized fractional order chaotic systems using state feedback control," *Chaos Solitons and Fractals*, vol. 22, pp. 141–150, 2004.
43. J. Wang, X. Xiong, and Y. Zhang, "Extending synchronization scheme to chaotic fractional-order chen systems," *Physica A: Statistical Mechanics and its Applications*, vol. 370, no. 2, pp. 279–285, 2006.
44. J. G. Lu and G. Chen, "A note on the fractional-order chen system," *Chaos Soliton & Fractal*, vol. 27:685–8, pp. 1044–1051, 2006.
45. M. M. Asheghan, M. T. Hamidi Beheshti, and M. S. Tavazoei, "Robust synchronization of perturbed chen's fractional-order chaotic systems," *Communications in Nonlinear Science and Numerical Simulations*, vol. 16, pp. 1044–1051, 2011.
46. M. S. Tavazoei and M. Haeri, "Synchronization of chaotic fractional-order systems via active sliding mode controller," *Physica A Statistical Mechanics and its Applications*, vol. 387, pp. 57–70, 2008.
47. A. Si Ammour, S. Djennoune, and M. Bettayeb, "A sliding mode control for linear fractional systems with input and state delays," *Communications in Nonlinear Science and Numerical Simulation*, vol. 14, no. 5, pp. 2310–2318, 2009.
48. S. Dadras and H. R. Momeni, "Control of a fractional-order economical system via sliding mode," *Physica A: Statistical Mechanics and its Applications*, vol. 389, no. 12, pp. 2434–2442, 2010.
49. H. Hosseinnia, R. Ghaderi, A. Ranjbar, M. Mahmoudian, and S. Momani, "Sliding mode synchronization of an uncertain fractional order chaotic system," *Computers & Mathematics with Applications*, vol. 59, pp. 1637–1643, 2010.

50. H. Delavari, R. Ghaderi, A. Ranjbar, and S. Momani, "Fuzzy fractional order sliding mode controller for nonlinear systems," *Communications in Nonlinear Science and Numerical Simulation*, vol. 15, pp. 963–978, 2010.
51. Z.-M. Ge and C.-Y. Ou, "Chaos synchronization of fractional order modified duffing systems with parameters excited by a chaotic signal," *Chaos, Solitons & Fractals*, vol. 35, no. 4, pp. 705–717, 2008.
52. A. Matouk, "Feedback control and synchronization of a fractional-order modified autonomous van der pol-duffing circuit," *Communications in Nonlinear Science and Numerical Simulation*, vol. 16, pp. 975–986, 2011.
53. X. Wu, H. Lu, and S. Shen, "Synchronization of a new fractional-order hyperchaotic system," *Physics Letter A*, vol. 373, pp. 2329–2337, 2009.
54. P. Zhou and W. Zhu, "Function projective synchronization for fractional-order chaotic systems," *Nonlinear Analysis: Real World Applications*, vol. 12, pp. 811–816, 2011.
55. J. Wang and Y. Zhou, "A class of fractional evolution equations and optimal controls," *Nonlinear Analysis: Real World Applications*, vol. 1, no. 12, pp. 262–272, 2011.
56. Z. Ruo-Xun and Y. Shi-Ping, "Designing synchronization schemes for a fractional-order hyperchaotic system," *Acta Physica Sinica*, vol. 57, no. 11, pp. 6837–6843, 2008.
57. G. M. Mophou, "Optimal control of fractional diffusion equation," *Computers & Mathematics with Applications*, vol. 61, no. 1, pp. 68–78, 2011.
58. Y. Zheng, Y. Nian, and D. Wang, "Controlling fractional order chaotic systems based on takagi–sugeno fuzzy model and adaptive adjustment mechanism," *Physics Letter A*, vol. 375, pp. 125–129, 2010.
59. X.-Y. Wang, Y.-J. He, and M.-J. Wang, "Chaos control of a fractional order modified coupled dynamos system," *Nonlinear Analysis*, vol. 12, no. 71, pp. 6126–6134, 2009.
60. J. Losada and J. J. Nieto, "Properties of a new fractional derivative without singular kernel," *Progress in Fractional Differentiation and Applications*, vol. 1, no. 2, pp. 87–92, 2015.
61. S. Dadras and H. R. Momeni, "A novel three-dimensional autonomous chaotic system generating two, three and four-scroll attractors," *Physics Letter A*, vol. 373, pp. 3637–3642, 2009.
62. M. S. Tavazoei and M. Haeri, "Chaotic attractors in incommensurate fractional order systems," *Physica D: Nonlinear Phenomena*, vol. 237, no. 20, pp. 2628–2637, 2008.
63. S. Roshan, H. Jafari, and D. Baleanu, "Solving FDEs with caputo-fabrizio derivative by operational matrix based on genocchi polynomials," *Mathematical Methods in the Applied Sciences*, vol. 41(18), pp. 9134–9141, 2018.
64. J. R. Loh, A. Isah, C. Phang, and Y. T. Toh, "On the new properties of caputo–fabrizio operator and its application in deriving shifted legendre operational matrix," *Applied Numerical Mathematics*, vol. 132, pp. 138–153, 2018.
65. S. Ahmad, A. Ullah, A. Akgül, and M. De la Sen, "A novel homotopy perturbation method with applications to nonlinear fractional order KdV and burger equation with exponential-decay kernel," *Journal of Function Spaces*, vol. 2021, 8770488, 2021.
66. Z. Körpinar, "On numerical solutions for the caputo-fabrizio fractional heat-like equation," *Thermal Science*, vol. 22, pp. 274–274, 2017.
67. H. Yépez-Martínez and J. F. Gómez-Aguilar, "Laplace variational iteration method for modified fractional derivatives with non-singular kernel," *Journal of Applied and Computational Mechanics*, vol. 6, no. 3, pp. 684–698, 2020.
68. H. G. Taher, "Adomian decomposition method for solving nonlinear fractional PDEs," *Journal of Research in Applied Mathematics*, vol. 7, pp. 21–27, 2021.

69. K. Diethelm, N. J. Ford, and A. D. Freed, "A predictor-corrector approach for the numerical solution of fractional differential equations," *Nonlinear Dynamics*, vol. 29, no. 1, pp. 3–22, 2002.
70. E. J. Moore, S. Sekson, and K. Sanoe, "A caputo–fabrizio fractional differential equation model for HIV/AIDS with treatment compartment," *Advances in Difference Equations*, 2019.
71. Y. Toh, C. Phang, and J. R. Loh, "New predictor-corrector scheme for solving nonlinear differential equations with caputo-fabrizio operator," *Mathematical Methods in the Applied Sciences*, vol. 42, pp. 175–185, 2019.
72. R. Douaifia, S. Bendoukha, and S. Abdelmalek, "A newton interpolation based predictor–corrector numerical method for fractional differential equations with an activator–inhibitor case study," *Mathematics and Computers in Simulation*, vol. 187, pp. 391–413, 2021.
73. S. J. Achar, C. Baishya, and M. K. Kaabar, "Dynamics of the worm transmission in wireless sensor network in the framework of fractional derivatives," *Mathematical Methods in the Applied Sciences*, vol. 45, pp. 4278–4294, 2021.

12 A Fractional Order Model with Non-Singular Mittag-Leffler Kernel

Ali Akgül
Siirt University
Near East University

CONTENTS

12.1 INTRODUCTION

Fractional calculus has taken much interest recently. There are many types of fractional derivatives in the literature. Many different kernels have been used to construct different fractional derivatives. Mittag-Leffler kernel is one of the kernels used to construct the fractional derivative. This kernel is very useful in many applications. Thus, this function is very important in the fractional calculus. In this work, we consider the following fractional model with the Mittag-Leffler kernel as [1–4]:

$$ {}^{ABC}D^{\alpha}\boldsymbol{S}_B(t) = -\beta_{TB}B^{\sigma}\frac{\boldsymbol{I}_T(t)}{N_T}\boldsymbol{S}_B(t) + (\mu_B^{\sigma} + \alpha_B^{\sigma})\,\boldsymbol{R}_B(t) $$

$$ {}^{ABC}D^{\alpha}\boldsymbol{E}_B(t) = \beta_{TB}B^{\sigma}\frac{\boldsymbol{I}_T(t)}{N_T}\boldsymbol{S}_B(t) - \nu_B^{\sigma}\boldsymbol{E}_B(t) $$

$$ {}^{ABC}D^{\alpha}\boldsymbol{I}_B(t) = \nu_B^{\sigma}\boldsymbol{E}_B(t) - \lambda_B^{\sigma}\boldsymbol{I}_B(t) $$

DOI: 10.1201/9781003368069-12

$$^{ABC}D^{\alpha}\boldsymbol{R}_B(t) = \lambda_B^{\sigma}\boldsymbol{I}_B(t) - (\mu_B^{\sigma} + \alpha_B^{\sigma})\boldsymbol{R}_B(t),$$

$$^{ABC}D^{\alpha}\boldsymbol{S}_T(t) = -\beta_{BT}B^{\sigma}\frac{\boldsymbol{I}_B(t)}{N_B}\boldsymbol{S}_T(t) + p\mu_T^{\sigma}(\boldsymbol{I}_T(t) + \boldsymbol{E}_T(t))$$

$$^{ABC}D^{\alpha}\boldsymbol{E}_T(t) = \beta_{BT}B^{\sigma}\frac{\boldsymbol{I}_B(t)}{N_B}\boldsymbol{S}_T(t) - (p\mu_T^{\sigma} + \nu_T^{\alpha})\boldsymbol{E}_T(t)$$

$$^{ABC}D^{\alpha}\boldsymbol{I}_T(t) = -p\mu_T^{\sigma}\boldsymbol{I}_T(t) + \nu_T^{\sigma}\boldsymbol{E}_T(t)$$

with the initial conditions

$$S_B(0), E_B(0), I_B(0), R_B(0), S_T(0), E_T(0), I_T(0) \geqslant 0.$$

For more details see [5–14].

Definition 1. *[15] Let $f \in H^1(a,b)$, $a < b, \alpha \in [0,1]$ then the definition of the fractional derivative in the sense of Atangana-Baleanu becomes:*

$$\left({}_a^{ABC}D^{\alpha} f\right)(t) = \frac{\mathscr{AB}(\alpha)}{1-\alpha}\int_a^t f'(x)E_{\alpha}\left(-\alpha\frac{(t-x)^{\alpha}}{1-\alpha}\right)dx.$$

and the associated fractional integral is:

$$\left({}_a^{AB}I^{\alpha} f\right)(t) = \frac{1-\alpha}{\mathscr{AB}(\alpha)}f(t) + \frac{\alpha}{\mathscr{AB}(\alpha)}\left({}_aI^{\alpha} f\right)(t).$$

where $\mathscr{AB}(\alpha)(\alpha) = 1 - \alpha + \frac{\alpha}{\Gamma(\alpha)}$.

We organize our work as follows: We give the existence and uniqueness of the solutions of the model in Section 12.2. We apply the numerical method in Section 12.3. We demonstrate the numerical simulations in Section 12.4. We give the conclusion in the last section.

12.2 EXISTENCE AND UNIQUENESS OF THE SOLUTIONS

In this section, we prove the existence and uniqueness of the solutions of the following model:

$$^{ABC}D^{\alpha}\boldsymbol{S}_B(t) = -\beta_{TB}B^{\sigma}\frac{\boldsymbol{I}_T(t)}{N_T}\boldsymbol{S}_B(t) + (\mu_B^{\sigma} + \alpha_B^{\sigma})\ \boldsymbol{R}_B(t)$$

$$^{ABC}D^{\alpha}\boldsymbol{E}_B(t) = \beta_{TB}B^{\sigma}\frac{\boldsymbol{I}_T(t)}{N_T}\boldsymbol{S}_B(t) - \nu_B^{\sigma}\boldsymbol{E}_B(t)$$

$$^{ABC}D^{\alpha}\boldsymbol{I}_B(t) = \nu_B^{\sigma}\boldsymbol{E}_B(t) - \lambda_B^{\sigma}\boldsymbol{I}_B(t)$$

$$^{ABC}D^{\alpha}\boldsymbol{R}_B(t) = \lambda_B^{\sigma}\boldsymbol{I}_B(t) - (\mu_B^{\sigma} + \alpha_B^{\sigma})\boldsymbol{R}_B(t),$$

$$
\begin{aligned}
{}^{ABC}D^{\alpha}\boldsymbol{S}_T(t) &= -\beta_{BT}B^{\sigma}\frac{\boldsymbol{I}_B(t)}{N_B}\boldsymbol{S}_T(t) + p\mu_T^{\sigma}(\boldsymbol{I}_T(t)+\boldsymbol{E}_T(t)) \\
{}^{ABC}D^{\alpha}\boldsymbol{E}_T(t) &= \beta_{BT}B^{\sigma}\frac{\boldsymbol{I}_B(t)}{N_B}\boldsymbol{S}_T(t) - (p\mu_T^{\sigma}+\nu_T^{\alpha})\boldsymbol{E}_T(t) \\
{}^{ABC}D^{\alpha}\boldsymbol{I}_T(t) &= -p\mu_T^{\sigma}\boldsymbol{I}_T(t) + \nu_T^{\sigma}\boldsymbol{E}_T(t)
\end{aligned}
$$

To prove the existence and uniqueness of the solutions of the above model, we verify the linear growth and Lipschitz condition.

12.2.1 LINEAR GROWTH

Let

$$
\begin{aligned}
&-\beta_{TB}B^{\sigma}\frac{\boldsymbol{I}_T(t)}{N_T}\boldsymbol{S}_B(t) + (\mu_B^{\sigma}+\alpha_B^{\sigma})\ \boldsymbol{R}_B(t) = A_1(t,S_B,E_B,I_B,R_B,S_T,E_T,I_T) \\
&\beta_{TB}B^{\sigma}\frac{\boldsymbol{I}_T(t)}{N_T}\boldsymbol{S}_B(t) - \nu_B^{\sigma}\boldsymbol{E}_B(t) = A_2(t,S_B,E_B,I_B,R_B,S_T,E_T,I_T) \\
&\nu_B^{\sigma}\boldsymbol{E}_B(t) - \lambda_B^{\sigma}\boldsymbol{I}_B(t) = A_3(t,S_B,E_B,I_B,R_B,S_T,E_T,I_T) \\
&\lambda_B^{\sigma}\boldsymbol{I}_B(t) - (\mu_B^{\sigma}+\alpha_B^{\sigma})\boldsymbol{R}_B(t) = A_4(t,S_B,E_B,I_B,R_B,S_T,E_T,I_T) \\
&-\beta_{BT}B^{\sigma}\frac{\boldsymbol{I}_B(t)}{N_B}\boldsymbol{S}_T(t) + p\mu_T^{\sigma}(\boldsymbol{I}_T(t)+\boldsymbol{E}_T(t)) = A_5(t,S_B,E_B,I_B,R_B,S_T,E_T,I_T) \\
&\beta_{BT}B^{\sigma}\frac{\boldsymbol{I}_B(t)}{N_B}\boldsymbol{S}_T(t) - (p\mu_T^{\sigma}+\nu_T^{\alpha})\boldsymbol{E}_T(t) = A_6(t,S_B,E_B,I_B,R_B,S_T,E_T,I_T) \\
&-p\mu_T^{\sigma}\boldsymbol{I}_T(t) + \nu_T^{\sigma}\boldsymbol{E}_T(t) = A_7(t,S_B,E_B,I_B,R_B,S_T,E_T,I_T)
\end{aligned}
$$

We have

$$
\begin{aligned}
&|A_1(t,S_B,E_B,I_B,R_B,S_T,E_T,I_T)|^2 \\
&= \left|-\beta_{TB}B^{\sigma}\frac{\boldsymbol{I}_T(t)}{N_T}\boldsymbol{S}_B(t) + (\mu_B^{\sigma}+\alpha_B^{\sigma})\ \boldsymbol{R}_B(t)\right|^2 \\
&\leq 2\left(-\beta_{TB}B^{\sigma}\frac{\boldsymbol{I}_T(t)}{N_T}\right)^2 |\boldsymbol{S}_B(t)|^2 + 2(\mu_B^{\sigma}+\alpha_B^{\sigma})^2\,|\boldsymbol{R}_B(t)|^2 \\
&\leq 2(\mu_B^{\sigma}+\alpha_B^{\sigma})^2\,|\boldsymbol{R}_B(t)|^2\left(1+\frac{\left(-\beta_{TB}B^{\sigma}\frac{I_T(t)}{N_T}\right)^2|\boldsymbol{S}_B(t)|^2}{(\mu_B^{\sigma}+\alpha_B^{\sigma})^2\,|\boldsymbol{R}_B(t)|^2}\right)
\end{aligned}
$$

If

$$
\frac{\left(-\beta_{TB}B^{\sigma}\frac{I_T(t)}{N_T}\right)^2}{(\mu_B^{\sigma}+\alpha_B^{\sigma})^2\,|\boldsymbol{R}_B(t)|^2} < 1,
$$

then, we get

$$|A_1(t,S_B,E_B,I_B,R_B,S_T,E_T,I_T)|^2 \quad \leq \quad K_1\left(1+|S_B|^2\right)$$

where

$$K_1 = 2\left(\mu_B^\sigma + \alpha_B^\sigma\right)^2 |\,\boldsymbol{R}_B(t)|^2.$$

$$\begin{aligned}
&|A_2(t,S_B,E_B,I_B,R_B,S_T,E_T,I_T)|^2 \\
&\quad = \left|\beta_{TB}B^\sigma \frac{\boldsymbol{I}_T(t)}{N_T}\boldsymbol{S}_B(t) - \nu_B^\sigma \boldsymbol{E}_B(t)\right|^2 \\
&\quad \leq 2\left(\beta_{TB}B^\sigma \frac{\boldsymbol{I}_T(t)}{N_T}\right)^2 |\boldsymbol{S}_B(t)|^2 + 2\left(-\nu_B^\sigma\right)^2 |\,\boldsymbol{E}_B(t)|^2 \\
&\quad \leq 2\left(\beta_{TB}B^\sigma \frac{\boldsymbol{I}_T(t)}{N_T}\right)^2 |\boldsymbol{S}_B(t)|^2 \left(1+\frac{\left(-\nu_B^\sigma\right)^2 |\,\boldsymbol{E}_B(t)|^2}{\left(\beta_{TB}B^\sigma \frac{\boldsymbol{I}_T(t)}{N_T}\right)^2 |\boldsymbol{S}_B(t)|^2}\right)
\end{aligned}$$

If

$$\frac{\left(-\nu_B^\sigma\right)^2}{\left(\beta_{TB}B^\sigma \frac{\boldsymbol{I}_T(t)}{N_T}\right)^2 |\boldsymbol{S}_B(t)|^2} < 1,$$

then, we get

$$|A_2(t,S_B,E_B,I_B,R_B,S_T,E_T,I_T)|^2 \quad \leq \quad K_2\left(1+|E_B|^2\right)$$

where

$$K_2 = 2\left(\beta_{TB}B^\sigma \frac{\boldsymbol{I}_T(t)}{N_T}\right)^2 |\boldsymbol{S}_B(t)|^2.$$

$$\begin{aligned}
|A_3(t,S_B,E_B,I_B,R_B,S_T,E_T,I_T)|^2 &= |\nu_B^\sigma \boldsymbol{E}_B(t) - \lambda_B^\sigma \boldsymbol{I}_B(t)|^2 \\
&\leq 2\left(\nu_B^\sigma\right)^2 |\boldsymbol{E}_B(t)|^2 + 2\left(-\lambda_B^\sigma\right)^2 |\,\boldsymbol{I}_B(t)|^2 \\
&\leq 2\left(\nu_B^\sigma\right)^2 |\boldsymbol{E}_B(t)|^2 \left(1+\frac{\left(-\lambda_B^\sigma\right)^2 |\,\boldsymbol{I}_B(t)|^2}{\left(\nu_B^\sigma\right)^2 |\boldsymbol{E}_B(t)|^2}\right)
\end{aligned}$$

If

$$\frac{\left(-\lambda_B^\sigma\right)^2}{\left(\nu_B^\sigma\right)^2 |\boldsymbol{E}_B(t)|^2} < 1,$$

then, we get

$$|A_3(t,S_B,E_B,I_B,R_B,S_T,E_T,I_T)|^2 \quad \leq \quad K_3\left(1+|I_B|^2\right)$$

where

$$K_3 = 2\left(\nu_B^{\sigma}\right)^2 |\boldsymbol{E}_B(t)|^2.$$

$$\begin{aligned}|A_4(t,S_B,E_B,I_B,R_B,S_T,E_T,I_T)|^2 &= |\lambda_B^{\sigma}\boldsymbol{I}_B(t)-(\mu_B^{\sigma}+\alpha_B^{\sigma})\boldsymbol{R}_B(t)|^2\\ &\leq 2\left(\lambda_B^{\sigma}\right)^2|\boldsymbol{I}_B(t)|^2+2\left(-\left(\mu_B^{\sigma}+\alpha_B^{\sigma}\right)\right)^2|\boldsymbol{R}_B(t)|^2\\ &\leq 2\left(\lambda_B^{\sigma}\right)^2|\boldsymbol{I}_B(t)|^2\left(1+\frac{\left(-\left(\mu_B^{\sigma}+\alpha_B^{\sigma}\right)\right)^2|\boldsymbol{R}_B(t)|^2}{\left(\lambda_B^{\sigma}\right)^2|\boldsymbol{I}_B(t)|^2}\right)\end{aligned}$$

If

$$\frac{\left(-\left(\mu_B^{\sigma}+\alpha_B^{\sigma}\right)\right)^2}{\left(\lambda_B^{\sigma}\right)^2|\boldsymbol{I}_B(t)|^2}<1,$$

then, we get

$$|A_4(t,S_B,E_B,I_B,R_B,S_T,E_T,I_T)|^2 \quad \leq \quad K_4\left(1+|R_B|^2\right)$$

where

$$K_4 = 2\left(\lambda_B^{\sigma}\right)^2|\boldsymbol{I}_B(t)|^2$$

$$\begin{aligned}&|A_5(t,S_B,E_B,I_B,R_B,S_T,E_T,I_T)|^2\\ &\quad= -\beta_{BT}B^{\sigma}\frac{\boldsymbol{I}_B(t)}{N_B}\boldsymbol{S}_T(t)+p\mu_T^{\sigma}\left(\boldsymbol{I}_T(t)+\boldsymbol{E}_T(t)\right)\\ &\quad\leq 2\left(-\beta_{BT}B^{\sigma}\frac{\boldsymbol{I}_B(t)}{N_B}\right)^2|\boldsymbol{S}_T(t)|^2+2\left(p\mu_T^{\sigma}\right)^2|\boldsymbol{I}_T(t)+\boldsymbol{E}_T(t)|^2\\ &\quad\leq 2\left(p\mu_T^{\sigma}\right)^2|\boldsymbol{I}_T(t)+\boldsymbol{E}_T(t)|^2\left(1+\frac{\left(-\beta_{BT}B^{\sigma}\frac{\boldsymbol{I}_B(t)}{N_B}\right)^2|\boldsymbol{S}_T(t)|^2}{\left(p\mu_T^{\sigma}\right)^2|\boldsymbol{I}_T(t)+\boldsymbol{E}_T(t)|^2}\right)\end{aligned}$$

If

$$\frac{\left(-\beta_{BT}B^{\sigma}\frac{\boldsymbol{I}_B(t)}{N_B}\right)^2}{\left(p\mu_T^{\sigma}\right)^2|\boldsymbol{I}_T(t)+\boldsymbol{E}_T(t)|^2}<1,$$

then, we get

$$|A_5(t,S_B,E_B,I_B,R_B,S_T,E_T,I_T)|^2 \quad \leq \quad K_5\left(1+|S_T|^2\right)$$

where

$$K_5 = 2(p\mu_T^\sigma)^2 \,|\, \boldsymbol{I}_T(t) + \boldsymbol{E}_T(t)|^2.$$

$$\begin{aligned}
&|A_6(t, S_B, E_B, I_B, R_B, S_T, E_T, I_T)|^2 \\
&= \left| \beta_{BT} B^\sigma \frac{\boldsymbol{I}_B(t)}{N_B} \boldsymbol{S}_T(t) - (p\mu_T^\sigma + \nu_T^\alpha) \boldsymbol{E}_T(t) \right|^2 \\
&\leq 2\left(\beta_{BT} B^\sigma \frac{\boldsymbol{I}_B(t)}{N_B}\right)^2 |\boldsymbol{S}_T(t)|^2 + 2(-(p\mu_T^\sigma + \nu_T^\alpha))^2 \,|\, \boldsymbol{E}_T(t)|^2 \\
&\leq 2\left(\beta_{BT} B^\sigma \frac{\boldsymbol{I}_B(t)}{N_B}\right)^2 |\boldsymbol{S}_T(t)|^2 \left(1 + \frac{(-(p\mu_T^\sigma + \nu_T^\alpha))^2 \,|\, \boldsymbol{E}_T(t)|^2}{\left(\beta_{BT} B^\sigma \frac{\boldsymbol{I}_B(t)}{N_B}\right)^2 |\boldsymbol{S}_T(t)|^2}\right)
\end{aligned}$$

If

$$\frac{(-(p\mu_T^\sigma + \nu_T^\alpha))^2}{\left(\beta_{BT} B^\sigma \frac{\boldsymbol{I}_B(t)}{N_B}\right)^2 |\boldsymbol{S}_T(t)|^2} < 1,$$

then, we get

$$|A_6(t, S_B, E_B, I_B, R_B, S_T, E_T, I_T)|^2 \quad \leq \quad K_6\left(1 + |E_T|^2\right)$$

where

$$K_6 = 2\left(\beta_{BT} B^\sigma \frac{\boldsymbol{I}_B(t)}{N_B}\right)^2 |\boldsymbol{S}_T(t)|^2.$$

$$\begin{aligned}
|A_7(t, S_B, E_B, I_B, R_B, S_T, E_T, I_T)|^2 &= |-p\mu_T^\sigma \boldsymbol{I}_T(t) + \nu_T^\sigma \boldsymbol{E}_T(t)|^2 \\
&\leq 2(-p\mu_T^\sigma)^2 |\boldsymbol{I}_T(t)|^2 + 2(\nu_T^\sigma)^2 \,|\, \boldsymbol{E}_T(t)|^2 \\
&\leq 2(\nu_T^\sigma)^2 \,|\, \boldsymbol{E}_T(t)|^2 \left(1 + \frac{(-p\mu_T^\sigma)^2 |\boldsymbol{I}_T(t)|^2}{(\nu_T^\sigma)^2 \,|\, \boldsymbol{E}_T(t)|^2}\right)
\end{aligned}$$

If

$$\frac{2(-p\mu_T^\sigma)^2}{2(\nu_T^\sigma)^2 \,|\, \boldsymbol{E}_T(t)|^2} < 1,$$

then, we get

$$|A_7(t, S_B, E_B, I_B, R_B, S_T, E_T, I_T)|^2 \quad \leq \quad K_7\left(1 + |I_T|^2\right)$$

where

$$K_7 = 2(\nu_T^\sigma)^2 \,|\, \boldsymbol{E}_T(t)|^2.$$

12.2.2 LIPSCHITZ CONDITION

Now, we verify the Lipschitz condition.

$$\begin{aligned}&|A_1(t,S_B,E_B,I_B,R_B,S_T,E_T,I_T)-A_1(t,S_{B_1},E_B,I_B,R_B,S_T,E_T,I_T)|^2\\&\quad=\left|-\beta_{TB}B^{\sigma}\frac{\boldsymbol{I}_T(t)}{N_T}(\boldsymbol{S}_B(t)-\boldsymbol{S}_{B_1}(t))\right|^2\end{aligned}$$

$$\begin{aligned}&\sup_{t\in[0,T]}|A_1(t,S_B,E_B,I_B,R_B,S_T,E_T,I_T)-A_1(t,S_{B_1},E_B,I_B,R_B,S_T,E_T,I_T)|^2\\&\quad=\left(-\beta_{TB}B^{\sigma}\frac{\boldsymbol{I}_T(t)}{N_T}\right)^2\sup_{t\in[0,T]}|\boldsymbol{S}_B(t)-\boldsymbol{S}_{B_1}(t)|^2\end{aligned}$$

$$\begin{aligned}&\|A_1(t,S_B,E_B,I_B,R_B,S_T,E_T,I_T)-A_1(t,S_{B_1},E_B,I_B,R_B,S_T,E_T,I_T)\|_\infty^2\\&\quad=\bar{K}_1\,\|\boldsymbol{S}_B(t)-\boldsymbol{S}_{B_1}(t)\|_\infty^2\end{aligned}$$

Similar results for A_2, A_3, A_4, A_5, A_6 and A_7 can be obtained easily. Therefore, under the condition

$$\begin{aligned}\max&\left(\frac{\left(-\beta_{TB}B^{\sigma}\frac{\boldsymbol{I}_T(t)}{N_T}\right)^2}{(\mu_B^{\sigma}+\alpha_B^{\sigma})^2\,|\,\boldsymbol{R}_B(t)|^2},\frac{(-\nu_B^{\sigma})^2}{\left(\beta_{TB}B^{\sigma}\frac{\boldsymbol{I}_T(t)}{N_T}\right)^2|\boldsymbol{S}_B(t)|^2},\frac{(-\lambda_B^{\sigma})^2}{(\nu_B^{\sigma})^2\,|\boldsymbol{E}_B(t)|^2}\right.\\&\frac{(-(\mu_B^{\sigma}+\alpha_B^{\sigma}))^2}{(\lambda_B^{\sigma})^2\,|\boldsymbol{I}_D(t)|^2},\frac{\left(-\beta_{BT}B^{\sigma}\frac{\boldsymbol{I}_B(t)}{N_B}\right)^2}{(p\mu_T^{\sigma})^2\,|\,\boldsymbol{I}_T(t)+\boldsymbol{E}_T(t)|^2},\frac{(-(p\mu_T^{\sigma}+\nu_T^{\alpha}))^2}{\left(\beta_{BT}B^{\sigma}\frac{\boldsymbol{I}_B(t)}{N_B}\right)^2|\boldsymbol{S}_T(t)|^2},\\&\left.\frac{2(-p\mu_T^{\sigma})^2}{2(\nu_T^{\sigma})^2\,|\,\boldsymbol{E}_T(t)|^2}\right)<1\end{aligned}$$

the system has unique solution.

12.3 NUMERICAL METHOD

We consider

$$\begin{aligned}{}^{ABC}D^{\alpha}\boldsymbol{S}_B(t)&=-\beta_{TB}B^{\sigma}\frac{\boldsymbol{I}_T(t)}{N_T}\boldsymbol{S}_B(t)+(\mu_B^{\sigma}+\alpha_B^{\alpha})\,\boldsymbol{R}_B(t)\\{}^{ABC}D^{\alpha}\boldsymbol{E}_B(t)&=\beta_{TB}B^{\sigma}\frac{\boldsymbol{I}_T(t)}{N_T}\boldsymbol{S}_B(t)-\nu_B^{\sigma}\boldsymbol{E}_B(t)\\{}^{ABC}D^{\alpha}\boldsymbol{I}_B(t)&=\nu_B^{\sigma}\boldsymbol{E}_B(t)-\lambda_B^{\sigma}\boldsymbol{I}_B(t)\\{}^{ABC}D^{\alpha}\boldsymbol{R}_B(t)&=\lambda_B^{\sigma}\boldsymbol{I}_B(t)-(\mu_B^{\sigma}+\alpha_B^{\sigma})\boldsymbol{R}_B(t),\end{aligned}$$

$$
\begin{aligned}
{}^{ABC}D^{\alpha}\boldsymbol{S}_T(t) &= -\beta_{BT}B^{\sigma}\frac{\boldsymbol{I}_B(t)}{N_B}\boldsymbol{S}_T(t) + p\mu_T^{\sigma}\left(\boldsymbol{I}_T(t)+\boldsymbol{E}_T(t)\right)\\
{}^{ABC}D^{\alpha}\boldsymbol{E}_T(t) &= \beta_{BT}B^{\sigma}\frac{\boldsymbol{I}_B(t)}{N_B}\boldsymbol{S}_T(t) - (p\mu_T^{\sigma}+\nu_T^{\alpha})\boldsymbol{E}_T(t)\\
{}^{ABC}D^{\alpha}\boldsymbol{I}_T(t) &= -p\mu_T^{\sigma}\boldsymbol{I}_T(t) + \nu_T^{\sigma}\boldsymbol{E}_T(t)
\end{aligned}
$$

For simplicity, we define

$$
\begin{aligned}
A(t,S_B,E_B,I_B,R_B,S_T,E_T,I_T) &= -\beta_{TB}B^{\sigma}\frac{\boldsymbol{I}_T(t)}{N_T}\boldsymbol{S}_B(t) + (\mu_B^{\sigma}+\alpha_B^{\alpha})\ \boldsymbol{R}_B(t)\\
B(t,S_B,E_B,I_B,R_B,S_T,E_T,I_T) &= \beta_{TB}B^{\sigma}\frac{\boldsymbol{I}_T(t)}{N_T}\boldsymbol{S}_B(t) - \nu_B^{\sigma}\boldsymbol{E}_B(t)\\
C(t,_B,E_B,I_B,R_B,S_T,E_T,I_T) &= \nu_B^{\sigma}\boldsymbol{E}_B(t) - \lambda_B^{\sigma}\boldsymbol{I}_B(t)\\
D(t,S_B,E_B,I_B,R_B,S_T,E_T,I_T) &= \lambda_B^{\sigma}\boldsymbol{I}_B(t) - (\mu_B^{\sigma}+\alpha_B^{\sigma})\boldsymbol{R}_B(t),\\
K(t,S_B,E_B,I_B,R_B,S_T,E_T,I_T) &= -\beta_{BT}B^{\sigma}\frac{\boldsymbol{I}_B(t)}{N_B}\boldsymbol{S}_T(t) + p\mu_T^{\sigma}\left(\boldsymbol{I}_T(t)+\boldsymbol{E}_T(t)\right)\\
L(t,S_B,E_B,I_B,R_B,S_T,E_T,I_T) &= \beta_{BT}B^{\sigma}\frac{\boldsymbol{I}_B(t)}{N_B}\boldsymbol{S}_T(t) - (p\mu_T^{\sigma}+\nu_T^{\alpha})\boldsymbol{E}_T(t)\\
M(t,S_B,E_B,I_B,R_B,S_T,E_T,I_T) &= -p\mu_T^{\sigma}\boldsymbol{I}_T(t) + \nu_T^{\sigma}\boldsymbol{E}_T(t)
\end{aligned}
$$

Then, we have

$$
\begin{aligned}
{}^{ABC}D^{\alpha}\boldsymbol{S}_B(t) &= A(t,S_B,E_B,I_B,R_B,S_T,E_T,I_T)\\
{}^{ABC}D^{\alpha}\boldsymbol{E}_B(t) &= B(t,S_B,E_B,I_B,R_B,S_T,E_T,I_T)\\
{}^{ABC}D^{\alpha}\boldsymbol{I}_B(t) &= C(t,S_B,E_B,I_B,R_B,S_T,E_T,I_T)\\
{}^{ABC}D^{\alpha}\boldsymbol{R}_B(t) &= D(t,S_B,E_B,I_B,R_B,S_T,E_T,I_T)\\
{}^{ABC}D^{\alpha}\boldsymbol{S}_T(t) &= K(t,S_B,E_B,I_B,R_B,S_T,E_T,I_T)\\
{}^{ABC}D^{\alpha}\boldsymbol{E}_T(t) &= L(t,S_B,E_B,I_B,R_B,S_T,E_T,I_T)\\
{}^{ABC}D^{\alpha}\boldsymbol{I}_T(t) &= M(t,S_B,E_B,I_B,R_B,S_T,E_T,I_T)
\end{aligned}
$$

Applying Atangana-Baleanu integral gives:

$$
\begin{aligned}
S_B(t)-S_B(0) &= \frac{1-\alpha}{AB(\alpha)}A(t,S_B,E_B,I_B,R_B,S_T,E_T,I_T)\\
&\quad+\frac{\alpha}{AB(\alpha)\Gamma(\alpha)}\int_0^t A(\tau,S_B,E_B,I_B,R_B,S_T,E_T,I_T)(t-\tau)^{\alpha-1}d\tau
\end{aligned}
$$

$$
\begin{aligned}
E_B(t)-E_B(0) &= \frac{1-\alpha}{AB(\alpha)}B(t,S_B,E_B,I_B,R_B,S_T,E_T,I_T)\\
&\quad+\frac{\alpha}{AB(\alpha)\Gamma(\alpha)}\int_0^t B(\tau,S_B,E_B,I_B,R_B,S_T,E_T,I_T)(t-\tau)^{\alpha-1}d\tau
\end{aligned}
$$

$$I_B(t)-I_B(0)=\frac{1-\alpha}{AB(\alpha)}C(t,S_B,E_B,I_B,R_B,S_T,E_T,I_T)$$
$$+\frac{\alpha}{AB(\alpha)\Gamma(\alpha)}\int_0^t C(\tau,S_B,E_B,I_B,R_B,S_T,E_T,I_T)(t-\tau)^{\alpha-1}d\tau$$

$$R_B(t)-R_B(0)=\frac{1-\alpha}{AB(\alpha)}D(t,S_B,E_B,I_B,R_B,S_T,E_T,I_T)$$
$$+\frac{\alpha}{AB(\alpha)\Gamma(\alpha)}\int_0^t D(\tau,S_B,E_B,I_B,R_B,S_T,E_T,I_T)(t-\tau)^{\alpha-1}d\tau$$

$$S_T(t)-S_T(0)=\frac{1-\alpha}{AB(\alpha)}K(t,S_B,E_B,I_B,R_B,S_T,E_T,I_T)$$
$$+\frac{\alpha}{AB(\alpha)\Gamma(\alpha)}\int_0^t K(\tau,S_B,E_B,I_B,R_B,S_T,E_T,I_T)(t-\tau)^{\alpha-1}d\tau$$

$$E_T(t)-E_T(0)=\frac{1-\alpha}{AB(\alpha)}L(t,S_B,E_B,I_B,R_B,S_T,E_T,I_T)$$
$$+\frac{\alpha}{AB(\alpha)\Gamma(\alpha)}\int_0^t L(\tau,S_B,E_B,I_B,R_B,S_T,E_T,I_T)(t-\tau)^{\alpha-1}d\tau$$

$$I_T(t)-I_T(0)=\frac{1-\alpha}{AB(\alpha)}M(t,S_B,E_B,I_B,R_B,S_T,E_T,I_T)$$
$$+\frac{\alpha}{AB(\alpha)\Gamma(\alpha)}\int_0^t M(\tau,S_B,E_B,I_B,R_B,S_T,E_T,I_T)(t-\tau)^{\alpha-1}d\tau$$

We discretize the above equations at t_{n+1} as:

$$S_B^{n+1}-S_B^0=\frac{1-\alpha}{AB(\alpha)}A(t_{n+1},S_B^n,E_B^n,I_B^n,R_B^n,S_T^n,E_T^n,I_T^n)$$
$$+\frac{\alpha}{AB(\alpha)\Gamma(\alpha)}\int_0^{t_{n+1}} A(\tau,S_B,E_B,I_B,R_B,S_T,E_T,I_T)(t_{n+1}-\tau)^{\alpha-1}d\tau$$

$$E_B^{n+1}-E_B^0=\frac{1-\alpha}{AB(\alpha)}B(t_{n+1},S_B^n,E_B^n,I_B^n,R_B^n,S_T^n,E_T^n,I_T^n)$$
$$+\frac{\alpha}{AB(\alpha)\Gamma(\alpha)}\int_0^{t_{n+1}} B(\tau,S_B,E_B,I_B,R_B,S_T,E_T,I_T)(t_{n+1}-\tau)^{\alpha-1}d\tau$$

$$I_B^{n+1}-I_B^0=\frac{1-\alpha}{AB(\alpha)}C(t_{n+1},S_B^n,E_B^n,I_B^n,R_B^n,S_T^n,E_T^n,I_T^n)$$
$$+\frac{\alpha}{AB(\alpha)\Gamma(\alpha)}\int_0^{t_{n+1}} C(\tau,S_B,E_B,I_B,R_B,S_T,E_T,I_T)(t_{n+1}-\tau)^{\alpha-1}d\tau$$

$$R_B^{n+1} - R_B^0 = \frac{1-\alpha}{AB(\alpha)} D(t_{n+1}, S_B^n, E_B^n, I_B^n, R_B^n, S_T^n, E_T^n, I_T^n) + \frac{\alpha}{AB(\alpha)\Gamma(\alpha)} \int_0^{t_{n+1}} D(\tau, S_B, E_B, I_B, R_B, S_T, E_T, I_T)(t_{n+1}-\tau)^{\alpha-1} d\tau$$

$$S_T^{n+1} - S_T^0 = \frac{1-\alpha}{AB(\alpha)} K(t_{n+1}, S_B^n, E_B^n, I_B^n, R_B^n, S_T^n, E_T^n, I_T^n) + \frac{\alpha}{AB(\alpha)\Gamma(\alpha)} \int_0^{t_{n+1}} K(\tau, S_B, E_B, I_B, R_B, S_T, E_T, I_T)(t_{n+1}-\tau)^{\alpha-1} d\tau$$

$$E_T^{n+1} - E_T^0 = \frac{1-\alpha}{AB(\alpha)} L(t_{n+1}, S_B^n, E_B^n, I_B^n, R_B^n, S_T^n, E_T^n, I_T^n) + \frac{\alpha}{AB(\alpha)\Gamma(\alpha)} \int_0^{t_{n+1}} L(\tau, S_B, E_B, I_B, R_B, S_T, E_T, I_T)(t_{n+1}-\tau)^{\alpha-1} d\tau$$

$$I_T^{n+1} - I_T^0 = \frac{1-\alpha}{AB(\alpha)} M(t_{n+1}, S_B^n, E_B^n, I_B^n, R_B^n, S_T^n, E_T^n, I_T^n) + \frac{\alpha}{AB(\alpha)\Gamma(\alpha)} \int_0^{t_{n+1}} M(\tau, S_B, E_B, I_B, R_B, S_T, E_T, I_T)(t_{n+1}-\tau)^{\alpha-1} d\tau$$

We use two-point Lagrange-polynomial and obtain [15]:

$$S_B^{n+1} = S_B^0 + \frac{1-\alpha}{AB(\alpha)} A(t_{n+1}, S_B^n, E_B^n, I_B^n, R_B^n, S_T^n, E_T^n, I_T^n) + \frac{\alpha}{AB(\alpha)} \sum_{j=0}^{n} \left[\frac{h^\alpha A(t_j, S_B^n, E_B^n, I_B^n, R_B^n, S_T^n, E_T^n, I_T^n)}{\Gamma(\alpha+2)} ((n+1-j)^\alpha (n-j+2+\alpha) - (n-j)^\alpha (n-j+2+2\alpha)) \right]$$

$$E_B^{n+1} = E_B^0 + \frac{1-\alpha}{AB(\alpha)} B(t_{n+1}, S_B^n, E_B^n, I_B^n, R_B^n, S_T^n, E_T^n, I_T^n) + \frac{\alpha}{AB(\alpha)} \sum_{j=0}^{n} \left[\frac{h^\alpha B(t_j, S_B^n, E_B^n, I_B^n, R_B^n, S_T^n, E_T^n, I_T^n)}{\Gamma(\alpha+2)} ((n+1-j)^\alpha (n-j+2+\alpha) - (n-j)^\alpha (n-j+2+2\alpha)) \right]$$

$$I_B^{n+1} = I_B^0 + \frac{1-\alpha}{AB(\alpha)} C(t_{n+1}, S_B^n, E_B^n, I_B^n, R_B^n, S_T^n, E_T^n, I_T^n) + \frac{\alpha}{AB(\alpha)} \sum_{j=0}^{n} \left[\frac{h^\alpha C(t_j, S_B^n, E_B^n, I_B^n, R_B^n, S_T^n, E_T^n, I_T^n)}{\Gamma(\alpha+2)} ((n+1-j)^\alpha (n-j+2+\alpha) - (n-j)^\alpha (n-j+2+2\alpha)) \right]$$

$$R_B^{n+1} = R_B^0 + \frac{1-\alpha}{AB(\alpha)} D(t_{n+1}, S_B^n, E_B^n, I_B^n, R_B^n, S_T^n, E_T^n, I_T^n)$$
$$+ \frac{\alpha}{AB(\alpha)} \sum_{j=0}^{n} \left[\frac{h^\alpha D(t_j, S_B^n, E_B^n, I_B^n, R_B^n, S_T^n, E_T^n, I_T^n)}{\Gamma(\alpha+2)} ((n+1-j)^\alpha (n-j+2+\alpha) \right.$$
$$\left. -(n-j)^\alpha (n-j+2+2\alpha)) \right]$$

$$S_T^{n+1} = S_T^0 + \frac{1-\alpha}{AB(\alpha)} K(t_{n+1}, S_B^n, E_B^n, I_B^n, R_B^n, S_T^n, E_T^n, I_T^n)$$
$$+ \frac{\alpha}{AB(\alpha)} \sum_{j=0}^{n} \left[\frac{h^\alpha K(t_j, S_B^n, E_B^n, I_B^n, R_B^n, S_T^n, E_T^n, I_T^n)}{\Gamma(\alpha+2)} ((n+1-j)^\alpha (n-j+2+\alpha) \right.$$
$$\left. -(n-j)^\alpha (n-j+2+2\alpha)) \right]$$

$$E_T^{n+1} = E_T^0 + \frac{1-\alpha}{AB(\alpha)} L(t_{n+1}, S_B^n, E_B^n, I_B^n, R_B^n, S_T^n, E_T^n, I_T^n)$$
$$+ \frac{\alpha}{AB(\alpha)} \sum_{j=0}^{n} \left[\frac{h^\alpha L(t_j, S_B^n, E_B^n, I_B^n, R_B^n, S_T^n, E_T^n, I_T^n)}{\Gamma(\alpha+2)} ((n+1-j)^\alpha (n-j+2+\alpha) \right.$$
$$\left. -(n-j)^\alpha (n-j+2+2\alpha)) \right]$$

$$I_T^{n+1} = I_T^0 + \frac{1-\alpha}{AB(\alpha)} M(t_{n+1}, S_B^n, E_B^n, I_B^n, R_B^n, S_T^n, E_T^n, I_T^n)$$
$$+ \frac{\alpha}{AB(\alpha)} \sum_{j=0}^{n} \left[\frac{h^\alpha M(t_j, S_B^n, E_B^n, I_B^n, R_B^n, S_T^n, E_T^n, I_T^n)}{\Gamma(\alpha+2)} ((n+1-j)^\alpha (n-j+2+\alpha) \right.$$
$$\left. -(n-j)^\alpha (n-j+2+2\alpha)) \right]$$

12.4 RESULTS OF THE SIMULATION

We use the parameters and initial conditions as [16] (Table 12.1):

$$\begin{cases} S_B(0) = 10000, E_B(0) = 100, I_B(0) = 1500, R_B(0) = 1000 \\ S_T(0) = 3000, E_T(0) = 350, I_T(0) = 40. \end{cases}$$

We obtain the following figures. In these figures, we can see the effect of fractional order α. In addition, we present the simulations for different values of σ. In Figures 12.1–12.8, we demonstrate the numerical simulations for $\sigma = 1$ and different values of fractional order. In these figures, we use fractional order $\alpha = 1$, $\alpha = 0.95$, $\alpha = 0.90$ and $\alpha = 0.85$. In Figures 12.9–12.15, we present the numerical simulations for $\sigma = 0.95$. In Figures 12.16–12.22, we show the numerical simulations for $\sigma = 0.90$. In Figures 12.23–12.28, we demonstrate the numerical simulations for $\sigma = 0.85$.

Table 12.1
Biological Parameters for Simulation [16]

Parameters	Values	Biological Meaning
β_{TB}	0.0017	The probability that a bite by an infectious tick will infect a bovine
β_{BT}	0.027	The probability that biting an infectious bovine will infect a tick
B	0.4162	The infestation rate
μ_B	0.00029	The proportion of the bovine's population that dies every day
μ_T	0.0016	The proportion of the tick's population that dies every day
λ_B	0.00265	The daily bovines rate that has been subjected to treatment against babesiosis
α_B	0.001	The proportion of the treated bovine may return to susceptible state
ν_B	0 .0667	The passage rate from the exposed state to the infected state for bovines
ν_T	0.035	The passage rate from the exposed state to the infected state for ticks
p	0.1	The probability that a susceptible tick was born from an infected one

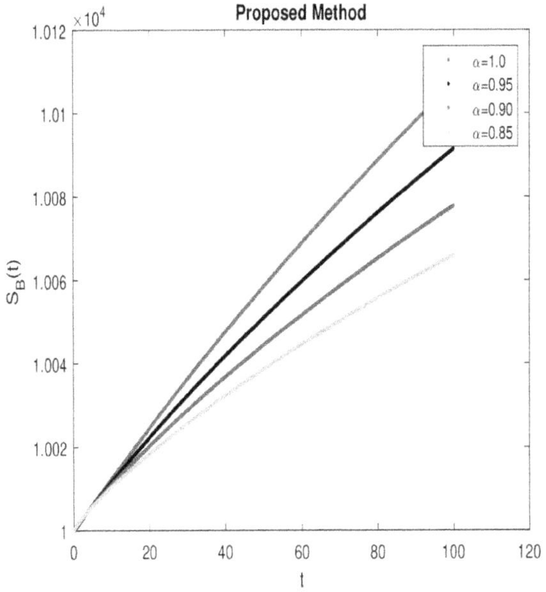

Figure 12.1 Numerical simulations for $\sigma = 1$.

12.5 CONCLUSION

In this chapter, we investigated the bovine Babesiosis epidemic model. We used fractional derivative in the model. We proved the existence and uniqueness of positive solution. We applied a very effective numerical method to the given fractional model. We demonstrated the numerical simulations to show the accuracy of the proposed technique.

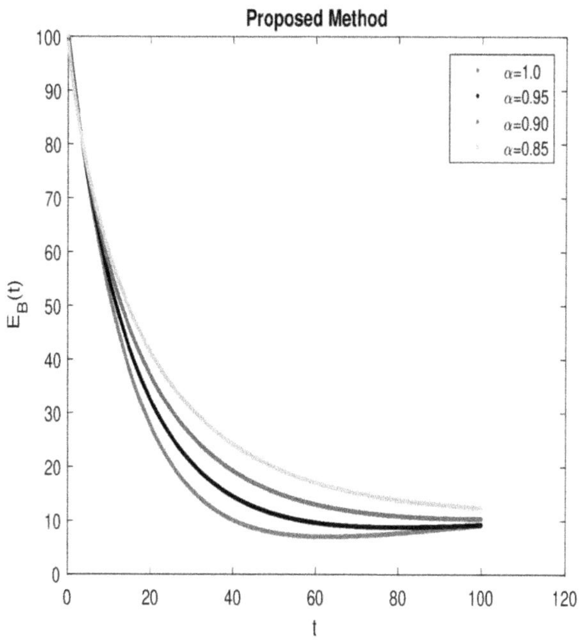

Figure 12.2 Numerical simulations for $\sigma = 1$.

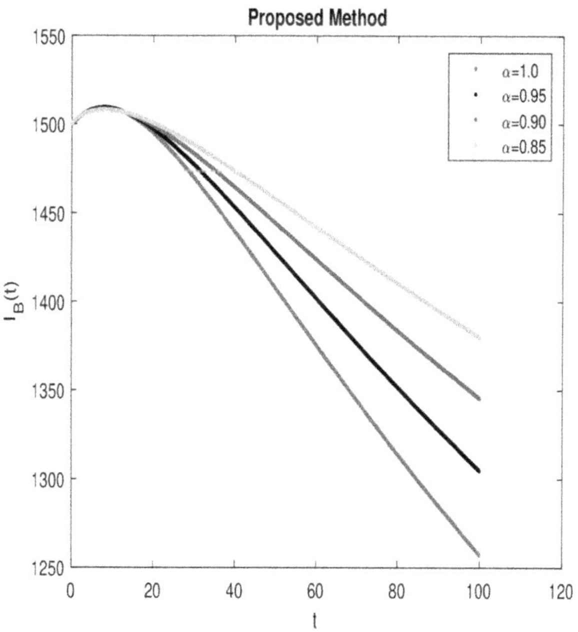

Figure 12.3 Numerical simulations for $\sigma = 1$.

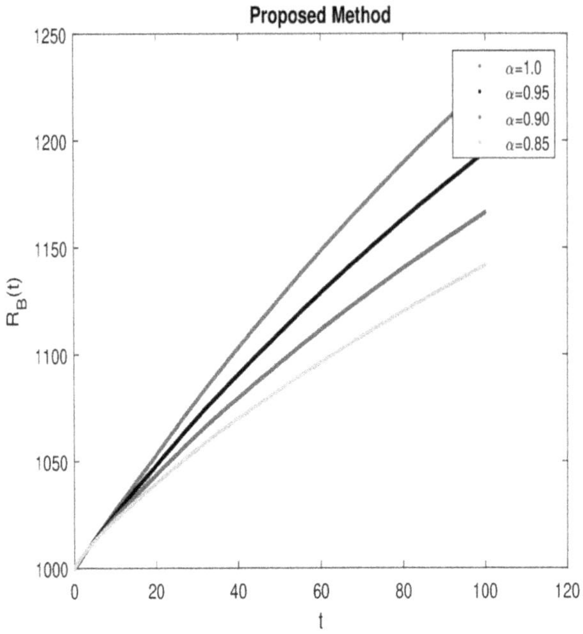

Figure 12.4 Numerical simulations for $\sigma = 1$.

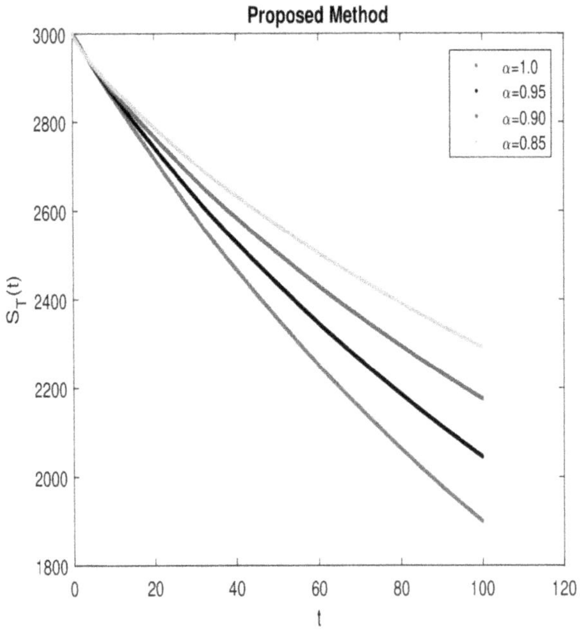

Figure 12.5 Numerical simulations for $\sigma = 1$.

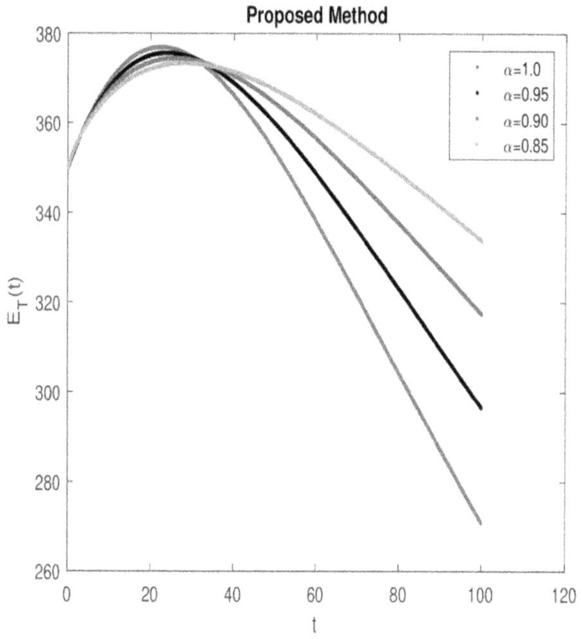

Figure 12.6 Numerical simulations for $\sigma = 1$.

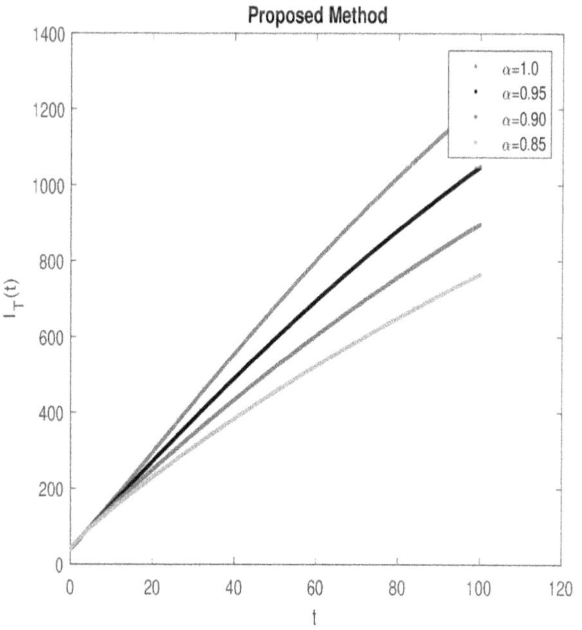

Figure 12.7 Numerical simulations for $\sigma = 1$.

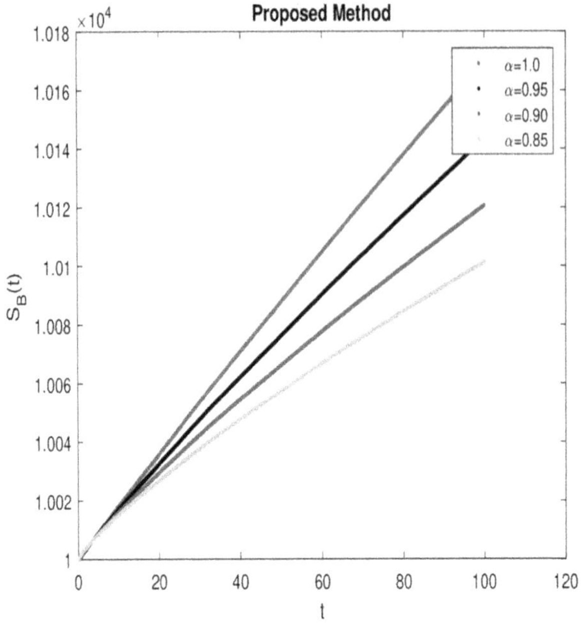

Figure 12.8 Numerical simulations for $\sigma = 0.95$.

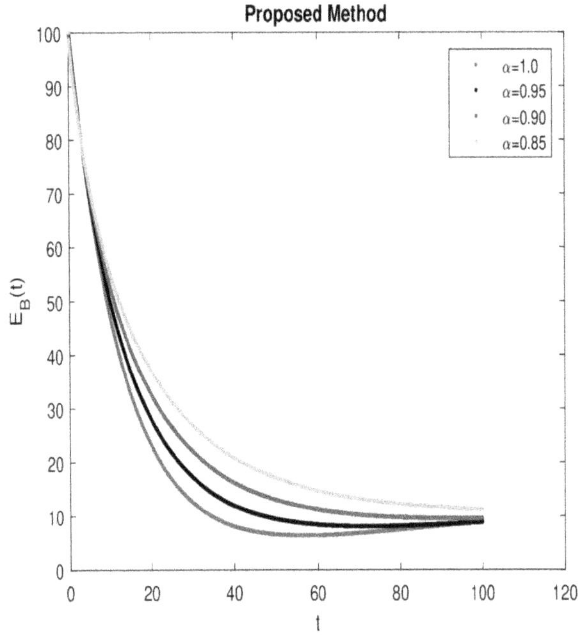

Figure 12.9 Numerical simulations for $\sigma = 0.95$.

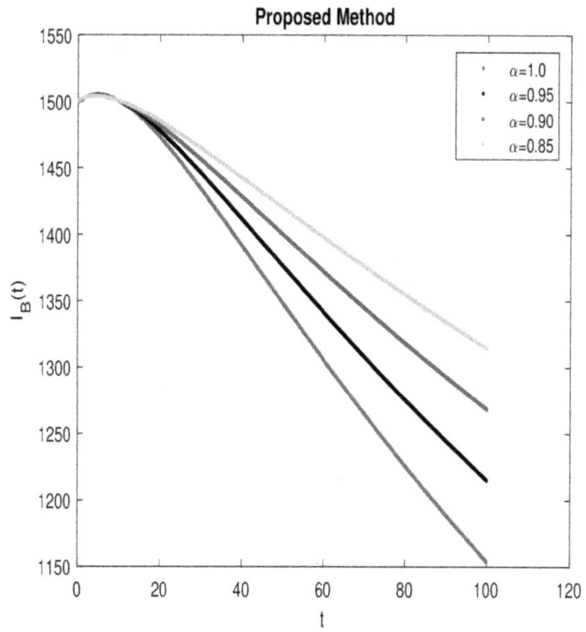

Figure 12.10 Numerical simulations for $\sigma = 0.95$.

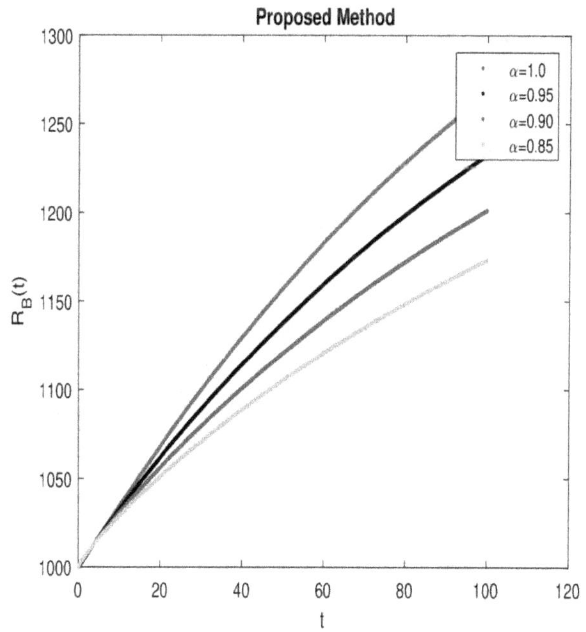

Figure 12.11 Numerical simulations for $\sigma = 0.95$.

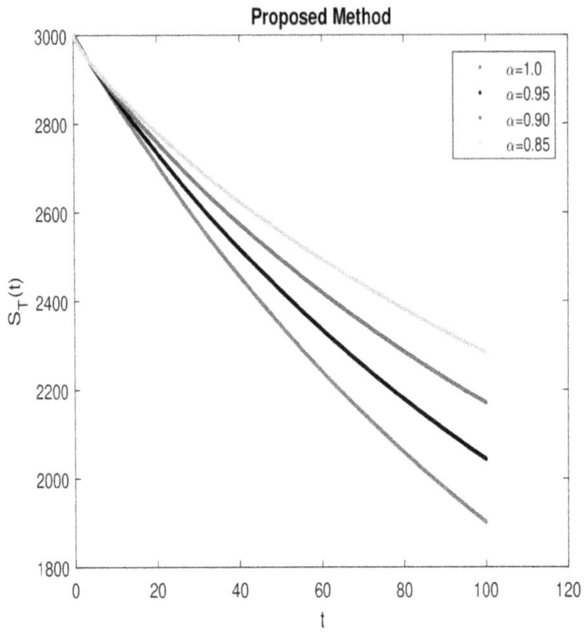

Figure 12.12 Numerical simulations for $\sigma = 0.95$.

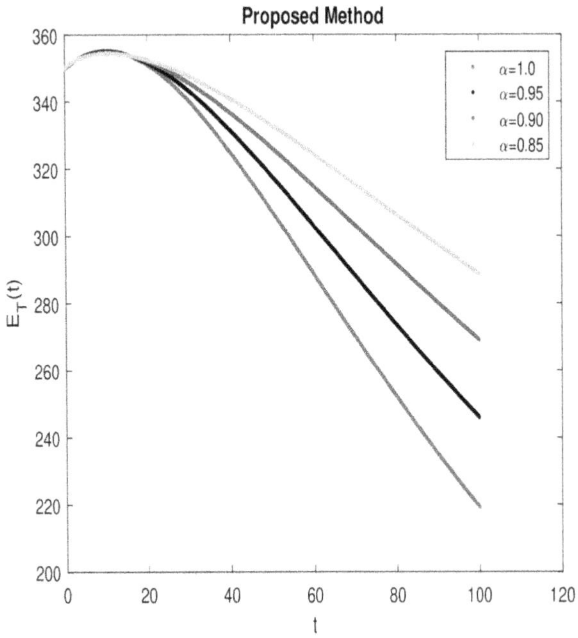

Figure 12.13 Numerical simulations for $\sigma = 0.95$.

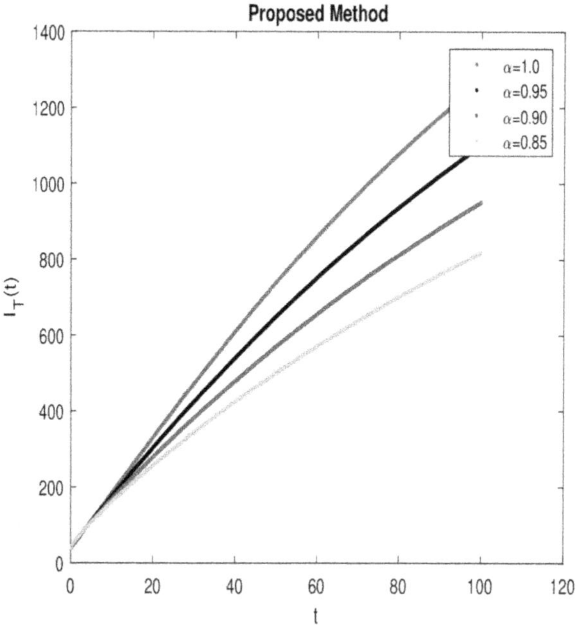

Figure 12.14 Numerical simulations for $\sigma = 0.95$.

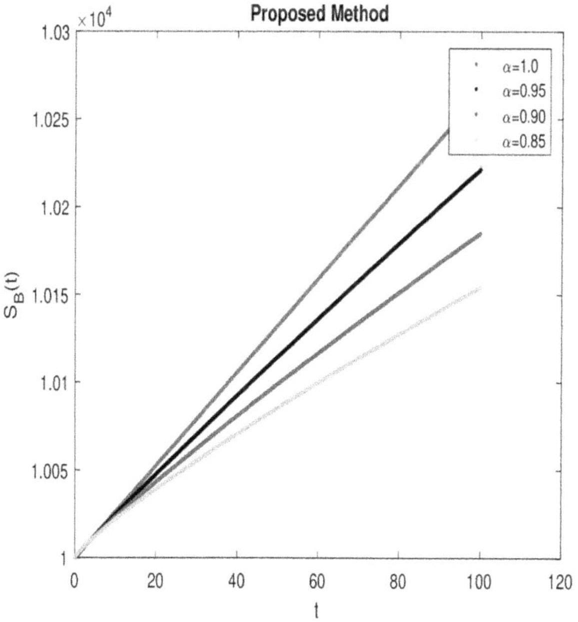

Figure 12.15 Numerical simulations for $\sigma = 0.9$.

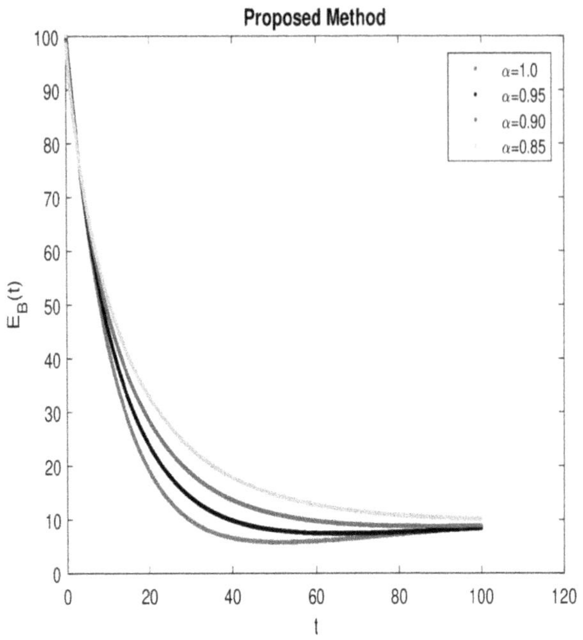

Figure 12.16 Numerical simulations for $\sigma = 0.9$.

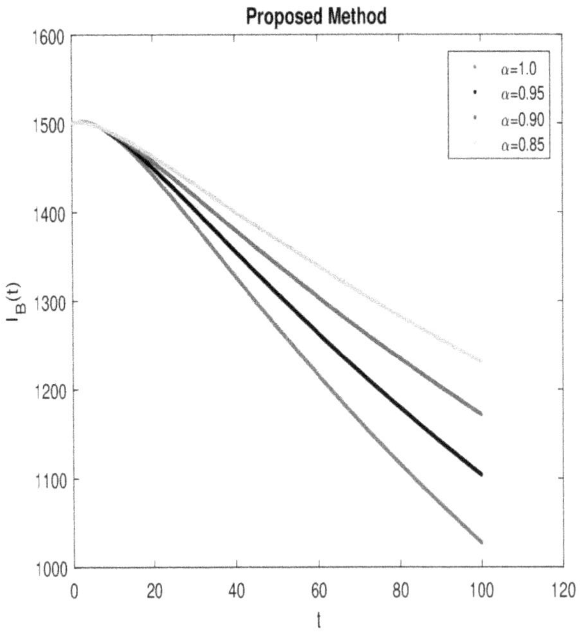

Figure 12.17 Numerical simulations for $\sigma = 0.9$.

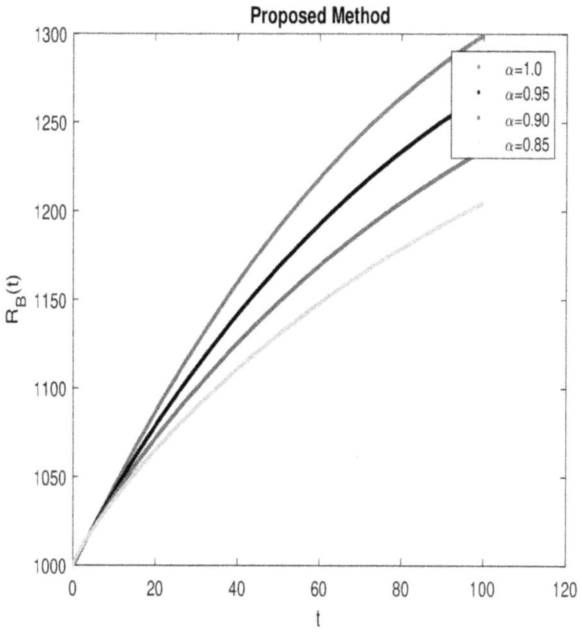

Figure 12.18 Numerical simulations for $\sigma = 0.9$.

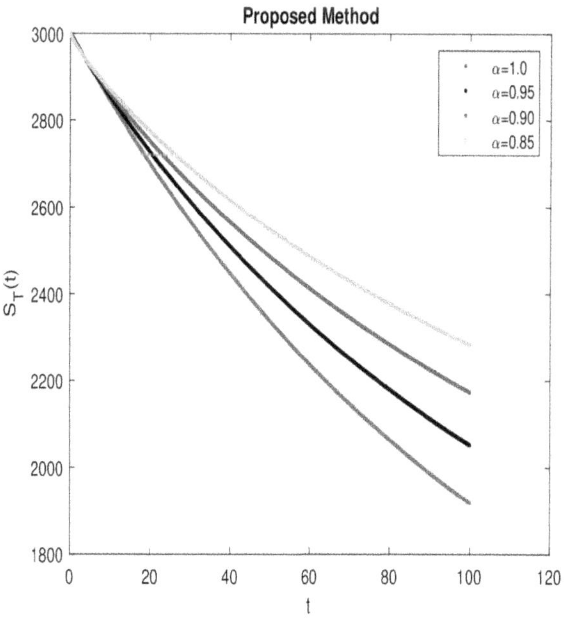

Figure 12.19 Numerical simulations for $\sigma = 0.9$.

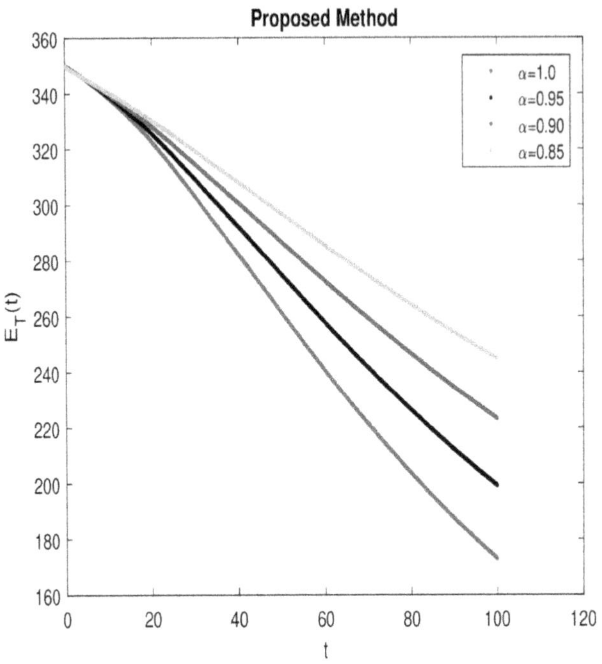

Figure 12.20 Numerical simulations for $\sigma = 0.9$.

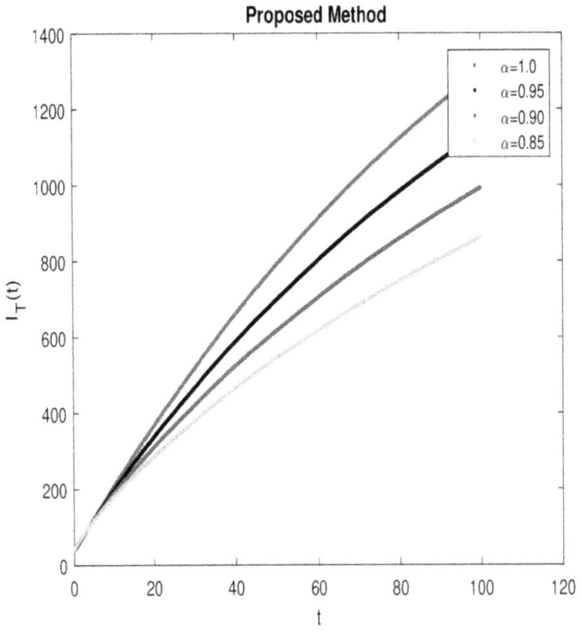

Figure 12.21 Numerical simulations for $\sigma = 0.9$.

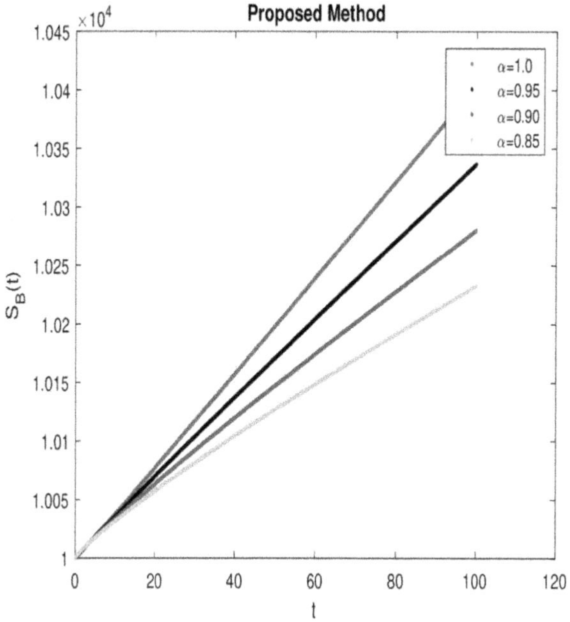

Figure 12.22 Numerical simulations for $\sigma = 0.85$.

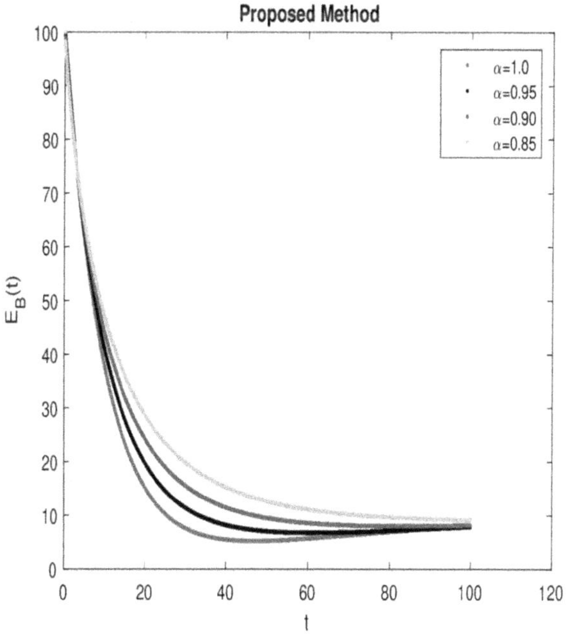

Figure 12.23 Numerical simulations for $\sigma = 0.85$.

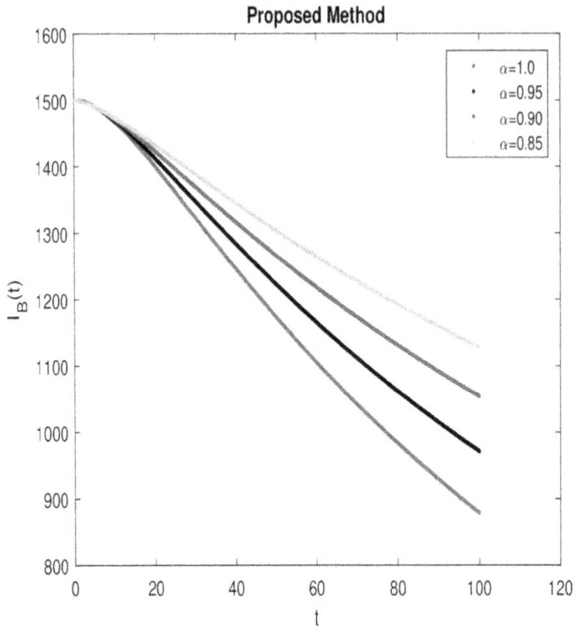

Figure 12.24 Numerical simulations for $\sigma = 0.85$.

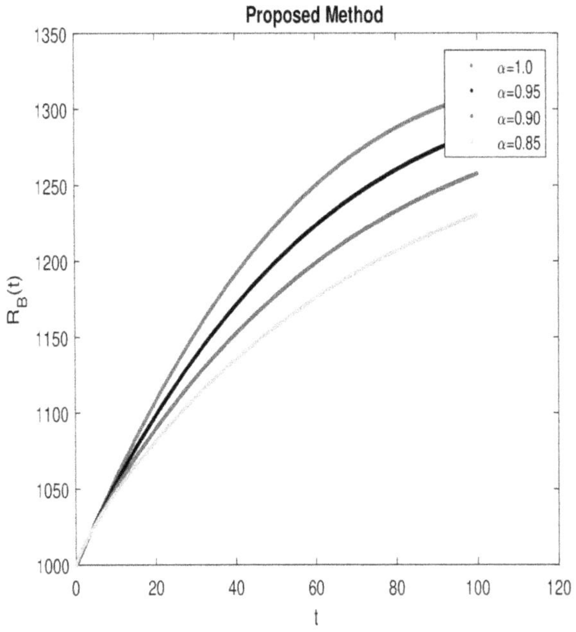

Figure 12.25 Numerical simulations for $\sigma = 0.85$.

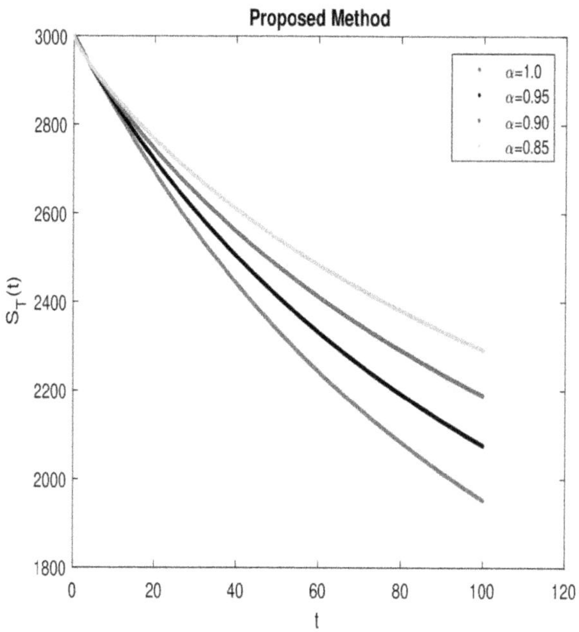

Figure 12.26 Numerical simulations for $\sigma = 0.85$.

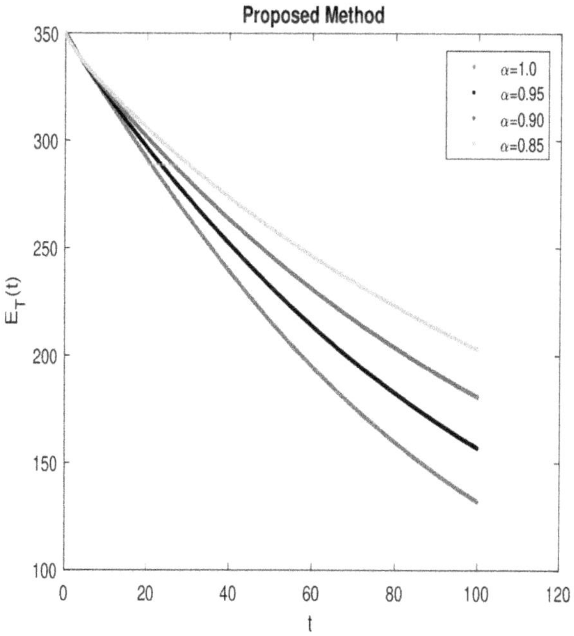

Figure 12.27 Numerical simulations for $\sigma = 0.85$.

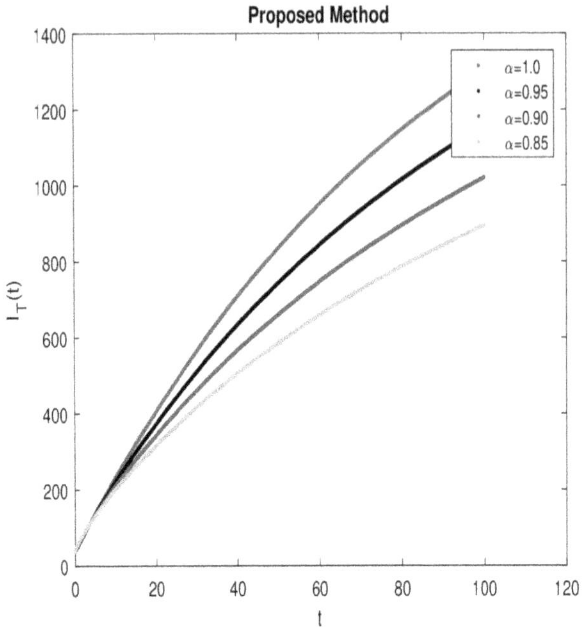

Figure 12.28 Numerical simulations for $\sigma = 0.85$.

References

1. A. Akgül, A novel method for a fractional derivative with non-local and non-singular kernel, *Chaos, Solitons and Fractals* 114, 478–482, 2018.
2. A. Atangana, A. Akgül, K. M. Owolabi, Analysis of fractal fractional differential equations, *Alexandria Engineering Journal* 59(3), 1117–1134, 2020.
3. D. Baleanu, A. Fernandez, A. Akgül, On a fractional operator combining proportional and classical differintegrals, *Mathematics* 8(3), 360, 2020.
4. K. M. Owolabi, A. Atangana, A. Akgül, Modelling and analysis of fractal-fractional partial differential equations: Application to reaction-diffusion model, *Alexandria Engineering Journal* 59(4), 2477–2490, 2020.
5. H. Singh, A. M. Wazwaz, Computational method for reaction diffusion-model arising in a spherical catalyst, *International Journal of Applied and Computational Mathematics* 7(3), 65, 2021.
6. H. Singh, Analysis for fractional dynamics of Ebola virus model, *Chaos Solitons and Fractals* 138, 109992, 2020.
7. H. Singh, Solving a class of local and nonlocal elliptic boundary value problems arising in heat transfer, *Heat Transfer* 51, 1524–1542, 2022.
8. V. M. Tripathi, H. M. Srivastava, H. Singh, C. Swarup, S. Aggarwal, Mathematical analysis of non-isothermal reaction-diffusion models arising in spherical catalyst and spherical biocatalyst, *Applied Sciences* 11, 10423, 2021.
9. H. Singh, An efficient computational method for non-linear fractional Lienard equation arising in oscillating circuits, In: H. Singh, D. Kumar, D. Baleanu (Eds.), *Methods of*

Mathematical Modelling: Fractional Differential Equations. CRC Press, Taylor & Francis Group: Boca Raton, FL, pp. 39–50, 2019.
10. H. Singh, Chebyshev spectral method for solving a class of local and nonlocal elliptic boundary value problems, *International Journal of Nonlinear Science and Numerical Simulations*, 000010151520200235, 2021.
11. H. Singh, H. M. Srivastava, Numerical investigation of the fractional-order Liénard and Duffing equations arising in oscillating circuit theory, *Frontiers in Physics* 8, 120, 2020.
12. H. Singh, H. M. Srivastava, J. J. Nieto, *Handbook of Fractional Calculus for Engineering and Science*. CRC Press, Taylor & Francis Group: Boca Raton, FL, 2022.
13. H. Singh, D. Kumar, D. Baleanu, *Methods of Mathematical Modelling: Fractional Differential Equations*. CRC Press, Taylor & Francis: Boca Raton, FL, 2019.
14. H. Singh, H. Srivastava, D. Baleanu, *Methods of Mathematical Modelling: Infectious Disease*. Elsevier Science: Amsterdam, Netherlands, 2022 (ISBN: 9780323998888).
15. M. Toufik, A. Atangana, New numerical approximation of fractional derivative with non-local and non-singular kernel: Application to chaotic models, *The European Physical Journal Plus* 132, 444, 2017.
16. M. Abdelheq et al., A predictive spatio-temporal model for bovine Babesiosis epidemic transmission, *Journal of Theoretical Biology* 480, 192–204, 2019.

Index

For Product Safety Concerns and Information please contact our EU
representative GPSR@taylorandfrancis.com
Taylor & Francis Verlag GmbH, Kaufingerstraße 24, 80331 München, Germany

www.ingramcontent.com/pod-product-compliance
Lightning Source LLC
Chambersburg PA
CBHW060333220326
41598CB00023B/2698
9781032436029